HACCP System Auditing for Food Safety

HACCP System Auditing for Food Safety

Principles and Techniques

Luis Couto Lorenzo

WILEY

Published by John Wiley & Sons, Inc., Hoboken, New Jersey.

Published simultaneously in Canada.

Originally published as *Auditoría del Sistema de APPCC*
© Diaz de Santos, 2019

Library of Congress Cataloging-in-Publication Data Applied for:

Hardback ISBN: 9781394254729

Cover Design: Wiley
Cover Image: © Nguyen Quang Ngoc Tonkin/Shutterstock

Set in 9.5/12.5pt STIXTwoText by Straive, Chennai, India

To my dear wife Lola, for her support through all these years.

Contents

Preface

Among the many subjects and disciplines covered by food safety, hazard analysis and critical control point (HACCP) is central. It provides the rational and scientific basis necessary to have the vision, the analysis, the ability to anticipate problems, and to learn from the experiences and situations that arise in the food industry. Unlike the application of general control measures, which are also essential to ensure food hygiene, HACCP provides the ability to think, rethink, and adapt resources and control strategies to achieve the desired level of food safety.

From professionals working in the food industry, who are the first and most involved in the application of the HACCP methodology in the very field for which it was created, to other practitioners such as consultants, auditors, official control agents, without forgetting the university teaching and research sector, all need to have a solid knowledge of the principles and steps to be followed to develop a consistent, feasible, and an effective HACCP system.

A great virtue of HACCP is that it is the same in all countries, so we can all speak the same language and food operators can adopt it as a system of mutual recognition to facilitate food trade.

In essence, HACCP is a simple, rational, and logical system that has remained largely unchanged since it was adopted by Codex as an indispensable tool for food safety, but at the same time it requires a thorough understanding of its concepts and the correct application of its principles to develop HACCP systems that are truly useful to the operators who have to apply them.

The comparative simplicity and also the consistency of the HACCP system have allowed it to reach our days respecting the original structure agreed in the 1990s. Thus, the modifications introduced in the latest version of the Codex document in 2022, although significant, concern nonessential aspects of the system and had already been assimilated by professionals in the previous years. Moreover, the experience gained over more than 30 years has enabled the system to mature and a general consensus has been reached in the most appropriate way to implement HACCP in the day-to-day work of the food industry.

The aim of this book is to explain in detail all the elements of the HACCP methodology, clarifying the related concepts, the phases of the HACCP study, and the application of the seven principles, respecting their internal logic and the way in which they should be interrelated to achieve a satisfactory result, adapted to the specificities of the food operator and truly effective. This is the only way to ensure that the operators will regard HACCP as an invaluable tool for the smooth running of their business and not as a burden imposed by law or customer requirements.

Once the HACCP system is operating in the food business, it needs to be periodically verified; this is not only a requirement of the HACCP system itself, in particular its principle 6, but is the way to ensure that it is working properly and is effective in achieving the food safety objectives.

In practice, HACCP should be understood as a management system, in which, starting from the commitment of the management or the company's board of directors, tasks, and responsibilities should be distributed among staff at different levels of the organization.

The best way to verify the HACCP system is to use management system auditing techniques. Auditing is a proven tool that has a wealth of experience in its use to verify the performance of management systems, whether for internal audits, supplier audits, or for certification purposes. This book follows the guidelines for auditing set out in ISO 19011:2018, the international reference standard for management system auditing.

There is a saying in the quality world that you learn to audit by auditing. But first it is necessary to have a good knowledge of the principles and techniques of auditing that must be practiced to achieve the competence required to perform this task effectively. The second part of this book is devoted to detailing the system to be followed and the skills and personal attributes that should inspire the auditor's behavior during his or her professional practice.

Luis Couto Lorenzo
Lalín, Spain
June 2024

Acknowledgement

I want to thank all the wonderful staff at Wiley, especially Jonathan Rose and Shwathi Srinivasan, for welcoming my proposal and guiding me through the whole process of publishing this book. Thanks to them, a dream has come true!

I also want to thank Estefanía Cuíña and Juan Castro for their invaluable contribution to case study 1.

And finally, I want to thank Manuel Araújo and Montserrat López for their essential contribution to including case study 2.

1

A Necessary Evolution

Since its conception, at the end of the 1950s, until today, the hazard analysis and critical control point (HACCP) system has gone through a series of phases and circumstances that have modified it until becoming a fundamental and irreplaceable tool to ensure food safety, as recognized by numerous international bodies and institutions.

The origin of HACCP is related to the failure mode and effects analysis (FMEA) system, which we can consider its predecessor and from which it borrows part of its methodology and the systematic approach to assessing risks in a given production, for example, using flow diagrams and probability risk matrices × gravity.

Based on the FMEA, the Pillsbury Company, NASA, and US Army Laboratories in Natick developed the original formulation of HACCP for its use in flights crewed from NASA's aerospace program. In 1971, the concept was presented publicly at the National Conference for Food Protection, co-sponsored by the Food and Drug Administration (FDA) and the American Public Health Association. At that time, the HACCP system included only three principles:

- hazard identification;
- determination of critical control points (CCPs); and
- monitoring of CCPs.

Its application in the food industry began to take shape in 1974 with the enactment by the FDA of the regulations that applied to canned food with low acidity. And already in the 1980s, its use gradually spread to other food industry sectors.

The major milestone in the application of the HACCP system occurred in the 1990s, specifically when the National Advisory Committee on Microbiological Criteria for Foods (NACMCF) defined the seven HACCP principles in 1992. One year later, the Codex Alimentarius Commission adopted these principles and published the Recommended International Code of Practice – General Principles of Food Hygiene, which included the annex Hazard Analysis and Critical Control Point – Guidelines for Their Application, describing the HACCP system. After several revisions, the HACCP has reached the current version: CXC 1-1969 (FAO and WHO 2022).

The importance of the publication of this General Principles of Food Hygiene is unquestionable, establishing the bases and principles of the system in a definitive way. Since then, we all have started speaking the same language and knowing what elements form the system, its methodology, and the concepts and terms that define it. This document is an essential reference for food safety auditors since it contains the requirements and criteria against which they can compare the HACCP systems designed by food operators.

With the Codex document was set up what we can call the "backbone" of HACCP: a basic structure that, since then, and with the experience acquired in the implementation of the system

HACCP System Auditing for Food Safety: Principles and Techniques, First Edition. Luis Couto Lorenzo.

in the food industries, has required some modifications and the intervention of other elements that are becoming necessary to overcome the difficulties involved in its practical implementation. Some of the most significant milestones in the evolution of the HACCP system since its origin could be the following:

- Publication of the Codex document.
- Universal acceptance.
- Need for prerequisite programs (PRPs).
- The emergence of the first HACCP standards with the elements of quality management systems included.
- Need for flexibility in its application to different types of operators.
- Use of the audit tool to verify its operation.
- Acceptance of HACCP as a food safety management system.
- Application of operational prerequisites.
- Need for validation of the system before its implementation.

What is not very evident in the annex Hazard Analysis and Critical Control Points – Guidelines for their Application, just mentioned in a sentence that occupies only one line, is the crucial role of the PRPs in the system's operation, as it was recognized years later. HACCP cannot work in a vacuum; it needs to be supported on the solid basis of properly implemented prerequisites (Sperber 1998).

Over time, the practical experience gained implementing HACCP has given rise to the strong international consensus on the primary role that codes of hygiene practice should play in controlling an essential part of the hazards identified in each case. Good hygiene practices or prerequisites are the foundations of the food safety system, must be implemented first, and are valid for all food operators, including primary production; HACCP is in the second place and may not be fully applicable in some cases (Figure 1.1).

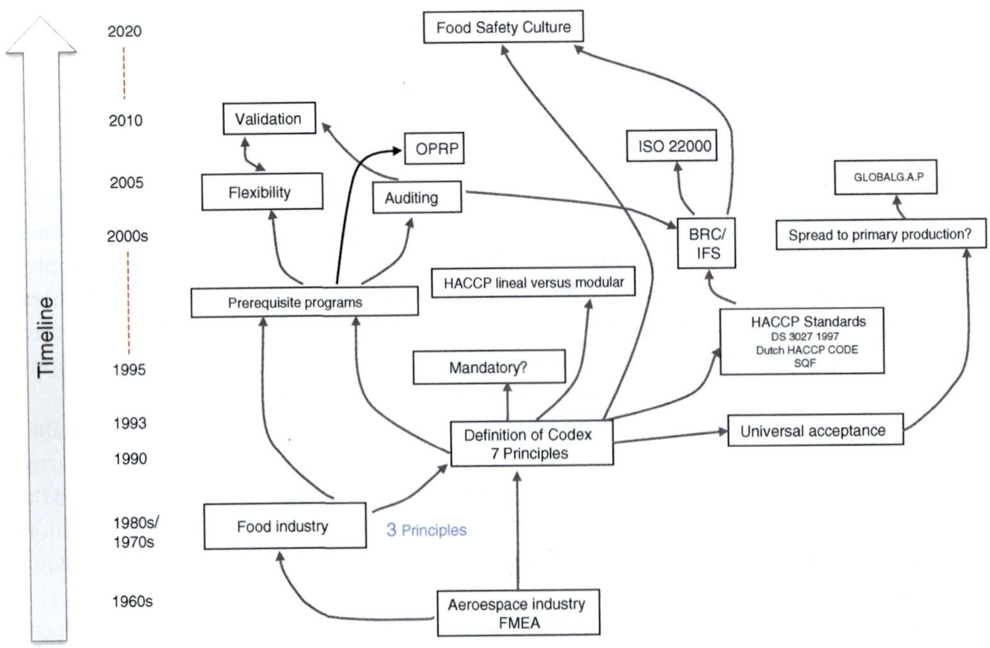

Figure 1.1 HACCP evolution. Source: Adapted from Griffith.

Another change in HACCP is the introduction of flexibility criteria. Since in the 1990s, the HACCP system became a mandatory requirement for operators in the different sectors of the food chain – except primary production – as is the case in the European Union with Directive 93/43, the difficulties that arose in some small and medium-sized companies to be able to implement the system as a whole effectively became evident.

At that time, it became necessary to set up provisions on the flexibility necessary to adapt the application of HACCP-based procedures in some establishments. Both documents CXC 1-1969, Regulation (EC) 852/2004 of the European Parliament and of the Council of 29 April 2004 on the hygiene of foodstuffs, and the guidance documents prepared by Directorate – General Health and Consumer Protection (DG SANCO) provide for the application of flexibility criteria for certain food operators. These provisions must be taken into account by food safety auditors when assessing self-control systems in these kinds of operators.

These changes and adaptations in the HACCP system have led some authors to lament the loss of its essence and original identity (Cliver 2010). There have indeed been changes, but we must also consider that HACCP was born for a very particular use and for conditions that we may describe as "in vitro" to achieve an objective of "zero defects" in the feeding of astronauts, necessary to ensure the success of the aerospace program. These conditions have little to do with the reality of the agri-food industries and with the application of the HACCP system in the different stages of the food chain, in a real food production environment, where it is not feasible to assume the zero-risk approach and in which food business operators must adjust their costs to be economically viable in a highly competitive market.

One issue that for years has been, and continues to be, the subject of debate is the possibility of extending the implementation of HACCP to all stages in the food chain, particularly primary production. Given the origin of the HACCP system, we know that its application was always more oriented to the central stages of the food chain, where there are companies that can process raw materials and carry out specific operations for the control or elimination of hazards; however, at the beginning and the end of the food chain, primary production and consumption, respectively, it is more challenging to implement.

Several HACCP experts (Sperber 2005; Cerf and Donnat 2011) recognize the difficulties in implementing the system in primary production due to the need for adequate infrastructure, technology, or even proper human resources. However, there is now a broad consensus on the need to promote good hygiene practices and good agricultural or veterinary practices as the best way to avoid or minimize the risk of hazards present at the primary production stage, which otherwise can reach the following steps in the supply chain.

In the history of HACCP, the appearance of the first standards or HACCP certification schemes was relatively early; Australia, the Netherlands, and Denmark were the most innovative and those who first saw the need for or usefulness of these standards in implementing HACCP. In 1994, the first food safety standard, the Safe Quality Food (SQF), appeared in Australia, and shortly after HACCP became mandatory in Europe, the second standard, the Dutch HACCP code "Requirements for a HACCP-Based Food Safety System" appeared in 1996, followed somewhat later by the Danish standard DS 3027:1997.

The change introduced by these standards is fundamental to understanding the subsequent HACCP evolution. Although what they do is add some aspects of quality systems, such as ISO 9000, to facilitate the management of the HACCP system, the important aspect is the qualitative leap that occurs when moving from the idea of the plan or HACCP study done on paper to a food safety management system, understood as a set of procedures to be applied by the staff of a specific organization in their daily activity – something that has ended up being entirely accepted and assumed by the bodies and institutions involved in HACCP management.

The assignment of roles and responsibilities both at management levels and between operators and supervisors, document control, implementation of documented procedures, notification and withdrawal of nonconforming products, and calibration of equipment, among others, are elements of quality systems that, over time, have demonstrated their compatibility with the HACCP system and synergy with its functioning and efficiency within the framework of a food safety management system.

In the first versions of the Danish and Dutch standards, there was little mention of PRPs, which has been corrected in their subsequent editions in conformity with the broad consensus achieved later regarding the role of prerequisites within the self-control system. Currently, safety standards or certification schemes typically include three components: the HACCP plan, PRPs, and elements of quality management; however, they may differ in the relative importance given to each of these components within the structure of the standard.

With the experience acquired in its implementation by food operators, HACCP goes from being understood as a mere document – a theoretical exercise on paper carried out by a team of people more or less competent, which ended with the delivery of the HACCP study – to become conceived as a set of provisions and procedures to implement in the different levels of business management, i.e. a food safety management system. Having a HACCP system is not having the HACCP study; the HACCP study alone does nothing for food safety. Food operators need to implement HACCP in their daily activity to see its results. As we will see later, one of the objectives of the audits is precisely to check the degree to which they comply with the provisions established in the documented system.

Despite the time that has passed, there are still people with difficulties in understanding what HACCP consists in, and this will continue as long as they approach this system without the appropriate training and guidance. HACCP is not a magic wand that alone solves the problems in a system, nor a fairy tale; it is a set of ideal arrangements, impeccable on paper but unfeasible in practice because of their cost or difficulty in implementation in a real food production environment. The basis for its effectiveness undoubtedly resides in a good study, adapted to the circumstances of each production in a specific organization. However, it will obtain the expected results only with constant work in its implementation and efforts to keep it operational and updated.

Other types of reactions to HACCP may come from skeptics, who think that HACCP contributes nothing since it is unnecessary and only a bureaucratic burden imposed from the official control – in countries where HACCP is mandatory – or a requirement to conclude a supply contract. Indeed, among those who nurture this group, there are many operators who, by the type of product or by the less exposed position they occupy in the food chain, are not aware of having lived or generated any food incident and also those who think that end-product testing is enough to ensure the safety of their food products, therefore they do not have to be worried. That is a misconception that can turn out very expensive, but it is possible to avoid this with appropriate training and awareness or, unfortunately, after experiencing an incident in their production. That is when many food operators begin to understand the false sense of safety caused by final product analyses and only then realize that the right way to ensure food safety is with the proper implementation of PRPs of hygiene – and HACCP when needed – as the most effective tools to achieve the required level of food hygiene.

For years, I have seen many times how managers or food safety technicians finally realized the real importance of the implementation of prerequisites only after overcoming a severe setback or how they started to recognize, for example, the irreplaceable value of a traceability program not only in achieving the food safety required but for the consistent performance of the business. In any case, to the most reluctant to adopt the HACCP system, we should ask: Do you know another better alternative?

Once HACCP was understood as a safety management system, it was easier to accept that the audit was the most appropriate way to check the functioning of the HACCP system in practice; a tool that, at that time, had already been tested enough in quality and environmental management systems, which facilitated the transfer of all that experience to its application in food safety. Therefore, we can assume that the guidance provided by ISO 19011, "Guidelines for auditing management systems," perfectly applies to the HACCP system.

Compared with food safety standards audits, the HACCP system audit entails an added degree of difficulty as it does not specify the requirements as explicitly or as detailed as they are in those certification schemes. This difficulty requires the HACCP auditor to have a solid knowledge of the system's principles and methodology and sufficient experience to determine whether the HACCP team applied the principles correctly and efficiently.

The International Organization for Standardization (ISO) published its standard for food safety management systems, ISO 22000, in September 2005, a standard that was born with the vocation of becoming the reference standard for all operators in the food chain at a time when there were already too many food standards in place. Although this objective has not been achieved since it remained with other certification schemes with great global implementation, such as the International Featured Standard (IFS) and Brand Reputation Compliance Global Standards (BRCGS), ISO 22000 became a reference model of the structure and elements a food safety management system should contain. ISO 22000 is based on the structure of ISO 9000 and integrates HACCP principles and Codex implementation, setting up auditable requirements for the HACCP plan in combination with PRPs and other aspects of management systems.

One of the concepts introduced by ISO 22000:2005 was that of operational PRPs, a novelty that caused some debate in the field of food hygiene on whether it was really necessary to add new concepts and whether it brought some advantages against the elements already defined. We will discuss these aspects later, but we may anticipate that their rationale can be justified by the need to control some hazards identified in the production process that does not fit with the characteristics required for CCPs. However, they require enhanced monitoring than prerequisites related to the control of the production environment, so they are a better choice for improving control in specific food processes.

In recent years, it has been noted that operators have difficulty understanding the concept of validation of the system. Although this concept has been clarified and consolidated, thanks to the work of some authors and the publication of guides explaining the differences between validation and verification (Ilsi 1999; FAO and WHO 2008), in practice, there continues to be some confusion and lack of experience among HACCP practitioners on how to validate control measures and the HACCP system as a whole.

This may be because, during this time, the validation has remained somewhat hidden among the verification activities within the HACCP Principle 6, as evidenced by the fact that, in practice, many manuals barely include validation activities documented. Another reason that may explain this lack of understanding is the need for sound scientific knowledge by HACCP team members to conduct HACCP validation properly. Some authors (Wallace et al. 2011) suggested the convenience of considering validation as an independent HACCP principle to make it more visible to HACCP professionals.

Thus, validation and maintenance of the system, also claimed as an independent principle for similar reasons, would be the two new principles of the HACCP system to add to the seven established since the 1990s, according to some authors. Without going into assessing the benefits

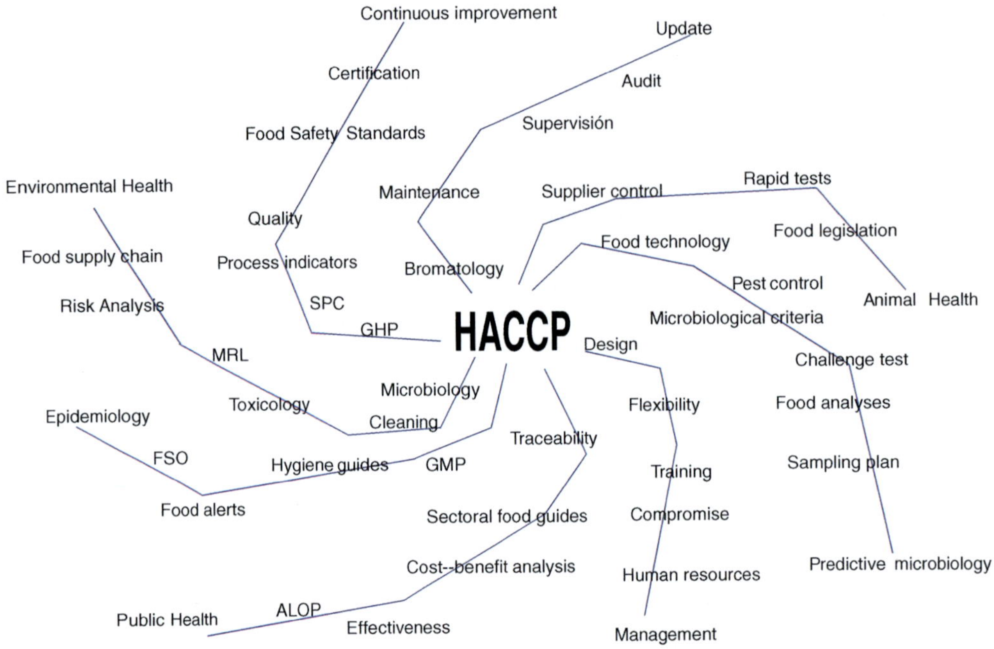

Figure 1.2 HACCPverse. HACCP Universe.

arising from this proposal, the truth is that these requirements are already part of the system, included in the "macro principle" 6, and in that sense, they must be applied with the same rigor as the rest, even if the operators are not aware of this in many cases.

Sometimes, a greater specificity in the system description may be missing, for example, in Codex documents, so that HACCP principles can be interpreted and applied in the same way by all interested parties: operators, technicians, consultants, official control agents, and auditors. Thus, some authors (Wallace et al. 2011) state the lack of guidance or guidelines available to HACCP team members on how to carry out the practical application of HACCP principles, such as conducting the hazard analysis, making informed decisions, and determining which hazards must be considered significant.

The fact that we refer to principles rather than more detailed requirements of food safety, at first glance, can be considered a deficiency or a limitation; however, it may be the key to maintaining the HACCP structure stable and recognizable by all for the past 30 years – a period characterized precisely by significant transformations in the food industry. What is certain is that this basic structure – the backbone of HACCP, described so concisely in the Codex document – has not stopped adding concepts, disciplines, techniques, and applications to constitute a whole set of subjects related to the implementation and management of self-control systems: this is the universe HACCP, or – imitating the concept "Twitterverse" coined for the former name of the well-known social network X – the HACCPverse (Figure 1.2).

References

Cerf, O. and Donnat, E. (2011). *Application of hazard analysis – critical control point (HACCP) principles to primary production: what is feasible and desirable? Food Control* 22: 1839–1843.

Cliver, D.O. (2010). *HACCP is dead. Food Technology* 64 (3): 104.

Danish Standards (2002). DS 3027:2002: Management of Food Safety based on HACCP (Hazard Analysis and Critical Control Points) – Requirements for a Management System for Food Producing Organizations and Their Suppliers. Danish Standards. Copenhaguen.

Europe, European Parliament and Council (2004). Regulation (EC) No 852/2004 of the European Parliament and of the Council of 29 April 2004 on the hygiene of foodstuffs. *Official Journal of the European Union* 139 (1): 1–54.

FAO and WHO (2022). *General Principles of Food Hygiene. Codex Alimentarius Code of Practice, No. CXC 1-1969*. Rome: Codex Alimentarius Commission.

FAO and WHO (2008). *Guidelines for the Validation of Food Safety Control Measures. Codex Alimentarius Guideline, No. CXG 69-2008*. Rome: Codex Alimentarius Commission.

International Life Science Institute (ILSI) Europe (1999). Validation and Verification of HACCP. http://www.ilsi.eu/publication/

National Advisory Committee on Microbiological Criteria for Food (NACMCF) (1997). *Hazards analysis and critical control point principles and application guidelines.* https://www.fda.gov/food/hazard-analysis-critical-control-point-haccp/haccp-principles-application-guidelines/ (accessed 25 November).

Sperber, W.H. (1998). *Auditing and verification of food safety and HACCP. Food Control* 9 (2–3): 157–162.

Sperber, W.H. (2005). *HACCP does not work from farm to table. Food Control* 16: 511–514.

Wallace, C.A., Sperber, W.H., and Mortimore, S.E. (2011). *Food Safety for the 21st Century*. Oxford, UK: Wiley-Blackwells.

2

The HACCP System

In the hazard analysis and critical control point (HACCP) audit, we may distinguish two parts: the object of the audit, i.e. "what" is audited; and the audit system, i.e. "how" we do this task. The versatility of this management tool allows us to check management systems based on different disciplines: quality, environment, safety, hygiene at work, etc. In the field of food safety, it is the HACCP system that determines the object and scope of audit methods and techniques.

Other questions that may arise are "what for," regarding the purpose and type of audit, and "against what," i.e. what are the criteria, requirements, or reference standards to use as elements "against" which compliance with the system is compared (Figure 2.1).

In the audit process of a HACCP system, we may distinguish two phases:

- Assess whether the HACCP plan is under the HACCP methodology and principles.
- Check whether the provisions of the HACCP plan are implemented as described; i.e. verify whether "they do what they say they do."

Most of the activities carried out by the audit team focus on verifying these two objectives. But we still have to add another, not the least important: to check the system's effectiveness and assess if the expected results are being achieved. Because if, after all the effort made for the design and implementation of the system, the results obtained are far from those expected, then there is something wrong with the whole system, which involves analyzing the cause of this ineffectiveness and adopting the appropriate solution (Figure 2.2).

To determine the system's effectiveness, we must use objective data related to production and business development or indicators such as the number of noncompliant products, incidents, reprocesses, customer returns, customer complaints. In addition, this is how the food operator can perceive the real added value of food safety management systems (FSMSs). In the first stage, the task of the auditor is to review the documentation of the HACCP system and, based on the information available of the company, such as food sector, size, organization, production lines, evaluate if the food operator correctly applied the HACCP principles and if they completed the HACCP study for each of the activities included. To carry out this assessment, the primary reference available to the auditor is the HACCP methodology and principles, as described in the document Guidelines for the Application of the Hazard Analysis and Critical Control Point System, approved in 1993 by the Food and Agriculture Organization (FAO) Codex Alimentarius Commission, or in the NACMCF HACCP Implementation Guide, adopted in 1997.

The importance of this Codex document for the application of the HACCP system is crucial, as it defines what the HACCP system is and the essential elements that make it up. In this way, from the very moment of its publication, it was possible to have a standard reference for all countries,

HACCP System Auditing for Food Safety: Principles and Techniques, First Edition. Luis Couto Lorenzo.

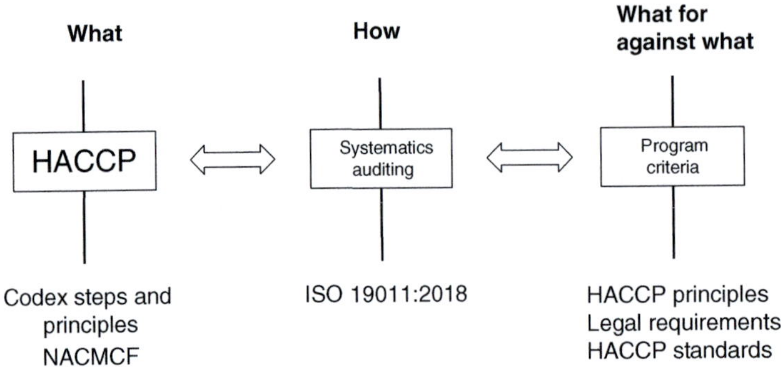

Figure 2.1 In audits we can distinguish the "what", the "how" and the "what for".

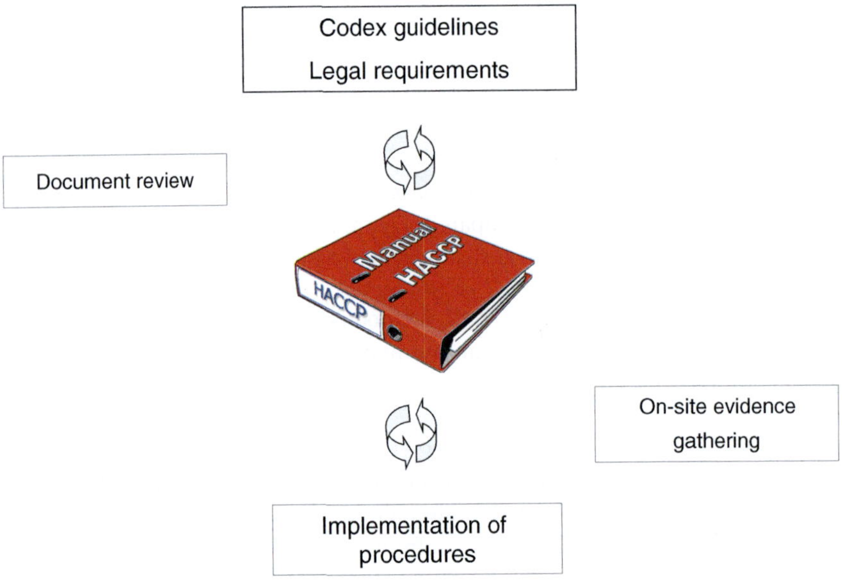

Figure 2.2 Two-step audit approach.

achieving that all the parties involved both in the implementation of HACCP and in its verification used the same concepts, spoke the same language, and referred to the same system unequivocally.

If we did not have this reference document, the HACCP system would have as many different contents as each user and take multiple forms, and the task of the auditors would be very complicated because they did not have the elements or criteria with which to compare the HACCP plans of food operators and thus determine their compliance.

Auditing against the principles of the HACCP system rather than against the requirements defined in the clauses and points of a standard or certification scheme can pose an added difficulty for food safety auditors. This lack of specificity of HACCP principles requires the auditor to have a deep knowledge of the methodology and principles of the system and to know how general principles apply to the particular situation of each food operator. That lack of specificity also requires auditors to have a margin of flexibility to assess the adaptation of HACCP requirements to the characteristics of each food operator, such as size and production orientation, provided that the result must be a consistent and effective system to ensure the safety of the food produced.

Because of the above, it is worth asking if, to facilitate the HACCP evaluation, it would be desirable for the HACCP system to have more development and specificity in its requirements, as is the case in HACCP standards. It may be an issue to debate; however, we should note that, presumably, the fact that the system is defined only from the steps and principles is what has given it the robustness and stability necessary to remain during all these years as the international reference and the basis of FSMSs for all operators in the food chain.

2.1 Characteristics of the HACCP System

Schematically, we may state that the HACCP system has the following characteristics:

- structured;
- logical;
- preventive;
- based on science;
- risk management;
- practical;
- common sense;
- cost-effective; and
- realist.

Like the failure mode and effects analysis (FMEA), its methodological antecedent, the HACCP study is structured through the sequential application of a series of steps. The HACCP team, in a reasoned and logical way, based on the process data and the available scientific information, must identify the hazards that may arise from each stage of a given production. The next step is to establish the likelihood that the hazards identified as potential will cause harm to consumers. Having determined the risk for each significant hazard, the HACCP team must establish what preventive measures can avoid them or reduce their presence to an acceptable level.

One concept that should be clear from the outset is that HACCP does not aim to control and manage all potential hazards as if achieving a zero-risk level were possible. This approach is neither realistic nor economically feasible; such control measures would entail an unbearable cost for food operators and make their productive activity practically impossible.

The generalization of the implementation of the HACCP system among food operators meant a paradigm shift in food safety, which we can describe as follows:

- The responsibility for providing safe food to consumers lies with food companies through their self-monitoring systems.
- Official control now has the role of supervision and verification of the self-control systems of food operators.
- Food safety assurance is no longer based on inspection and end-product testing but on the application of preventive measures.
- Food legislation no longer specifies most of the particular requirements, but food operators decide based on HACCP principles and procedures.
- Both official control and food business operators (FBOs) adopt audit techniques to check the functioning and effectiveness of self-control systems.

2.2 The Limitations of Final Product Analyses

The traditional food control system was based on inspection and analysis of the final product; despite the widespread implementation of HACCP, there is still some predisposition, both in food operators and in official control agents and other stakeholders, to overestimate the usefulness of food safety inspection and analysis for food safety control.

Therefore, we must remember that the final product analysis has certain limitations that we should not overlook:

- **They are corrective, not preventive**: they detect problems after they have occurred; this means having to act on the noncompliant product: reprocess or dispose of it while it is still in the possession of the manufacturer; or, in some cases, recover it from the market when it has already been distributed, with the higher costs that this entails, as well as the loss of image and consumer confidence.
- **Slow**: Some tests may take several days to obtain results; this means a longer time until the problem is detected and implies keeping the merchandise retained or, worse, the risk of having to withdraw it from the market when it has already left the food operator's premises.
- **Expensive**: although it depends on the number of samples analyzed, in general, laboratory analysis has a high cost for the operator, to which must be added the units of product that are lost when the analyses are destructive.
- The probability of detecting defective products is often low. Undoubtedly, the most crucial limitation from the point of view of food safety, since, despite the false sense of security that sampling plans can provide to the operator, the calculation of probabilities explains how often it can be to consider acceptable batches of product with a certain number of defective units that end up reaching the consumer.

Setting the goal of inspecting 100% of production units is not always realistic or feasible under the conditions in which people typically work in the food industry. Even in cases where the operators can have a system that allows them to inspect all the units, human errors will eventually occur, mainly if we refer to monotonous tasks, in which it is normal for the attention to decline over time or by the distractions that occur during the working day.

In addition, not all defects are equally obvious or easy to detect on visual inspection, and then there are all those that are not noticeable to the naked eye. Therefore, inspecting the final product is not a method to ensure the absence of defects and food safety.

The same goes for final product analyses; it is not possible to analyze 100% of the production, first because, in the case of destructive analyses, we would leave nothing to sell and second because that would mean an unaffordable cost for FBOs. The only solution is to take a sample of the set of units produced and infer the state of the whole production from the result of that sample. The problem, which many people need to realize, is that sampling is always associated with the probability of detection of defective or contaminated units. The number, size, and nature of the collected samples greatly influence the results' representativeness. However, even in the most comprehensive sampling plans, increasing the number and size of the samples, it must always be borne in mind that there is a probability of detecting defective units and that there are going to be cases where the result of the analysis of the batch sample is acceptable when the batch is not acceptable.

To illustrate this concept of probability associated with sampling plans, we can start with a simple case. Imagine that we take a single sample ($n = 1$) to determine the acceptability of a batch in which the presence of a contaminated unit would lead to the rejection of the batch ($c = 0$). In this case, the probability of detecting the contaminated units coincides precisely with the percentage of contaminated units in that batch. For example, if 50% of the units are contaminated, with a single sample half the time, we can detect these contaminated units. The bad news is that batches with 50% of their units contaminated would be distributed to the market an equal number of times. If the percentage of contaminated units drops to 10%, then only one in ten times would we be able to detect contaminated batches, which means that 90% of batches with 10% of their units contaminated would come to market. As we can see, this approach is unacceptable from the point of view of food safety. Suppose we want to improve the detection capacity. In that case, we can increase the number of samples, so if we take two samples ($n = 2$), the probability of accepting a batch with 10% of the contaminated units would be $0.9 \times 0.9 = 0.81$ (the probability of two independent facts is the product of their probabilities); as we can see, this probability decreases. However, we still have only a 19% chance of detecting the defective batch.

To understand better the effect of increasing the number of samples, in the following graph, we can compare the probabilities of acceptance between two sampling plans $n = 1$ and $n = 5$ for different percentages of defective units (Figure 2.3).

When we look at the column of 5% defective units, with the one-sample plan, the probability of accepting the lot is 0.95, and with the five-sample plan, the probability drops to 0.77, although it is still too high.

In addition to the number of samples, the sample's representativeness depends on the nature of the food and the homogeneity of the distribution of contamination among the units of the lot, more homogeneous in the case of liquids.

In the following table, we can see the joint effect of the number of samples, the concentration of contamination, and the homogeneous or heterogeneous distribution of contamination among all the lot units on the probabilities of detection of the contaminated units (Figure 2.4).

Sampling plans are used to decide whether a batch of product is acceptable; for this, it is necessary to have a criterion to compare the results of the units of samples analyzed. A sampling plan, therefore, consists of a sampling procedure and a decision criterion.

Each sampling plan is defined by n and c; n is the number of sample units collected independently and separately, and c is the maximum number of sample units to obtain positive results to accept the batch. For example, in a plan $n = 10$ and $c = 2$, we take ten sample units, and we can accept the lot when two of ten, or less, test positive. In other cases, the sampling plan includes the term defined by the letter m (in a two-class attribute plan), which refers to the maximum concentration of a microorganism for the result of the unit analyzed to be acceptable. Thus, the plan $n = 10$, $c = 2$, and limit m means that the lot is acceptable even if two samples of the ten collected exceed the concentration m. The attribute plans of three classes still define one more term: M, a concentration greater than m, which cannot be exceeded in any case.

% Nonconforming units	**0**	**5**	**10**	**20**	**30**	**50**
P(a) (1 sample)	1	0.95	0.90	0.80	0.70	0.50
P(a) (5 samples)	1	0.77	0.59	0.33	0.17	0.03

Figure 2.3 Probability of accepting a lot P(a) as a function of the actual number of defective units for two sampling plans $n = 1$ and $n = 5$ (ICMSF 2011).

	Concentration of contamination	Number of random samples	Probability of detection (1)
Homogeneously contaminated	5 cells/kg	10	71%
	1 cells/kg	10	22%
Heterogeneously contaminated	5 cells/kg in 1% of the batch	10	<2%
	10^4 cells/kg in 1% of the batch	10	<15%

Figure 2.4 Probability of detection in final product analysis of Salmonella contaminated milk powder (1) assumes 100% detection capability of the test (most are 90%).

Predictably, the higher the term n and the smaller the c, i.e. $c = 0$, the plan's ability to discriminate defective batches increases. However, unless the entire batch is analyzed, which we have already seen as an unfeasible approach, there is no way to avoid some degree of risk in each acceptance and rejection of a lot.

Therefore, in practice, we are looking for a compromise or balance solution between the number of samples to be analyzed and the risk we can assume. The most severe plans are used for the most serious risks to the population.

In all sampling plans, it is possible to determine what is known as the consumer risk, i.e. the probability of accepting a batch whose microbial load exceeds the limit specified or with a proportion of defective units higher than established. Reciprocally, there is also the producer's risk, i.e. the probability of rejecting an acceptable lot. No matter how much we try to increase sampling plans' rigor or discrimination capacity, these two types of risk will always be present.

The following table shows how different sampling plans influence the probability of accepting a batch with a known percentage of defective units. This table again clearly shows how final product analysis is not an effective way to ensure the safety of food produced when the percentages of defective units are low. It would be necessary to analyze more than 3000 or 5000 sample units to detect batches with 1% of their units defective with 95% or 99% probability. It should also be borne in mind that a defect rate of 1% is intolerable for most food operators; if we consider the average batch sizes produced in the food industry, this would mean sending hundreds of defective products to the market every day. No company could work in such conditions, and it is not acceptable from the point of view of food safety (Figure 2.5).

After looking at these examples, it is easy to conclude that food safety cannot be based solely on the analysis of the final product; food operators must focus their resources on establishing control systems that prevent or minimize the occurrence of problems in a given production or activity. This is, therefore, one of the most straightforward reasons for justifying the need for HACCP.

What we say does not mean that final product analyses are unnecessary, but that we must recognize their limitations, determine the most appropriate uses, and see at which points in the food chain the information they provide is most helpful. For example, in the case of imported products, where there is no information on the conditions of the preparation of the consignment received, the only way to decide on its destination is by analyzing an appropriate number of samples, depending on the characteristics of the lot and the associated risks.

% of defective units in the batch	Probability of acceptance (%) for each plan defined by *n* and *c*			
	$n = 1, c = 0$	$n = 5, c = 0$	$n = 10, c = 0$	$n = 60, c = 0$
1	99.0	95.1	90.4	54.7
2	98.0	90.4	81.7	30.0
5	95.0	77.4	59.9	4.6
10	90.0	59.1	34.9	0.18
20	80.0	32.8	10.7	0.00015

Figure 2.5 Effect of the quality of the lot (% of nonconforming units) on the probability of acceptance for different sampling plans.

2.3 Reasons to Implement HACCP

Many executives and managers of the food industry are unaware of the usefulness of the HACCP system to guarantee the safety of the food they market, nor of the benefits that its implementation can bring to the proper functioning of their business as a food company. The lack of knowledge of the system, the lack of experience, the biased perception of food risks, and preconceived ideas about what they can get from HACCP or how to manage it usually mean that the reasons why a company should implement the HACCP system are often not well understood. Among the reasons that we can expose to justify the implementation of HACCP are the following:

- **Problems can be avoided before they occur**: the HACCP preventive approach focuses on acting on the root causes of food hazards and production problems rather than on the control of processed foods.
- **We can act in time**: continuous or frequent surveillance measures on the processes allow immediate action, or after short periods, to correct the deviations detected in production and thus limit the amount of food processed under inadequate conditions. On the other hand, if we rely on the final product analysis, we always have to wait a while, sometimes several days, until the results are known and then be able to act accordingly.
- **Cost savings**: when the system is operating correctly, the reduction of expenses is achieved thanks to the fewer failures and nonconforming food that occur, the fewer reprocesses, the decrease in the amount of food that must be destroyed because it is not adequate, and the reduction in the number of product returns and claims, as well as recalls; with the added costs that this entails, since to the loss of raw material and/or product already prepared, we must add the expenses generated in the collection, transport, management, and disposal, where appropriate, of the food distributed.
- **Focus resources**: hazard analysis and risk assessment, attained with the HACCP methodology, allow the food operator to set priorities and focus available resources on controlling aspects that are truly important for food safety.

- **Protection of public health**: compared with the limitations of other strategies, such as inspection and analysis of the final product, HACCP allows to systematically and scientifically address safe food production.
- **Due diligence**: in case of litigation and as a recognized manner in some countries, HACCP can assist in the defense of the food operator by serving as a means of proof that it had taken all the necessary measures and all the precautions available to it to prevent the incident for which it is being tried. By extension, this concept applies to other situations that occur more frequently than the cases that become the subject of litigation; for example, in the event of customer complaints or official control actions motivated by complaints, alerts, or the investigation of a food outbreak, HACCP can be used to demonstrate that the food operator has fulfilled its responsibility to provide safe food. On the contrary, if what is evident is a poor implementation of HACCP or the lack of documentation related to the food or the batch in question, the position of the food operator worsens if the health administration initiates a disciplinary procedure.
- **Trademark protection**: hardly a commercial brand can emerge unscathed from any severe food alert or incident in which it is involved. The deterioration of the image, the loss of customer confidence, and the consequent reduction in sales often make the continuity of this brand in the market unsustainable. Due to their position in the food chain, not all FBOs are equally exposed to risks that can affect their trademark; for example, suppliers of intermediate products that do not reach consumers are less vulnerable to this effect. In any case, and for any food operator, implementing a preventive management system such as HACCP is the most effective way to prevent these setbacks and/or minimize their consequences.
- **Development of organizational culture**: the effort involved in implementing the HACCP system in an organization requires the commitment and responsibility of all parties involved in its management. When carried out correctly, the result is the development of a new organizational culture, more oriented to prevention, process control, and continuous improvement, leading to greater compliance with customer requirements and increased business expectations.
- **Comply with legal or customer requirements**: ideally, food operators should convince themselves of the benefits of HACCP without the need for it to be mandatory; in any case, in many countries, the implementation of the HACCP system is mandatory, and therefore, food operators are responsible for satisfying this legal requirement. Other times, customers require their suppliers to have a FSMS based on the HACCP system as a precondition for signing a supply contract or accepting them as approved suppliers.
- **Facilitates inspection by official control**: the documentation of the HACCP system provides much information on the activities, processes, and the most significant points for inspection, among other aspects, which helps the official control agent better prepare his performance.
- **Promotes international food trade**: being recognized as an international reference element, the HACCP system facilitates the assessment of the equivalence between the control and inspection systems of different countries; this increases confidence among operators and makes the exchange of food between economic agents more fluid.

Finally, it should be noted that, despite all these potential benefits, we can still find many food industry managers who are pretty skeptical about the usefulness of the HACCP system for their business. Given the problems of continuing with the classic approach, based on end-product testing and strictly regulated food production, these people should be asked if they know a better alternative, because so far, despite the effort it requires, the difficulties it poses, and the limitations in applying it to all food sectors, HACCP remains the most consistent system available to FBOs to ensure the safety of the food they produce.

2.4 Legal Basis of HACCP

At the EU level, the establishment of HACCP as a mandatory requirement for food businesses began with the publication of Council Directive 93/43/EEC on the hygiene of foodstuffs on June 14, 1993, which entered into force in December 1995.

This directive, taking into account the codes of international hygiene practice and the general principles of food hygiene of the Codex Alimentarius – as described in its second edition in 1985 – established in its article 3.2: FBOs shall identify any step in their activities which is critical to ensuring food safety and ensure that adequate safety procedures are identified, implemented, maintained, and reviewed based on the following principles, used to develop the system of HACCP (Figure 2.6).

In 2004, Regulation 852/2004 on food hygiene replaced Directive 93/43/EEC and consolidated the HACCP obligation for all FBOs after primary production.

Codex document CAC/RCP 1-1969 General Principles of Hygiene, the first version of which dates from 1969, provides the basis of Regulation 852/2004; in 1993, the annex containing the principles of HACCP and guidelines for their application was incorporated. After being adopted in 1969, the CXC 1-1969 (its code number now) was amended in 1999 and revised in 1997, 2003, 2020, and 2022.

The Codex General Principles of Hygiene document has the following content:

1. Introduction
2. Objectives
3. Scope
4. Use
5. General principles
6. Definitions

Part 1
Good Hygiene Practices

7. Introduction and control of food hazards
8. Primary production
9. Establishment – design of facilities and equipment
10. Training and competence

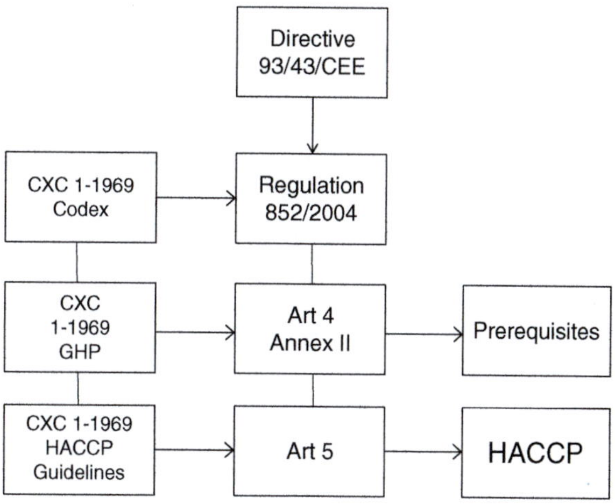

Figure 2.6 Legal basis of HACCP and prerequisites in Europe.

11. Establishment maintenance, cleaning, and disinfection
12. Personal hygiene
13. Control of operation
14. Product information and consumer awareness
15. Transportation

Part 2
Hazard Analysis and Critical Control Point (HACCP) System and Guidelines for its Application

16. Introduction to HACCP
17. Principles of the HACCP system
18. General guidelines for the application of the HACCP system
19. Application

Annex I: HACCP measures, logic sequence, and example
Annex II: Figure 1 – Logic sequence for application of HACCP
Annex III: Table 1 – Example of hazard analysis worksheet
Annex IV: Tools to determine the critical control points (CCPs)

As stated in the objectives set out in point 2 of this standard, the general principles of food hygiene: good hygiene practices (GHPs) and the HACCP system, aim to:

- provide principles and guidance on the application of GHPs applicable throughout the food chain to provide food that is safe and suitable for consumption;
- provide guidance on the application of HACCP principles;
- clarify the relationship between GHPs and HACCP; and
- provide the basis on which sector- and product-specific codes of practice can be established.

The codes of hygiene practice contained in this Codex document constitute the reference elements for developing Annex II to Regulation (EC) 852/2004. Article 4, "General and specific hygiene requirements," of this regulation states that FBOs operating in any of the stages after primary production shall comply with the general hygiene requirements laid down in Annex II and any specific requirements provided for in Regulation (EC) 853/2004.

Article 5 of Regulation (EC) 852/2004 "Hazard Analysis System and Critical Control Points" adopts the HACCP principles described in the annex to the Codex document as follows:

Article 5

Hazard Analysis and Critical Control Points

(1) Food business operators shall put in place, implement and maintain a permanent procedure or procedures based on the HACCP principles.
(2) The HACCP principles referred to in paragraph 1 consist of the following:
    (a) identifying any hazards that must be prevented, eliminated or reduced to acceptable levels;
    (b) identifying the critical control points at the step or steps at which control is essential to prevent or eliminate a hazard or to reduce it to acceptable levels;
    (c) establishing critical limits at critical control points which separate acceptability from unacceptability for the prevention, elimination or reduction of identified hazards;
    (d) establishing and implementing effective monitoring procedures at critical control points;
    (e) establishing corrective actions when monitoring indicates that a critical control point is not under control;

(f) establishing procedures, which shall be carried out regularly, to verify that the measures outlined in subparagraphs (a) to (e) are working effectively;
and

(g) establishing documents and records commensurate with the nature and size of the food business to demonstrate the effective application of the measures outlined in subparagraphs (a) to (f).

When any modification is made in the product, process, or any step, food business operators shall review the procedure and make the necessary changes to it.

(3) Paragraph 1 shall apply only to food business operators carrying out any stage of production, processing and distribution of food after primary production and those associated operations listed in Annex I.

(4) Food business operators shall:
(a) provide the competent authority with evidence of their compliance with paragraph 1 in the manner that the competent authority requires, taking account of the nature and size of the food business;
(b) ensure that any documents describing the procedures developed in accordance with this Article are up-to-date at all times;
(c) retain any other documents and records for an appropriate period.

These two articles of Regulation (EC) 852/2004 are essential to apprehend the legal basis of HACCP and its implementation by food business operators.

2.5 HACCP Pillars

Once the HACCP system became a legal and customer requirement, as its application began to spread in different countries, it was possible to verify that the level of implementation achieved by many food operators was not what was expected, especially in the case of small and medium-sized enterprises and for some food sectors. Something that draws the attention of professionals related to HACCP is to see how a system that on paper seems so logical, reasonable, and straightforward then, in practice, causes a lot of problems or difficulties that end up hindering its regular operation, according to the planned way. In the 1990s, several researchers (Griffith, Wallace, Panisello, Mortlock, Gilling, Taylor, Yiannas, etc.) began to analyze the reasons why this happened, being able to identify the presence of various technical barriers that complicate the implementation of HACCP by food operators. In many cases, these barriers relate to the knowledge, perceptions, attitudes, and behaviors of the people who have to implement the HACCP system. Although HACCP is initially conceived as a systematic, scientifically based study on a given food production line, in the end, it is the people who are responsible for putting it into practice; to give life to life in the day-to-day activity of that organization in order to achieve the established food safety objectives. The effectiveness of the HACCP system ultimately depends on the human factor; we can design an ideal HACCP with an impeccable design,

Figure 2.7 Essential elements for HACCP.

but if the people who have to apply it do not feel motivated or involved in its implementation, the results will hardly be as expected (Figure 2.7).

In their article "Technical Barriers to Hazard Analysis Critical Control Point (HACCP)," which is an obligatory reference when addressing this issue, Panisello and Quantick (2001) identified the four pillars on which the HACCP system must be based for its implementation to be effective:

- Commitment.
- Education and training.
- Availability of resources.
- External pressures.

Commitment: The need for a clear commitment from all levels of the organization involved in HACCP, from management to plant workers, is already recognized in the 1997 Codex document on HACCP and the NACMCF document (1998). Specifically, the Codex HACCP guidelines state, "For all types of food businesses, management awareness and commitment to food safety are necessary for implementation of an effective HACCP system." If the organization's commitment is not sure, allocating technical resources or those necessary for personnel training will not be adequate.

Education and training: In a people-based system, training is essential to acquire the knowledge and skills necessary for performing the tasks of each job and achieve the awareness and motivation required for the correct implementation of FSMSs.

Availability of resources: The HACCP system needs to have a series of resources so that its application can be guaranteed: facilities, equipment, training, analysis, etc. Those responsible for managing HACCP must identify material, technical, and personnel needs. They must provide the necessary resources to cover the corresponding costs as long as they are committed to the system.

External pressures: FBOs are pressured by governments, customers, and consumers to implement HACCP. In many countries, HACCP is a legal requirement, so in these cases, there is no choice; the operator must adopt the HACCP system on a mandatory basis, and the health administration, through the official control agents, will verify that the system is implemented and that it works properly. The food operator's customers exert another external pressure; many food companies require their suppliers to have a HACCP system or other standards or certification schemes to accept them as suppliers. In this case, economic reasons weigh heavily; if a FBO wants to sell to a customer or a specific market, he must meet the required conditions. Although somewhat different, there are also pressures from consumers; the food company is always exposed to the consumer's judgment, who in every purchase assesses whether the product satisfies him or not and who at any time can stop buying it or look for other options. In addition, the consumer can also submit complaints or claims that must be duly addressed. Finally, the media also exert pressure on the food industries, food news always arouses great interest in society, and any issue or incident in which food operators are involved can be quickly disseminated to the public and significantly affect their image.

For Panisello and Quantick, there is a priority or an order of importance in the action of these four pillars of HACCP; the first pillar – the one at the base of the pyramid – is the organization's commitment to the implementation of the system; the second is the education and qualification of all those responsible for the development and management of HACCP in the organization; in the next place, they put the availability of the material and human resources necessary for its practical application, and in last position, the external pressures exerted on the food operator, a kind of continuing source of tension that pushes the company, together with the other three pillars, for the maintenance and continuous improvement of the FSMS.

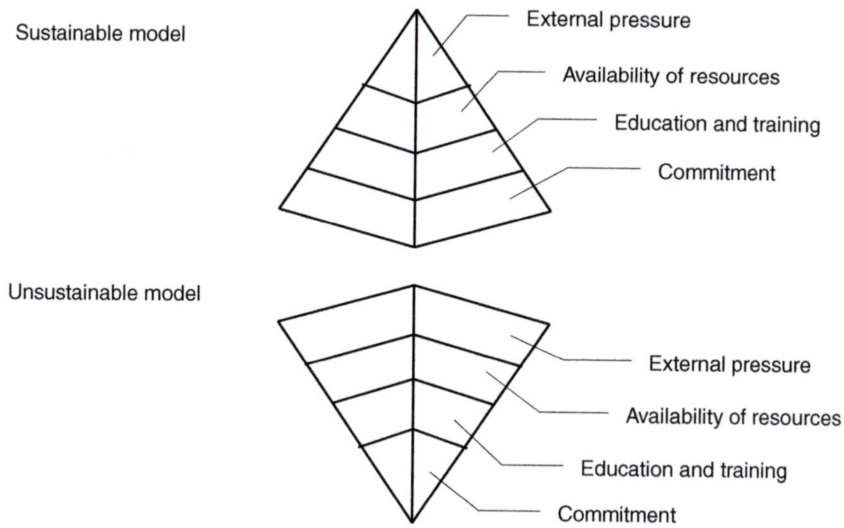

Figure 2.8 Pillars of HACCP.

In the scheme of Panisello and Quantick, the importance of each factor is represented by its size in the pyramid; thus, a sustainable HACCP management model is one in which operator commitment is the most important pillar, followed by education and training, while the effect of external pressures is the last. By contrast, the opposite model is one in which the operator receives many external pressures to implement HACCP but is not committed to its implementation and does not have adequate training for its personnel. This model, dependent on external pressures, is unsustainable over time because there is no conviction or commitment in the organization that is required for its management and continuous maintenance (Figure 2.8).

2.6 HACCP Barriers

Many authors, observing how the implementation of the HACCP system is much more complicated than one might suppose, seeing its principles and methodology, have investigated the barriers that could hinder its effective implementation. These barriers are often caused by the attitudes, perceptions, and behaviors that people show toward HACCP, which significantly affect their way of adopting it, both before its implementation and during its management and maintenance. In other cases, the reasons are objective, i.e. based on the existence of structural barriers or the availability of resources, which may complicate the application of HACCP in certain types of operators or food sectors.

Among the barriers identified by various authors are the following:

- **HACCP is not necessary**: many operators believe so; because of the way they work, they could live perfectly well without the HACCP system. Those operators consider that they know their processes perfectly and do not perceive any significant risk in their activity; in their opinion, they are very familiar with their activity, have always done so, and are unaware that they have caused any food incidents.
- **Small food companies**: it is generally accepted that smaller companies have more significant difficulties in developing and implementing HACCP. With less economic capacity, this means

less availability of material and human resources; almost all of the company's staff is devoted to production and marketing tasks, and they can rarely afford to have a technical department or technical director take on the task of forming and leading the HACCP team. Although they can request the help of an external consultant, the possibility of forming a sufficiently operational HACCP team is quite limited in this type of company. In any case, this limitation does not prevent small companies with acceptable HACCP management thanks to the workers' commitment, training, and experience.

- **Workload**: many FBOs perceive HACCP as an increase in workload. The obligations derived from the system's procedures become an additional task to those they already have for their productive activity. As the staff is adjusted not to increase personnel costs more than necessary, the management of HACCP implies the need to dedicate extra time to follow the procedures and fill in the corresponding records. Consequently, many workers associate HACCP with the paperwork it generates and see it as a bureaucratic burden for which they do not feel compensated.
- **Type of production**: depending on the number of lines and type of products, HACCP can be more complex to implement in some food companies than others. Thus, in the HACCP study in a company with a single production line and a single type of product, as may be the case of a ultra high temperature (UHT) milk packaging machine with a single flow diagram, even if they produce thousands of units per day, it will always be more straightforward than in the case of a food operator with many processes and different products, often at the same time, as in mass catering establishments, where the HACCP study would have to be completed for each of the products.
- **Inadequate facilities**: the design, size, and layout of the facilities determine the hygienic and sanitary conditions in which the operator's activity is carried out and mostly condition the manufacturing and handling practices of the workers. Establishments with structural deficiencies, low capacity, and poor design pose an obstacle to implementing HACCP due to increased risks derived from deficiencies in the facilities: more contamination from the environment, the crossing of manufacturing lines, inadequate temperatures, the presence of pests, etc.
- **Deficiencies in equipment**: within the food sector, there are notable differences in the degree of mechanization of processes, from those based on manual operations to those practically mechanized or automated. The existence of directly observable control parameters, the possibility of including online monitoring devices (with automated recording and/or alarm systems), and the availability of appropriate measuring instruments and equipment may affect operators' ability to comply with HACCP monitoring, verification, and recording requirements.
- **Lack of technical knowledge**: developing the steps of the HACCP study requires knowledge of food hygiene, technology, microbiology, toxicology, etc. The lack of sufficient knowledge and experience in these areas among the members of the HACCP team will be an obstacle to the development of HACCP, leading to increased difficulties in, for example, carrying out a proper hazard analysis or documenting the validation of the HACCP plan with the necessary scientific basis.
- **Lack of leadership**: the implementation of HACCP depends on the intervention of people with sufficient vision and leadership to involve the entire organization in developing and managing the HACCP system. In many cases, this person is at an intermediate level of the company's organizational chart, often performing the functions of a technical director or quality manager. For some authors, having a professional profile of these characteristics is a critical success factor for the implementation of HACCP; in addition to being a core member of the HACCP team, his role is to obtain the commitment of the management to meet the various needs of infrastructure, personnel, training, etc., generated by HACCP, as well as to raise awareness and motivate plant personnel to be involved in complying with system procedures. In smaller companies, where it

Figure 2.9 Some of the human factors involved in implementing HACCP.

is quite burdensome to keep a technical position of these characteristics on the payroll, they can turn to the help of an external advisor. However, their effect will always be more limited than when they are part of the company.

- **Inertia of workers**: introducing HACCP procedures in an organization requires overcoming people's natural resistance to changing their usual way of working for many years. This inertia to change among workers is also motivated because they do not understand why and because, in principle, they perceive HACCP as something unnecessary, complicated, and one more task to perform. In order to overcome the barrier represented by this negative attitude toward HACCP, challenging work on awareness and training of the staff is necessary, which helps them understand the reasons and the benefits of implementing HACCP in the food business.
- **Other factors**: working conditions like long hours, fast-paced production, job insecurity, low professionalization in some sectors, and work environment, among others, also affect the attitude and behavior of workers in the food industry when they assume responsibility for compliance with the procedures of the HACCP system (Figure 2.9).

2.7 HACCP Principles and Methodology

HACCP's principles and methodology were established internationally with the publication of two documents by the Codex Alimentarius and by the National Advisory Committee on Microbiological Criteria for Foods (NACMCF) in the early 1990s. The first version of the NACMCF HACCP System dates back to 1992; the following year, Codex Alimentarius published its Hazard Analysis and Critical Control Point (HACCP) System – Guidelines for its application as an annex to document CAC/RCP 1 – 1969 General Principles of Hygiene. In 1997, the NACMCF revised its HACCP System, comparing it with the Codex guide, resulting in a more concise drafting of HACCP principles and the inclusion of new definitions and other sections on prerequisites, training, implementation, and maintenance of the system.

Since then, these documents constitute the international reference for the HACCP System and contain the main criteria against which we can assess whether the HACCP is well documented and fit for purpose.

It is important to note that the Codex Alimentarius and NACMCF include the same principles and steps, describe the same methodology, and share most definitions. There are some differences

HACCP	Codex Alimentarius	NACMCF
Editions	1993, 1997, 2003, 2022	1992, 1997
Principles	7	7
Steps	12	12
Decision tree	1	2
Sections	2	3
Appendices	4 annexes and 2 tables	8

Figure 2.10 Comparing Codex and NACMCF.

between the two documents. However, these differences refer to nonessential aspects of the system, such as those related to the guidelines for its implementation, the examples provided as a guide, and the number of appendices or sections included. Therefore, the Codex Alimentarius and NACMCF describe the same HACCP system unequivocally and without contradictions. Both are equally valid reference documents for professionals in the implementation and management of HACCP and its supervision and audit (Figure 2.10).

At the EU level, the European Commission published two other guidance documents, which are very useful for the application of HACCP principles by food operators and competent authorities:

- 2020/C 199/01 Commission Notice providing guidance on FSMSs for food retail activities, including food donations. The purpose of this guidance is to facilitate and harmonize the implementation of the EU requirements for an FSMS with a central role of the hazard analysis for the following retailers: butcher, grocery, bakery, fish and ice cream shops, distribution centers, supermarkets, restaurants, catering services, and pubs. This notice, therefore, provides guidance on how these retailers, within their activity, can implement Regulation (EC) 852/2004, in particular, its Article 4 and Annex II as regards general hygiene requirements and its Article 5 as regards procedures based on the HACCP principles. It provides at the same time an example of how the retail sector can specifically implement the guidelines, including those on facilitation/flexibility, recommended in the 2016 Commission Notice, specifically at the retail level. Competent authorities may also use this guidance to verify the implementation of the EU requirements by these food businesses.
- 2022/C 355/01 Commission Notice on the implementation of FSMSs covering GHPs and procedures based on the HACCP principles, including the facilitation/flexibility of the implementation in certain food businesses.

The purpose of this guidance is to facilitate and harmonize the implementation of the EU requirements on GHPs and procedures based on the HACCP-based procedures as parts of FSMSs by providing practical guidance on:

- the relevant legislation, the link between GHP, prerequisite programs (PRP), operational PRP (OPRP) and HACCP-based procedures within a FSMS, the relationship with international standards and training, and the use of guides to GHPs;

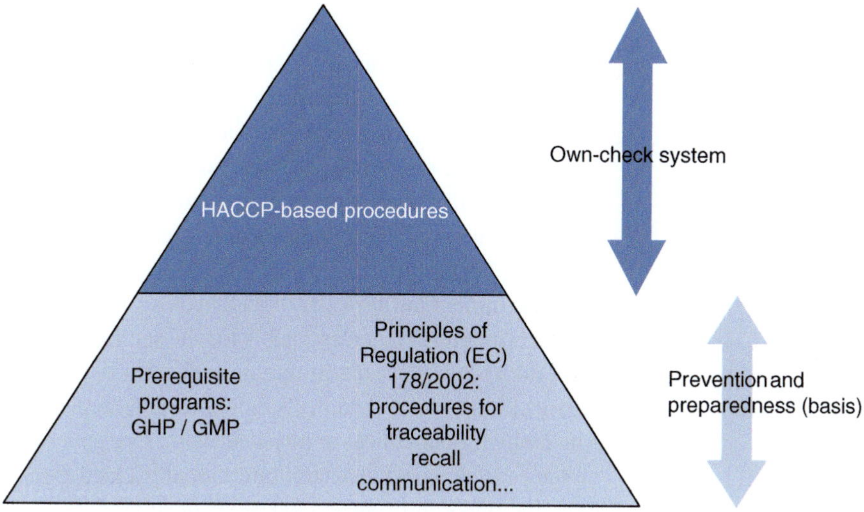

Figure 2.11 Elements of an FSMS. European Commission.

- the implementation of GHPs, including the flexibility provided for certain food establishments by EU legislation related to their implementation (Annex I);
- the implementation of HACCP-based procedures, including the flexibility provided for certain food establishments by EU legislation related to their implementation (Annex II);
- the auditing of FSMS (Annex III) (Figure 2.11).

2.7.1 The HACCP Document of the Codex

The principles and system of HACCP are described in the second part of document CXC 1-1969 General Principles of Hygiene, entitled Hazard Analysis and Critical Control Point (HACCP) and Guidelines for its Application. In the second part of this document, section 17 sets out the seven principles of the HACCP system. Section 18 provides general guidance for applying the HACCP system, and section 19 describes its application in 12 successive stages.

The introduction to this second part of CXC 1-1969 recognizes that implementation may vary and that a more flexible approach is needed depending on the circumstances and the capabilities of each FBO. The HACCP system, which is science based and systematic, identifies hazards and measures for their control to ensure food safety. HACCP is a tool to assess hazards and establish control systems that focus on control measures for significant hazards along the food chain rather than relying mainly on end-product testing.

This Codex document assumes that HACCP principles can be considered throughout the food chain, from primary production to final consumption, and its implementation should be guided by scientific evidence of risks to human health. However, applying HACCP in primary production is not always feasible; some of the principles can be applied and incorporated into good practice programs (e.g. good agricultural practices (GAP), etc.).

The first part of the Codex document includes in section 6 a list of definitions essential for understanding the principles of HACCP and its application, allowing us all to speak the same language and know exactly what we mean.

Definitions

Acceptable level	A level of hazard in a food at or below which the food is considered to be safe according to its intended use.
Allergen cross-contact	The unintentional incorporation of an allergenic food, or ingredient, into another food that is not intended to contain that allergenic food or ingredient.
Cleaning	The removal of soil, food residues, dirt, grease, or other objectionable matter.
Competent authority	The government authority or official body authorized by the government that is responsible for the setting of regulatory food safety requirements and/or for the organization of official controls including enforcement.
Contaminant	Any biological, chemical or physical agent, foreign matter or other substances not intentionally added to food that may compromise food safety or suitability.
Contamination	The introduction or occurrence of a contaminant in the food or food environment.
Control	• when used as a noun: the state wherein correct procedures are being followed and any established criteria are being met; and • when used as a verb: to take all necessary actions to ensure and maintain compliance with established criteria and procedures.
Control measure	Any action or activity that can be used to prevent or eliminate a hazard or reduce it to an acceptable level.
Corrective action	Any action taken when a deviation occurs in order to re-establish control, segregate and determine the disposition of the affected product if any and prevent or minimize reoccurrence of the deviation.
Critical control point (CCP)	A step at which a control measure or control measures, essential to control a significant hazard, is/are applied in a HACCP system.
Critical limit	A criterion, observable or measurable, relating to a control measure at a CCP which separates acceptability from unacceptability of the food.
Deviation	Failure to meet a critical limit or to follow a GHP procedure.
Disinfection	Reduction by means of biological or chemical agents and/or physical methods in the number of viable microorganisms on surfaces, in water or air to a level that does not compromise food safety and/or suitability.
Flow diagram	A systematic representation of the sequence of steps used in the production or manufacture of food.
Food business operator (FBO)	The entity responsible for operating a business at any step in the food chain.

Food handler Any person who directly handles packaged or unpackaged food, equipment and utensils used for food, or surfaces that come into contact with food and that is expected, therefore, to comply with food hygiene requirements.

Food hygiene All conditions and measures necessary to ensure the safety and suitability of food at all stages of the food chain.

Food hygiene system Prerequisite programmes, supplemented with control measures at CCPs, as appropriate, that when taken as a whole, ensure that food is safe and suitable for its intended use.

Food safety Assurance that food will not cause adverse health effects to the consumer when it is prepared and/or eaten according to its intended use.

Food suitability Assurance that food is acceptable for human consumption according to its intended use.

Good hygiene practices (GHPs) Fundamental measures and conditions applied at any step within the food chain to provide safe and suitable food.

HACCP plan Documentation or set of documents, prepared in accordance with the principles of HACCP to ensure control of significant hazards in the food business.

HACCP system The development of a HACCP plan and the implementation of the procedures in accordance with that plan.

Hazard A biological, chemical or physical agent in food with the potential to cause an adverse health effect.

Hazard analysis The process of collecting and evaluating information on hazards identified in raw materials and other ingredients, the environment, in the process or in the food, and conditions leading to their presence to decide whether or not these are significant hazards.

Monitor The act of conducting a planned sequence of observations or measurements of control parameters to assess whether a control measure is under control.

Primary production Those steps in the food chain up to and including storage and, where appropriate, transport of outputs of farming. This would include growing crops, raising fish and animals, and the harvesting of plants, animals or animal products from a farm or their natural habitat.

Prerequisite program Programmes including good hygiene practices, good agricultural practices and good manufacturing practices, as well as other practices and procedures such as training and traceability, that establish the basic

	environmental and operating conditions that set the foundation for implementation of a HACCP system.
Significant hazard	A hazard identified by a hazard analysis, as reasonably likely to occur at an unacceptable level in the absence of control, and for which control is essential given the intended use of the food.
Step	A point, procedure, operation, or stage in the food chain, including raw materials, from primary production to final consumption.
Validation of control measures	Obtaining evidence that a control measure or combination of control measures, if properly implemented, is capable of controlling the hazard to a specified outcome.
Verification	The application of methods, procedures, tests, and other evaluations, in addition to monitoring, to determine whether a control measure is or has been operating as intended.
NACMCF	The HACCP Principles and Application Guidelines document of the NACMCF shares many of the Codex definitions; however, it includes different ones, and some that it defines in a slightly different way, but that it is helpful to know as well.
CCP Decision Tree	A sequence of questions to assist in determining whether a control point is a CCP.
Control	(a) To manage the conditions of an operation to maintain compliance with established criteria. (b) The state where correct procedures are being followed and criteria are being met.
Control Measure	Any action or activity that can be used to prevent, eliminate or reduce a significant hazard.
Control Point	Any step at which biological, chemical, or physical factors can be controlled.
Corrective Action	Procedures followed when a deviation occurs.
Criterion	A requirement on which a judgement or decision can be based.
Critical Control Point	A step at which control can be applied and is essential to prevent or eliminate a food safety hazard or reduce it to an acceptable level.
Critical Limit	A maximum and/or minimum value to which a biological, chemical or physical parameter must be controlled at a CCP to prevent, eliminate or reduce to an acceptable level the occurrence of a food safety hazard.
Deviation	Failure to meet a critical limit.
HACCP	A systematic approach to the identification, evaluation, and control of food safety hazards.

HACCP Plan	The written document which is based upon the principles of HACCP and which delineates the procedures to be followed.
HACCP System	The result of the implementation of the HACCP plan.
HACCP Team	The group of people who are responsible for developing, implementing and maintaining the HACCP system.
Hazard	A biological, chemical, or physical agent that is reasonably likely to cause illness or injury in the absence of its control.
Hazard Analysis	The process of collecting and evaluating information on hazards associated with the food under consideration to decide which are significant and must be addressed in the HACCP plan.
Monitor	To conduct a planned sequence of observations or measurements to assess whether a CCP is under control and to produce an accurate record for future use in verification.
Prerequisite Programs	Procedures, including Good Manufacturing Practices, that address operational conditions providing the foundation for the HACCP system.
Severity	The seriousness of the effect(s) of a hazard.
Step	A point, procedure, operation or stage in the food system from primary production to final consumption.
Validation	That element of verification focused on collecting and evaluating scientific and technical information to determine if the HACCP plan, when properly implemented, will effectively control the hazards.
Verification	Those activities, other than monitoring, that determine the validity of the HACCP plan and that the system is operating according to the plan.

2.7.2 Principles of the HACCP System

The HACCP system is designed, validated, and implemented in accordance with the following seven principles:

Principle 1
Conduct a hazard analysis and identify control measures.

Principle 2
Determine the CCPs.

Principle 3
Establish validated critical limits.

Principle 4
Establish a system to monitor control of CCPs.

Principle 5
Establish the corrective actions to be taken when monitoring indicates a deviation from a critical limit at a CCP has occurred.

Principle 6

Validate the HACCP plan and then establish procedures for verification to confirm that the HACCP system is working as intended.

Principle 7

Establish documentation concerning all procedures and records appropriate to these principles and their application.

2.8 The Four Stages for the Implementation of HACCP

As stated in the Codex guidelines, the HACCP system should be applied to each specific operation separately; in that sense, the Codex document contains the precise indications to complete the HACCP study in a specific manufacturing line and some guidelines for its implementation. However, it should be borne in mind that this is only part of what the overall process of implementing HACCP in a food operator entails; some tasks need to be addressed before starting the HACCP study and also other subsequent ones, which are necessary to achieve the continuous maintenance of the HACCP system in the food business. Thus, four stages can be distinguished in implementing HACCP: preparation and planning, HACCP study, implementation, and system maintenance (Mortimore and Wallace 2013) (Figure 2.12).

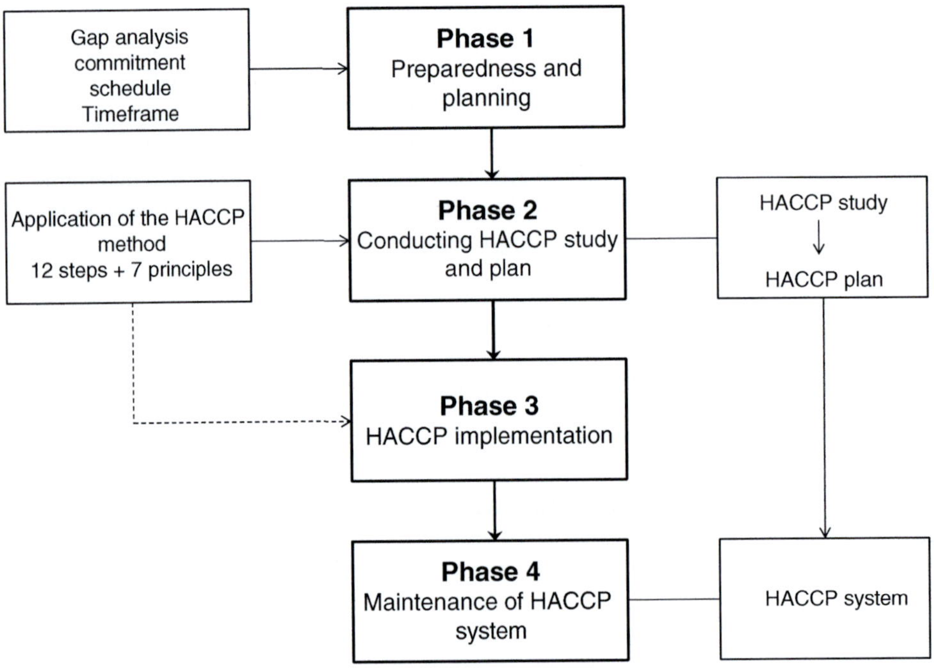

Figure 2.12 The four stages for HACCP implementation.

2.9 Preparation and Planning

Some tasks must be addressed before starting the HACCP study for a particular operation. The first aspect that needs to be considered is ensuring the commitment of those responsible for the food business.

As reflected in the scheme of Panisello and Quantick that we saw earlier, the basis to carry out the entire process that involves the implementation of HACCP is the commitment, which must start from the management and extend to all levels of the organization.

Without the conviction and support of the highest level of the company, it is not possible to finance or promote the changes in the organization that are necessary to ensure the effective implementation of the HACCP system. Panisello and Quantick consider commitment as the basis that should sustain the system. At the same time, Robert Gaze, in his book *HACCP: a practical guide* (Campden BRI 2009), argues that it should be the first phase of the HACCP study, the second determine the terms of reference and then apply the 12 phases of Codex.

In this preparation phase, those responsible for promoting the first steps to implement HACCP should begin to consider some issues prior to the HACCP study itself, and either themselves or with the HACCP team, once it has been formed, specify the following aspects:

- Set the terms of reference and scope of the HACCP study.
- Gap analysis.
- Timeframe.

2.9.1 Establish the Terms of Reference and Scope of the HACCP Study

Although it is not listed as an early stage in the Codex or NACMCF guidelines, everyone assumes it is necessary to define the terms of reference and the scope of the HACCP study before starting the HACCP study. This stage includes, on the one hand, specifying the types of hazards to include in the study: microbiological, chemical, and physical, and/or whether they are all to be addressed at the same time or in phases; on the other hand, delimiting the scope means identifying the activities that are under the responsibility of the FBO and where exactly their action within the food chain begins and ends. This delimitation of the scope is necessary because often, the food operator does not limit its control to the activities within its facilities but extends them to other stages in the food supply chain, both backward and forward. For example, in the dairy industry, the scope of HACCP may start at the milk reception phase, from the arrival of the milk tank, while in other cases, it may extend to transport and the milk collection route or even to some control over the livestock farms that are suppliers of the dairy center. Similarly, we may refer to the following stages in the food chain. In that case, the self-control manual must specify whether it ends with the dispatch of the product from the loading dock or whether it continues with the transport and distribution of the product.

Other aspects to consider to establish the scope are the type of HACCP and the organization's structure. In this way, it is necessary to consider if the company uses a HACCP of linear type, in which there is a flow diagram for each product, or if it chooses a modular approach, widely used in the catering industry, in which HACCP focuses on analyzing the shared processes that follow to elaborate on several products or groups of products, and also when the activities of the food operator are broken down into modules that are studied independently. In this case, knowing where each module begins and ends is necessary so that no activities are left unanalyzed when we change from one module to another.

In enterprises consisting of several units or headquarters, or with a head office and several satellite or peripheral units, such as large retail chains, it is necessary to delimit management responsibilities between the different units and to establish which activities are included in the self-control system of the central unit and which in those of peripheral units, so that there are no aspects out of control.

2.9.2 Gap Analysis

In order to ensure that the foundations of the HACCP system are sufficiently established, food operators should carry out a prior needs analysis before starting the HACCP study. This gap analysis consists of checking, in a more or less systematized way, the situation of the prerequisites of the HACCP system of the operator and comparing them with the optimal condition to start the HACCP study in the most effective way possible. This difference, or gap, between the current situation and the desired one must be solved through a specific action plan to correct the deficiencies detected (Figure 2.13).

The preliminary analysis of the needs may be structured around three main areas:

- Infrastructures and equipment.
- Control programs.
- Personnel.

Infrastructures include aspects such as:

- State of the facilities.
- Distribution of processes.
- Process capacity.
- Equipment.
- Measuring instruments.
- Water supply.

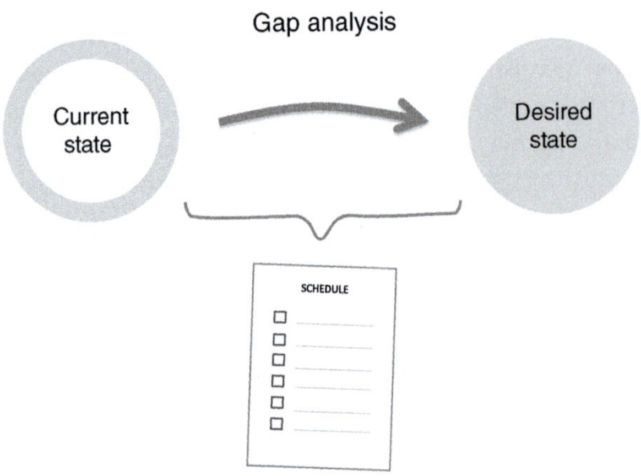

Figure 2.13 Gap analysis.

If the food operator has control programs prior to the implementation of HACCP, the effectiveness of elements such as the following must be checked:

- Authorized suppliers.
- Raw materials specifications.
- Cleaning and disinfection procedures.
- Pest control.
- Allergen control.
- Maintenance programs for facilities and equipment.
- Traceability system.

Concerning the personnel required to carry out the implementation of the HACCP system, the FBO must take into account the following issues:

- Total staff available.
- Organizational chart.
- Training needs at different levels of the organization.

2.9.3 Timeframe

Once the FBO has detected the needs, it must draw an action plan or schedule with each of the elements it considers necessary to solve, indicating the person responsible and the expected deadline for its completion in each case. When there are too many items or it is not easy to decide where it is most convenient to start the action plan, some authors recommend using the HACCP methodology to help determine the most priority elements. Although it is always stressed that PRPs must be developed before HACCP implementation, this does not contradict that already in these previous steps; it is possible to use the HACCP system, specifically hazard analysis, to help determine where the most significant risks are and thus focus the available resources on ensuring their control as quickly and effectively as possible (Mortimore and Wallace 2013).

2.10 The 12 Steps of HACCP

The application of the principles of the HACCP system involves carrying out the following tasks, as identified in this logical sequence composed of 12 steps (Figure 2.14):

In this sequence of twelve steps, two parts can be differentiated:

- Steps 1 to 5: preliminary phases.
- Steps 6 to 12: HACCP principles.

The preliminary phases include the tasks necessary to determine the specific activities and characteristics of the products and processes on which the specific hazard analysis needs to be performed and then to apply the HACCP principles (Figure 2.15).

As can be seen, the scheme and tasks included in the phases before applying HACCP show considerable similarity with the phases of the FMEA study, from which HACCP adopted part of its methodology. In both cases, the starting point is a multidisciplinary team that studies a specific process to analyze what failures may be, their probability of occurrence, and the consequences of these failures. The difference is that the FMEA focuses on failures in the functions of objects and HACCP on the occurrence of hazards in food.

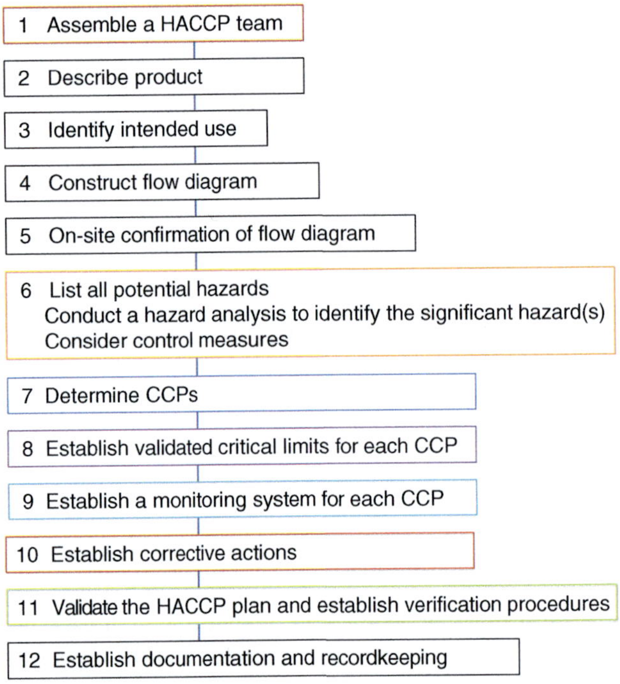

Figure 2.14 Twelve HACCP steps.

1. Assemble a cross-functional team
2. Identify the scope of the FMEA
3. Identify the functions of the scope
4. For each function, identify all the ways failure could happen
5. For each failure mode, identify all the consequences
6. Determine how serious each effect is, the severity rating S
7. For each failure mode, determine all the potential root causes
8. For each cause, determine the occurrence rating, O
9. For each cause, identify current process controls
10. For each control, determine the detection rating, D
11. Is this failure mode associated with a critical characteristic?
12. Calculate the risk priority number (RPN) $S \times O \times D$
13. Identify actions to reduce RPN
14. Recalculate RPN

Source: ASQ (American Society for Quality).

1. Select a process and draw its diagram
2. Assemble a cross-functional team
3. List potential failures and their consequences
4. Identify the potential causes of failures
5. Analyze current controls
6. Determine the likelihood, severity and detection rating (S, O, and D)
7. Calculate RPN (risk priority number) $S \times O \times D$
8. Establish actions to reduce RPN
9. Recalculate RPN

Source: Adapted from Michel Akiki.

Figure 2.15 Examples of models with the phases of an FMEA study.

2.11 Assemble the HACCP Team

The first step in HACCP systematics is to assemble a team of people to carry out the HACCP study. In a system whose success relies on people, it is essential to have, from the beginning, a team with sufficient knowledge and experience to develop an effective HACCP plan adapted to the characteristics of each process and the particular conditions of each organization. A HACCP system is only as good as the team that made it.

The HACCP team must be multidisciplinary, composed of people with different training and experience to cover all the fields and subjects that it is necessary to know in order to carry out the HACCP study, such as food hygiene, microbiology, process technology, and production, among others.

The size of the HACCP team that is considered most suitable – easy to manage and sufficient to carry out the tasks – typically ranges from four to six people, for whom they are permanent members. Occasionally, they can count on other people or turn to specialists to solve specific issues for which they do not have experience or knowledge within the usual team.

When assembling the HACCP team, consider whether there is enough staff within the food business and whether these people have the knowledge and experience necessary to participate in the team competently. As we have seen above, this needs analysis may have already been conducted, and measures may have already been taken to meet staffing and training needs.

All team members must understand and apply HACCP principles and methodology fluently. If this is not the case, it is necessary to ensure that everyone receives HACCP training of sufficient duration to acquire the required competence. The recommended course (Gaze 2009) for HACCP team members should include at least the following subjects:

- The principles and terminology of HACCP.
- The role of PRPs.
- How to conduct the HACCP logically, systematically, and in sufficient detail.
- The role of HACCP in food safety.
- The benefits of HACCP (Figure 2.16).

In medium-sized or large food companies, it is common to have professionals with the technical profile needed to constitute the HACCP team. They are usually technical managers and those

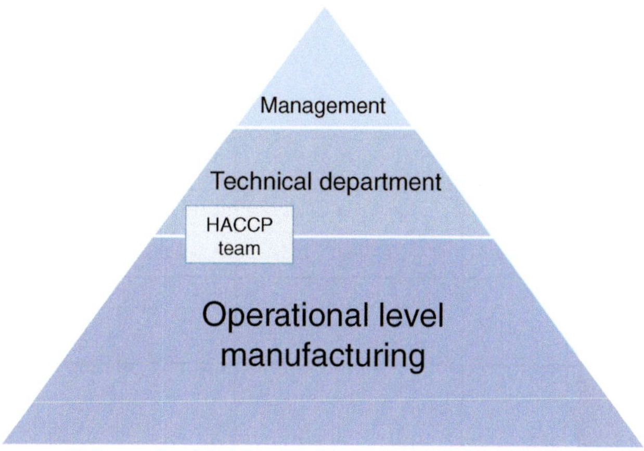

Figure 2.16 HACCP team position in an organization.

responsible for quality and production. These types of positions are common in food companies, and it is normal for them to play a predominant or leadership role within the team. In addition to the members of this intermediate or technical level of the company, the HACCP team must also include people more linked to the production processes. Their input to the team is significant because they are the ones who know what happens during manufacturing and are familiar with the failures and problems that can arise.

Within the HACCP team, we can define some roles; the first is the team leader, who promotes the tasks and coordinates the rest of the members, typically played by the quality manager or production manager. Some authors also point out the usefulness of appointing a team secretary or scribe whose function is to take notes and write the documents of the working sessions and meetings of the team and keep track of discussions, deadlines, and the dates for meetings (Wareing 2010).

The HACCP team leader is a member appointed by the company's management who must have sufficient responsibility and authority to lead the HACCP team and ensure that the organization implements the HACCP system correctly. Among their functions are to serve as a liaison between management, the HACCP team, and, in general, the rest of the organization, as well as to obtain from the management the support and resources necessary for implementing and maintaining the FSMS.

Another important aspect is the role that the leader can play, although not only him, to achieve that the work team is task oriented, cohesive, and with a favorable environment for participation in which all members feel equally involved and in which, in the end, the work obtained is the result of the contribution of all its members, from their different visions and experiences (Figure 2.17).

Like the FMEA, the HACCP methodology was created so that the system begins to be built "from below" based on knowledge of the processes and the concrete way each organization carries out its activities. That is where the contribution of the team members most linked to production tasks can be fundamental because they are the ones who best know the failures that usually occur, the problems that arise, and their consequences. This teamwork to analyze the hazards that may occur in a specific flow of operations is what adds value to HACCP to become a specific system, adapted

HACCP team member	Charge	Education and experience	Comments
Document date___________			

Figure 2.17 HACCP team members.

to the needs of each company, and is the basis from which we can add the rest of the elements of the system and the higher levels of control, from bottom to top: preventive measures, monitoring system, corrective actions, and verification.

An issue to take into account when assessing the competence of the HACCP team is that, in some food companies, even though the HACCP team has been formally constituted, in practice, it may happen that, due to poor planning of the team's work or other constraints, in the end, most of the tasks end up falling on a single person, who has to write the procedures and make most of the decisions about the design and structure of HACCP alone. When this situation occurs, the work of the HACCP team becomes something more symbolic and not so much a true teamwork, in which decisions are made jointly with the participation of all its members. Eventually, this ends up affecting the result obtained.

Small and/or less developed companies find it challenging to have enough staff with the technical or specialized knowledge required to form a multidisciplinary HACCP team, so they usually use an external consultant's help. In these cases, it may be the external advisor who assumes the leading role of carrying out the tasks of elaboration and implementation of the HACCP plan; however, this should not be to the detriment of the work of the HACCP team. The reasons are several: firstly, because the Codex systematics states so; secondly, because the HACCP plan will always be much more adapted to the characteristics of the company and its production processes when it has the participation of the people who apply them and who are the ones who know them best, and finally because it is always more manageable for the operator to seize the procedures and feel more motivated to apply them when the company's staff has participated in their preparation.

In the case of small companies that request the help of an external consultant to develop HACCP, it should not be the consultant who writes the entire HACCP plan alone (Gaze 2009). When the consultant develops the entire HACCP system with hardly any participation of the process owners, i.e. line managers and/or operators responsible for the processes, this will be a significant obstacle to the implementation of HACCP by workers since not only will it be necessary to overcome inertia to change their usual way of working and carrying out controls, but also operators will perceive these procedures as something alien to them, imposed from outside, so it is more difficult for them to understand them well or to adopt them as their own. These difficulties in implementing HACCP in small companies also stem from the fact that, in many cases, the external consultants who prepare HACCP deliver the manual to those responsible for the food operator without carrying out real work of advising and monitoring the implementation among the company's staff, in order to explain what the new procedures consist of, what changes may entail in the way of working and to help correct problems that might arise when the staff start applying the system.

The lower availability of human resources to constitute the HACCP team in small companies should not be understood as an impediment to the implementation of the HACCP system to be carried out effectively in this type of company; more than the number of people, what really counts are the preparation and awareness of those responsible for implementing the system; and in many cases, the shortage of personnel can be compensated with a greater involvement of system managers in the control of processes and product safety, who are aware of the economic repercussions that the appearance of any failure or problem in production can have for their business.

The leader and other members of the HACCP team will surely be the main interlocutors of the food company during the audit, so from the beginning, the auditors will have the opportunity to evaluate the team's competence, the preparation of its members, and the work they do. To do this, in addition to reviewing their training and professional experience, the audit team may want to check their knowledge of the HACCP system, their participation in the development of the HACCP plan, and their responsibilities in the management of the system, or if they are aware of the modifications and updates of their HACCP system. Another aspect of interest to auditors is the details related to the activity of the HACCP team, such as HACCP plan preparation documents, hazard lists, working documents, meeting minutes, internal communications, system reviews, in order to assess to what extent the work team maintains its activity over time.

2.12 Describe the Product

The HACCP team must complete a description of each product the food company markets. This description should include all finished food products, ready for distribution or sale to the final consumer, and intermediate products destined for other processing plants for food use. For example, in a cheese factory, in addition to the different types of cheese, butter, or cream, the HACCP plan must also contain a description of whey as an intermediate product when the whey is destined for another processing plant to obtain products intended for human consumption, such as infant formulas or nutritional supplements, among other uses.

The auditors will surely want to check for a complete description in the HACCP plan for every one of the products marketed by the food operator. Besides being a requirement of the HACCP system, this information is essential for planning activities and calculating the duration of the audit, along with the number of staff and the number and complexity of the processes.

According to HACCP guidelines, each food produced must have its own HACCP study, which begins with describing the product and analyzing the steps that make up its elaboration process. However, in some companies where the number of products is high, it is possible to facilitate this task by grouping the products whose manufacturing process goes through the same operations. For example, in the case of mass catering companies, where the number of dishes prepared can be substantial, it is common to use a modular or process grouping approach to simplify the preparation and management of the HACCP plan. In this way, HACCP is applied, not for each meal but for the different groups of meals that go through the same operations or modules included in the overall operation process.

From a global flow diagram, we can distinguish different food groups according to the sequence of operations or modules through which the preparation passes in each case. This simplified approach allows the use of the same flow diagram for the different types of processes of the food operator. It helps to focus the control measures on each operation or module comprising the manufacturing lines: reception, food preparation, cooking, cooling, reheating, and packaging, among others.

In its Managing Food Safety: A Manual for the Voluntary Use of HACCP Principles for Operators of Food Service and Retail Establishment, published in 2006, the Food and Drug Administration (FDA) proposes a classification of foods according to the thermal profile that results from the operations carried out in each preparation. According to this criterion, the meals are divided into three groups: uncooked, cooked and served on the same day, and complex, from the number of times that its elaboration passes through what they call the "dangerous zone," between 41 and 135 °F (5–57.2 °C), which is the one with the highest risk of microbial multiplication (Figure 2.18).

The following table shows examples of foods classified by the FDA into three groups according to the number of processes included in each preparation (Figure 2.19).

The safety of a food product depends both on intrinsic factors such as raw materials, ingredients, a_w or pH, among others, and on extrinsic factors such as packaging, storage conditions, or

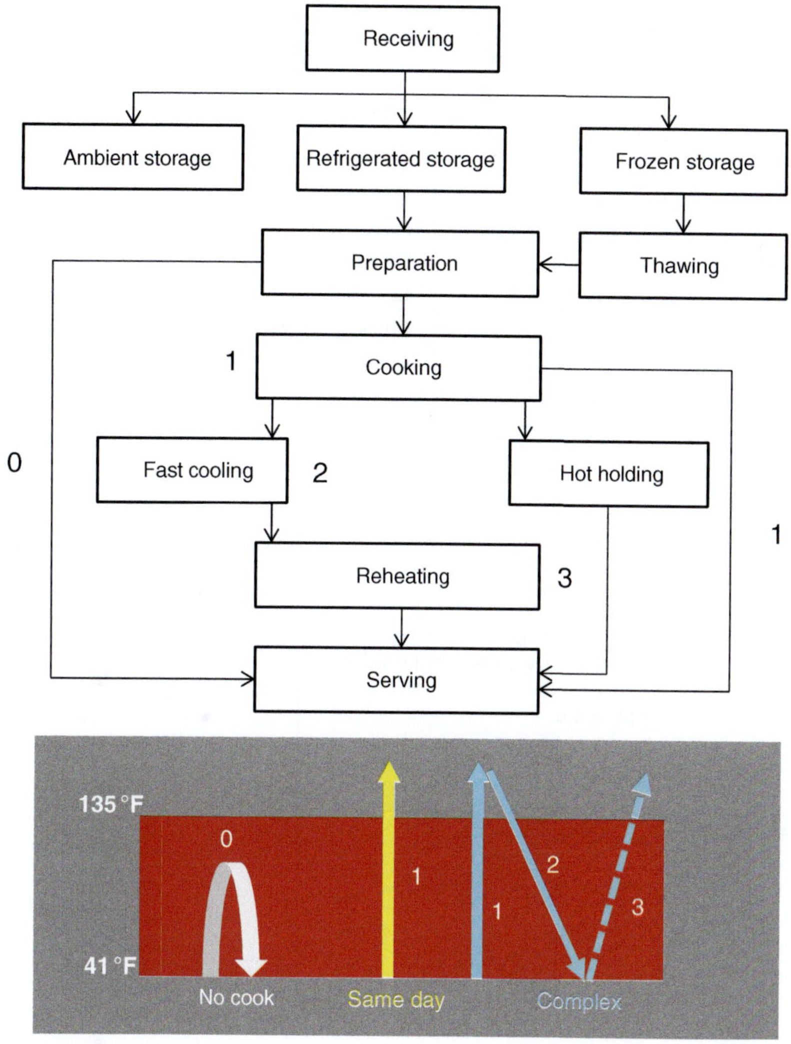

Figure 2.18 Flow diagram of food retail. Meals groups according to elaboration processes.

intended use; for this reason, the description of a food must be complete and contain all the necessary details so that the HACCP team can analyze what the hazards associated with each food are. The characteristics of a foodstuff are directly related to the processes followed to reach that state or the desired conditions. Thus, the product and process descriptions are linked, and they can be considered equivalent for the HACCP study purposes in many cases (Wallace et al. 2011).

For each food produced by the food operator, the HACCP team must consider what factors make the food safe for the consumer and how long it is estimated to retain its properties or guarantee its safety. Most foods on the market base their safety and/or conservation on the intervention of several factors, which can act additively or synergistically to avoid alteration or deterioration by microbial growth. Factors such as a_w, pH, salt content, heat treatment received usually act together, and rarely, the preservation of a food is based on the intervention of a single factor, as could be the case of food that the FBO intends to preserve solely by the action of salt; the salt content would have to be very high and that could have a sensory impact or on the characteristics of the food unwanted

Food preparation with no cook step	Food preparation for same day service	Complex food preparation
Raw meat and seafood (to be cooked by consumer)	Fried chicken	Soups
Salad greens	Broiled fish	Gravies
Fish for raw consumption	Fried oysters	Sauces
Fresh vegetables	Hamburgers	Large roasts
Oysters or clams served	Soup du jour	Chili
Raw tuna salad	Hot vegetables	Taco filling
Caesar salad dressing	Cooked eggs	Egg rolls
Cole slaw		Chicken salad (made from raw chicken)
Sliced sandwich meats		
Sliced cheese		
Chicken salad (made from canned chicken)		

Figure 2.19 Different groups of meals.

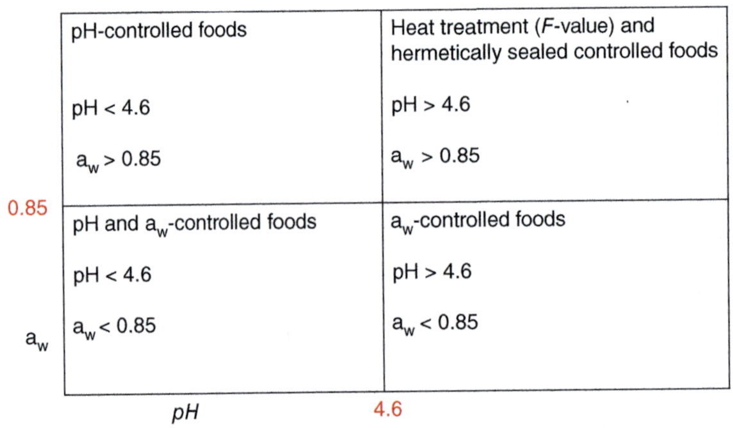

Figure 2.20 Types of foods according to control factors.

by consumers, for example by those who want to reduce the salt content in their diet. However, we must bear in mind that when analyzing the stability of the product, aspects related to the chemical and microbiological safety of food should have priority over the organoleptic properties. The pH, a_w, and heat treatment received are usually the main factors in establishing the preservation criteria of a food product (Rahman 2007). When the FBO plans a new food or changes the composition, ingredients, or characteristics of one already on the market, the HACCP team must consider how these variations may influence its safety characteristics, the hazards that may arise, and its new shelf life or best-before date (Figures 2.20 and 2.21).

Food category (examples of food to evaluate)	Pathogens to consider	Control types (alone or in combination)
Meat and poultry meat (sausages and cured meats)	*Clostridium botulinum*[a] and *Clostridium perfringens*, *Salmonella* spp., *Escherichia coli enterohemorrhagic*, *Camplylobacter jejuni*, *Yersinia enterocolitica*, *Staphylococcus aureus*, *Listeria monocytogenes*	Time/temperature, pH, a_W, preservatives, moisture: protein ratio, fermentation, heat treatment
Fish and seafood (smoked fish)	*Vibrio vulnificus*, *Vibrio parahaemolyticus*, *Vibrio cholerae*, *C. botulinum*[a], *L. monocytogenes*, *Salmonella* spp., *Shigella* spp., *S. aureus*	Time/temperature, origin control, fermentation, pH, a_W, salt in aqueous phase, preservatives, drying, salting
Fats, oils and dressings	*S. aureus*[b], *Salmonella* spp[b]., *B. cereus*[b], *C. botulinum*[b] (garlic in oil)	pH, a_W, salt
Sugar and syrups (light maple syrup)	*C. botulinum*[c]	a_W, acidification (light syrups)
Milk and dairy products (yoghurt)	*Salmonella* spp.[d], *L. monocytogenes*[d], *E. coli enterohemorrhagic*[d], *S. aureus*[d], *B. cereus* (cells[d] and spores[a]), *C. botulinum* (cells[d] and spores[a]), *Campylobacter jejuni*[d]	Production control, time/temperature, cooking/pasteurization, a_W, preservatives

[a]Only in anoxic environments.
[b]pH > 4.0 and $a_W \sim 0.92$ in salad dressings and mayonnaise would exclude the growth of pathogens to be considered.
[c]Only of concern in light syrups and can be controlled by acidification.
[d]In pasteurized products all preprocess vegetative pathogens would be controlled.

Figure 2.21 Examples of pathogens and methods for their control in different foods.

The product description must include the following information:

- Name of the product.
- Composition: raw materials, ingredients, additives, etc.
- Structure and physicochemical characteristics: solid, liquid, gel, emulsion, moisture content, a_W, pH.
- Process or transformation applied: frozen, dried, smoked, salted, etc.
- Packaging: type of container, hermetic, vacuum, modified atmosphere, active packaging, etc.
- Storage and distribution conditions.

- Shelf life.
- Instructions for use.
- Any applicable microbiological or chemical criteria.

To these previous points, the FBO can add information related to the labels with which it will market the product. In the case of products intended for delivery to the final consumer without further processing, the label must comply with the requirements of the rules on labeling and/or information provided to the consumer. The auditor may want to know the label templates used for different products and check if they include information such as the following:

- Indications regarding the ingredients or processing aids found in the finished product that may cause allergies or intolerances in the consumer.
- Special conditions of storage and/or use.
- How to use if it would be difficult to use the food correctly without this information.
- The minimum durability date or expiration date.
- Nutritional information.

On many occasions, information on changes in the composition of foods is not communicated to the HACCP team or updated correctly in the system documentation.

The description of the characteristics of a foodstuff and/or the processes followed in its preparation is also essential to determine which category of product the foodstuff falls into and, therefore, what regulatory criteria apply to it.

Thus, if we refer to microbiological criteria, Regulation 2073/2005 establishes the criteria that apply to the different categories of foodstuffs; for example, in the case of *Listeria monocytogenes*, the food safety criterion varies between absence at 25 g and 100 cfu/g, depending on the physicochemical characteristics of the food, its intended use, its shelf life, and whether or not the food may promote the growth of *L. monocytogenes* during its shelf life. Foods with pH $\leq$ 4.4, or $a_w \leq 0.92$, with pH $\leq$ 5, and $a_w \leq 0.94$, or with a shelf life of less than five days, according to this regulation are automatically considered not to favor the growth of *L. monocytogenes*, and therefore, the criterion is 100 cfu/g. With other values, it is the food operator who must demonstrate, to the satisfaction of the competent authority, that the product will not exceed the limit of 100 cfu/g during its shelf life, through a specific study. When they are ready-to-eat foods for infants or special medical purposes, the criterion is always absence in 25 g.

We find another example of the importance of the correct description of the product by the food operator in meat derivatives; thus, according to Regulation 1333/2008 on food additives, the main criterion for determining which additives the FBO may use in their manufacture is whether they are classified as meat preparations or as meat products. The key in this type of product is to clarify whether the transformation process that the meat has undergone has been sufficient to alter the internal structure of the muscle fiber, in which case it is classified as a meat product, or whether it has not been sufficient and hence cutting surface retains the characteristics of fresh meat and is considered a meat preparation. Taking into account that the food operator is the one who best knows how the elaboration process has been and the degree of transformation that the meat processing in question has been able to undergo, the regulations establish that the manufacturer is responsible for determining which category the FBO must market the product. This information must be part of the product description and remain available to the competent authority upon request (Figure 2.22).

Product description	
Brand	Campo de Muras
PDO	San Simón da Costa
Type	Matured high-fat pressed paste
Composition	Pasteurized cow's milk, lactic ferments, calcium chloride,rennet and salt
Size	1 kg
Country of origin	Spain
Sanitary registry	15.XXXXX/LU
Characteristics	
Fat content	45%
Milk solids	58%
pH	5.2–5.5
Ripening time	Minimum 45 days
Shape	Bullet or spinning top
Dimensions	16 cm × 11 cm
Organoleptic characteristics	
Cheese paste	Fine-textured, semi-hard, white to yellow, may have somesmall holes
Rind	Inedible, not surface-treated
Texture	Buttery and creamy
Color	Cream
Flavor	Smoked and dried fruits flavor
Aroma	Smokey, woody
Microbiological characteristics	
Listeria monocytogenes	Absence in 25 g
Salmonella	Absence in 25 g
E. coli	$m = 10^2$; $M = 10^3$; $n = 5$, $c = 2$
Coagulase-positive staphylococci	$m = 10^2$; $M = 10^3$; $n = 5$, $c = 2$
Shelf life	Best before 12 months after manufacture
Recommendations	Keep refrigerated at 2–8 °C
Target population	The whole population, except lactose intolerants
Intended use	Direct consumption, culinary preparations and desserts
Allergens	Milk and dairy products, lactose
Palletization and packing	
Type of pallet	European
Packaging	Carton of four units
Box dimensions	282 cm (length), 282 cm (height) and 160 (width)
Boxes per pallet	64
Nutrition facts per 100 g	
Calories	429 kcal/1.780 kJ
Fat	36.3 g
Saturated fats	23.3 g
Carbohydrate	3.7 g
Sugars	<0.1 g
Fiber	1.72 g
Protein	23.7 g
Salt	0.20 g
Applicable regulations: Regulation (EC) No. 852/2004 on the hygiene of foodstuffs, Regulation (EC) No. 853/2004 on specific hygiene rules for products of animal origin, Regulation (EC) No. 2073/2005 on microbiological criteria for foodstuffs, Regulation (EC) No. 1169/2011 on consumer information, and Royal Decree 1113/2006 approving the quality standard for cheeses and processed cheeses.	

Figure 2.22 Description of the product.

2.13 Determining Intended Use

In the next step of the HACCP study, Codex systematics requires the HACCP team to identify the intended or expected use of the product. This task includes, on the one hand, determining whether the product is ready for consumption or whether it requires any prior treatment by the consumer and, on the other, establishing whether the product is intended for the general population and/or whether certain vulnerable groups of the population can consume it.

Concerning the first aspect, the HACCP equipment must identify any misuse or incorrect handling of the product that may pose a risk to the consumer. In some cases, information regarding possible misuse by the consumer can be obtained from complaints related to the consumption of a particular product (ASQ 2014). For many foods, culinary tradition or habitual consumption practices do not usually raise doubts about whether the food can be eaten directly or the type of treatment or preparation required. However, the increasing development of fourth- and fifth-range products, i.e. ready-to-eat products or ready-to-eat meals, requires the manufacturer to indicate on the labeling which preparation is necessary to avoid the emergence of risks due to its incorrect use.

In some foods, the intended use is not easy to determine; for example, in the case of cooked sausages type Frankfurt, many consumers consider that this product can be consumed directly, while others usually give some precooking. If we check the labels of some manufacturers, we can see that in some cases, the FBO recommends a previous treatment or cooking, and in other cases, it indicates nothing. An International Commission on Microbiological Specifications for Food (ICMSF) study on this type of product concludes that there may be a risk of *L. monocytogenes* in the packaged product, which is usually reduced to acceptable levels when these sausages undergo some cooking or reheating prior to consumption. Therefore, each manufacturer, depending on the characteristics of the sausage production process, is the one who must decide if it is necessary to include in the labeling the recommendation of some treatment by the consumer.

The HACCP team should also define whether the food product is intended for the general population or whether it can be consumed by certain groups of people that are considered more vulnerable, such as:

- children;
- elderly;
- immunosuppressed;
- pregnant; and
- allergic.

When formulating food for infants and young children, the HACCP team should consider that the legal requirements for these foods are stricter regarding microbiological criteria, contaminant limits, and more restrictive use of food additives.

It is generally accepted that people included in these groups: elderly, pregnant women, infants and children, and people with weakened immune systems, are more likely to suffer from food-borne illnesses and food poisoning, not only because the infective dose is significantly lower in many cases, but also because the consequences of the disease are also usually more severe. Despite the recognition of these most vulnerable groups, the doctrine established in the food industry is that food must be equally safe for the entire population, including the most sensitive people.

Another important group of consumers to pay attention to is those who have some allergy or intolerance to certain food ingredients. The management of allergens in the food business requires a comprehensive approach within the HACCP system since it affects several procedures within the self-control system, such as the supply of raw materials, control of the separation between manufacturing lines, good handling practices, cleaning and disinfection programs, waste management, food packaging, and labeling. The food safety team, based on the information available on the raw materials and ingredients used in the manufacture of the different products, must carry out a risk analysis of the presence of these substances in the finished products, either because they are part of their formulation or because there is a possibility of cross contamination between production lines that it is not able to avoid with the means at its disposal: efficient separation between production lines or processes, effective cleaning systems, etc. Risk control for individuals with food allergies or intolerances lies in providing them, through labeling, with exact information on which substances that cause allergy or intolerance may be present in the finished product. Regulation 1169/2011 on the provision of food information to consumers obliges the food operator to include in the labeling any ingredient or technological adjuvant – or that derives from a substance or product – from one of the fourteen groups of substances that cause allergy or intolerance listed in Annex II of aforementioned European regulation.

A frequent cause of consumer complaints and alerts in the Rapid Alert System for Food and Feed (RASFF) is precisely the presence of undeclared allergens on the labeling of food products. The auditor may be interested in checking whether labeling food products declares all allergens present in processed foods or whether FBO has correctly analyzed the risk of cross contamination between different products and manufacturing lines (Figure 2.23).

Substances or products that cause allergies or intolerances Annex II to Regulation (EC) 1169/2011
Cereals containing gluten, namely wheat (such as spelt and Khorasan wheat), rye, barley, oats or their hybridized strains, and products thereof
Crustaceans and products thereof
Eggs and products thereof
Fish
Peanuts and products thereof
Soybeans and products thereof
Milk and products thereof (including lactose)
Nuts, namely almonds (*Amygdalus communis* L.), hazelnuts (*Corylus avellana*), walnuts (*Juglans regia*), cashews (*Anacardium occidentale*), pecan nuts (*Carya illinoinensis* (Wangenh.) K. Koch), Brazil nuts (*Bertholletia excelsa*), pistachio nuts (*Pistacia vera*), macadamia or Queensland nuts (*Macadamiaternifolia*), and products thereof, except for nuts used for making alcoholic distillates including ethyl alcohol of agricultural origin
Celery and products thereof
Mustard and products thereof
Sesame seeds and products thereof
Sulfur dioxide and sulfites at concentrations of more than 10mg/kg or 10 mg/l in terms of the total SO_2 which are to be calculated for products as proposed ready for consumption or as reconstituted according to the instructions of the manufacturers
Lupin and products thereof
Mollusks and products thereof

Figure 2.23 Allergens listed in Regulation 1169/2011.

2.14 Construct a Flow Diagram

The following preliminary step of the HACCP study consists of constructing a flow diagram for each production line. A flow diagram is a schematic representation of the operations through which the production process of a given food passes, from the reception of raw materials and ingredients to the finished product. However, it is also possible to include some steps before and after the manufacturing process, such as the transport of raw materials and the distribution of the final product, so that there is correspondence between the steps of the flow diagram and the scope of the activities of the food operator within the food chain.

The primary purpose of the flow diagram is to facilitate the task of the HACCP team to conduct the hazard analysis and determination of CCPs. The hazard analysis must be conducted for all the operations that make up the production process; for this reason, when these operations are correctly identified in the flow diagram, the work of the HACCP team can be better systematized, which is the best way to avoid overlooking some activities. When the team omits a stage in the flow diagram, this omission may cause an incomplete hazard analysis or unforeseen hazards, reducing the effectiveness of the entire HACCP system.

The flow diagram allows the logical and structured approach required for the determination of CCPs because it lists all the steps of the process in which the possible presence of CCPs must be considered and also because it provides the overview necessary for the team to check if there are subsequent steps that can eliminate or control the hazards analyzed in each specific stage. The stages identified in the flow diagram should correspond to those analyzed in the subsequent stages of the HACCP study, i.e. those included in the HACCP worksheet and on the team working documents related to the determination of CCPs.

The auditor may want to check these aspects to see if the HACCP team has followed a structured approach and if the HACCP study is consistent; when this is not evident, he may want to investigate the reasons why this mismatch occurs between the phases of the flow diagram and the systematic determination of the CCPs.

There are no rules on how to construct flow diagrams. The format is the choice of each food operator. What a flow diagram must comply with is that the information it provides is truthful and accurate. In its simplest version, the minimum information that a flow diagram must include is the operational steps that make up the food production process, ordered in the actual way it happens. From there, the FBO can add other information such as input and output of raw materials, ingredients, by-products or intermediates, data on parameters related to stages or operations such as times and temperatures, stages with CCPs or control points, the type of hazards considered in each stage, the phases where loops or rework can occur.

Codex guidelines state it as follows:

Flow diagrams should, as appropriate, include but not be limited to the following:

- the sequence and interaction of the steps in the operation;
- where raw materials, ingredients, processing aids, packaging materials, utilities, and intermediate products enter the flow;
- any outsourced processes;
- where applicable, reworking and recycling take place;
- where end products, intermediate products, waste, and by-products are released or removed.

However, it is not advisable to overload the flow diagram with so many details or so much information that it makes it more difficult to understand at first glance and too messy and loses its function of a schematic representation of the production process. The HACCP team will consider

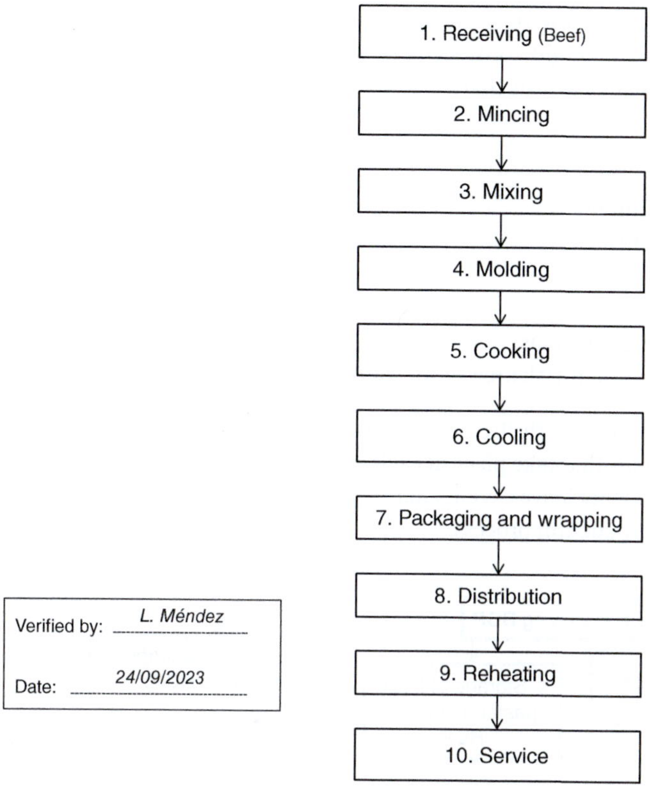

Figure 2.24 Flow diagram for the production of frozen meat pies.

much more information and other details not represented in the flow diagram, but that must be available, and to a greater extent, in other documents and other sources of information.

A widely used type of flow diagram is one in which rectangular text boxes joined by arrows represent the stages of the process; at the top of the page is the initial phase or the input of raw materials, and at the bottom is the final stage or the finished product; some companies find it helpful to use other shapes in text boxes such as rhombus, in places where it is necessary to make a decision, an oval shape for the final phase, a rectangle with a wavy bottom edge to indicate that a document or record is generated. Usually, it is advisable to number the steps in the text boxes (Figure 2.24).

The content and type of operations represented in the flow diagram vary depending on whether we refer to a linear HACCP or another with a modular approach; thus, we can find different types of flow diagrams:

- **Linear**: represents the specific manufacturing process of a single product or group of similar products.
- **Modular**: used for more complex food operations or when several different foods produced by the same business share certain parts of the production process, as in the case of mass catering establishments.
- **Modular set**: a master flow diagram that depicts the set of all the modules that make up the production process of a food operator.
- **Individual modular**: contains the specific operations carried out in a specific module of those that are part of the global module of the food operator.

In addition to the information provided by the flow diagram, variable according to the type and level of detail included, the HACCP team might consider other aspects that are also of interest to analyze in greater depth the processes and hazards that possibly originate in each of the operations such as:

- Drawings of the premises.
- Layout and characteristics of the equipment.
- Flows of raw materials and by-products.
- Routes followed by staff.
- Existence of line crossings.
- Separation between dirty areas and clean areas.
- Existence, or need, of clean rooms or high-risk areas with controlled environment.

With this complementary information and with the help of the flow diagram, the HACCP team can get a complete idea of the operations that take place at each step and thus be able to determine in a more systematic way what enters, what leaves, what changes, and what remains in each of the operational steps that makes up the production process.

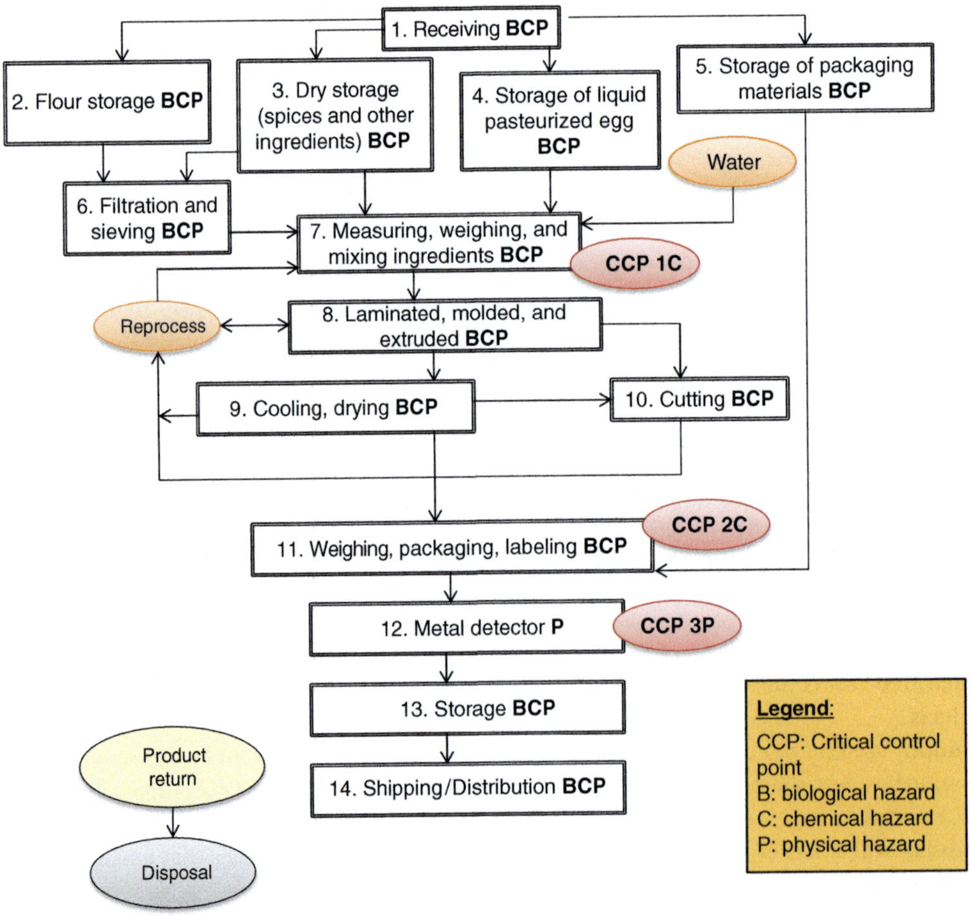

Figure 2.25 Flow diagram for the production of fresh unfilled pasta.

In many cases, to make the flow diagram manageable, various basic operations performed at a particular place or at a specific time can be grouped under the name of a single phase. So, in the flow diagrams representing food preparation processes, we can find stages such as food preparation, cooking, assembling composed of several specific culinary practices. For example, if we choose the stage "food preparation," this could be decomposed into cleaning, washing, boning, evisceration, peeling, chopping, mincing, marinating, etc. The HACCP team must consider which phases contain grouped processes and know the specific operations that comprise each to carry out the hazard analysis effectively.

Some food operators consider it appropriate to have several flow diagram models of their processes: a complete model containing all the information required for the HACCP study and which the FBO can retain as HACCP supporting documentation, and another abridged version, with little detail, which would be shared with customers and other stakeholders requesting information about the company (Figure 2.25).

For the audit of the self-control system, the flow diagram greatly helps plan the audit, so it is one of the documents usually requested previously from the operator. With the information it contains, the auditor can get an idea of the company's activities, the number of production lines, and the type and complexity of the processes, and, in this way, can better prepare the on-site audit. Then, when performing the audit activities, the auditor will want to verify that the FBO meets the requirement to document the flow diagrams, that there is one for each product or group of products, that they are up to date, and that the information they contain is correct.

2.15 On-site Flow Diagram Verification

Verifying the flow diagram is the last of the preparatory steps of the HACCP study before applying the principles. The systematic and exhaustive nature of HACCP aims to ensure that the study is carried out on all the operations and phases that make up the production process, with the ultimate objective that there are no aspects left unanalyzed or hazards left unprevented.

On-site flow diagram verification is a system requirement for HACCP equipment to ensure that the stages and data it contains are correct. This previous phase of the system is like a plus of guarantee so that the HACCP study is developed from a sound and reliable base. Owing to the interrelationship between the phases of the process and the application of HACCP principles, a failure or the omission of some operation at this time of the study would be unfailingly carried over to the following phases, leading to a deficient HACCP study.

Those responsible for verifying the flow diagram are usually one or more members of the HACCP team. However, they may also be assisted by others more directly involved in the production tasks. To confirm the flow diagram, we must "walk the line," i.e. walk throughout the factory floor to follow the different operations that make up the process. This verification is a complex task in many cases and may need some preparation or the help of checklists. Those responsible for conducting this verification should take their time to observe the activities closely, asking the operators or the owners of the processes, when necessary, to provide more information about the activities they carry out and about the circumstances that may affect the way of working, such as production shifts, changes in work rhythms, incidents or seasonal variations of the activity.

Not all flow diagrams present the same difficulty to be verified in situ. There are notable differences between the processes depending on the types of companies and the food sectors that we consider; hence, the strategy and the way of preparing the verification by the team have to adapt to each case. It is very different to verify the processes of a catering establishment or a cattle slaughterhouse, where most operations are manual and can be followed just by looking at them, or the flow diagram of a dairy milk packaging industry, where we only see interconnected tubes and equipment, or a wine cellar, in which there is little manual work and the only thing we can observe most of the time are deposits.

Another difference to consider for verification planning is the duration of the processes, from those that last an hour or less to others that last months or several years, as is the case with foods subjected to more or less prolonged stages of fermentation or maturation. Finally, we must also take into account how the production cycles are. There are short production cycles that are repeated continuously and that we can see any day of the year and others that vary according to the seasons or that are seasonal, such as the reception of the grapes in a winery; in this case, specific operations such as destemming, crushing, or pressing of grapes only can be checked during the time that the harvest lasts and the entry of the grapes into the wineries. The HACCP team can take into account all these factors when planning how they will carry out the verification of the flow diagrams in situ.

To verify compliance with this HACCP requirement, the auditor will examine the evidence of the confirmation of the flow diagram, such as the name of the responsible and the date of the verification or whether a procedure defines this responsibility.

Another issue the auditor can check is whether the diagram is updated. The audit team is likely to be interested in investigating whether there have been changes to products or manufacturing lines since the date of the last edition of the flow diagram and whether these are included in the diagram or whether there is any procedure defining the manner, frequency, and responsibilities for updating flow diagrams.

Similar to how the HACCP team had to proceed with the HACCP study, the auditor may also want to confirm the accuracy of the flow diagram by examining the sequence of plant operations. For this task, he has the same techniques and sources of information that have been discussed before for the team: direct observation of the processes, interviews with the workers and with those responsible for the process, review of records, etc. The fundamental difference is that while the HACCP team has to verify in situ all the phases and all the manufacturing lines included in the scope of the HACCP study, the auditor team will probably do so by sampling and choosing only some of the phases or processes or, based on the risks associated with particular operations, the observations they make during the visit to the plant or their professional criteria.

References

ASQ Food, Drug and Cosmetic Division (2014). *The Certified HACCP Auditor*. Wisconsin: ASQ Quality Press.

Europe, European Parliament and Council (2004a). *Regulation (EC) No 852/2004 of the European Parliament and of the Council of 29 April on the hygiene of foodstuffs. Official Journal of the European Union.* 139: 1–54.

Europe, European Parliament and Council (2004b). *Regulation (EC) No 853/2004 of the European Parliament and of the Council of 29 April laying down specific hygiene rules for food of animal origin. Official Journal of the European Union.* 139: 55–205.

Europe, European Parliament and Council (2008). *Regulation (EC) No 1333/2008 of the European Parliament and of the Council of 16 December on food additives. Official Journal of the European Union.* 354: 16–33.

European Commission (2022). *Commission Notice on the implementation of food safety management systems covering Good Hygiene Practices and procedures based on the HACCP principles, including the facilitation/flexibility of the implementation in certain food businesses* (2022/C 355/01).

FAO and WHO (2022). *General Principles of Food Hygiene. Codex Alimentarius Code of Practice, No. CXC 1-1969*. Rome: Codex Alimentarius Commission.

Food and Drug Administration. *Managing Food Safety: A Manual for the Voluntary Use of HACCP Principles for Operators of Food Service and Retail Establishments*. 2006.

Gaze, R. (2009). *HACCP – A practical Guide*, 4e, Guideline No. 42 . Chipping Campden, UK: Campden Bri.

ICMSF (International Commission on Microbiological Specifications for Food) (2011). *Microorganisms in Foods 8. Use of Data for Assessing Process Control and Product Acceptance*. Springer US.

Mortimore, S. and Wallace, C. (2013). *HACCP: A Practical Approach*, 3e. Springer US.

National Advisory Committee on Microbiological Criteria for Food (NACMCF) (1997). *Hazards analysis and critical control point principles and application guidelines*. https://www.fda.gov/food/hazard-analysis-critical-control-point-haccp/haccp-principles-application-guidelines/ (accessed 25 November).

Panisello, J.P. and Quantick, P.C. (2001). *Technical barriers to Hazard Analysis Critical Control Point (HACCP). Food Control* 12: 165–173.

Rahman, M.S. (2007). *Handbook of Food Preservation*, 2e. CRC Press.

Wareing, P. (2010). *HACCP: A Toolkit for Implementation*. Leatherhead Publishing.

Wallace, C.A., Sperber, W.H., and Mortimore, S.E. (2011). *Food Safety for the 21st Century*. Oxford, UK: Wiley-Blackwells.

3

The Seven Principles of HACCP

3.1 Principle 1

Hazard analysis is a crucial principle of hazard analysis and critical control point (HACCP), decisive for the result of the HACCP study to be a genuinely helpful and effective instrument for the food business operator (FBO). Although its application can be more flexible, we must understand the HACCP system as a whole, in which each of the principles fulfills its function to the extent that it complements the others, acting jointly. Therefore, it does not make much sense to compare the value of each HACCP principle separately; however, due to the breadth and nature of the tasks covered by principle 1, it is presumably the most important HACCP principle.

The application of principle 1 is a challenge for the work of the HACCP team because it tests the competence, knowledge, and experience of its members to carry out a structured and well-founded hazard analysis based on the thorough analysis of the production process and the handling of different sources of external information, necessary to identify all the potential hazards that can reasonably be expected in the preparation of a given food. A good part of the scientific basis required for the development of HACCP is used at this stage in the analysis of hazards and in determining effective control measures to reduce these hazards to an acceptable level. In any HACCP study, hazard analysis is fundamental. It makes the difference between a specific self-control system, focused on the failures that may occur and adapted to the needs of the food operator, and what is nothing more than a mere paper exercise to meet a requirement.

At this stage, the HACCP team must define "what" to control; the remainder of the study will set out "how" to control and how to demonstrate and document it. Therefore, it is easy to understand that any omission or deficiency in this phase will be transmitted to the rest of the study, conditioning the final result. Principle 1 is the "analysis principle." We understand by analysis the procedure of separating any material or abstract entity into its constituent elements or examining something, especially by separating it into its parts, to discover what it is or contains. Following Codex's guidelines, this is a task where true teamwork acquires crucial importance, where the participation of all its members must prevail, and not of a single person, no matter how much professional value it has.

In this phase of the HACCP study, we can distinguish the following three steps:

List all potential hazards
↓
Determine significant hazards
↓
Establish control measures for each significant hazard

HACCP System Auditing for Food Safety: Principles and Techniques, First Edition. Luis Couto Lorenzo.

Before starting this task, all HACCP team members must be familiar with hazard and hazard analysis concepts. As we saw on page 27, these definitions are included in the HACCP document of the Codex (FAO and WHO 2022), as follows:

> **Hazard**: A biological, chemical or physical agent in food with the potential to cause an adverse health effect.
> **Hazard analysis**: The process of collecting and evaluating information on hazards identified in raw materials and other ingredients, the environment, in the process or in the food, and conditions leading to their presence to decide whether or not these are significant hazards.

The HACCP Guide of the National Advisory Committee on Microbiological Criteria for Foods (NACMCF) (1997) contains a definition of hazard in terms very similar to that of the Codex, although it presents a slight difference, as we can see:

> **Hazard**: A biological, chemical, or physical agent that is reasonably likely to cause illness or injury in the absence of its control.

This definition of hazard contains a small difference, but that can be of great help to guide the work of the HACCP team because it refers to the need to consider the probability of occurrence of the hazard and that such evaluation is made "in the absence of control," i.e. without taking into account the effect of control measures and prerequisite programs (PRPs) on the probability of occurrence (Mortimore and Wallace 2013).

When assessing the risks associated with the identified hazards, HACCP equipment must take into account the following factors:

- the probability of occurrence of the hazard;
- the severity of adverse health effects;
- the multiplication or survival of pathogenic microorganisms and the unacceptable generation of chemicals in intermediate products, final products, the production chain, or their environment;
- the production or persistence in food of toxins and other undesirable products of microbial metabolism, chemicals, physical agents, or allergens;
- the causes of each of the hazards and the conditions that favor these risk factors;
- identified acceptable levels of the hazards in the food, e.g. based on regulation, intended use, and scientific information;
- the nature of the facility and the equipment used in making the food product;
- the intended use and/or probability of product mishandling by potential consumers that could render the food unsafe;
- the target population and vulnerability of exposed consumers; and
- conditions leading to the above.

Depending on the origin or nature of the hazards, and according to the Codex and NACMCF definitions just discussed, HACCP distinguishes three types of hazards:

- physical
- chemical, and
- microbiological.

In some sources or references, other types of hazards are added, such as allergens, treated as a separate category from chemical hazards, and radiological hazards, such as radium-226, 228, uranium-235, 238, strontium-90, iodine-131, and cesium-137, as advocated by some authors; this categorization began to arouse more interest since the publication of the Food and Drug Administration (FDA) Food Safety Modernization Act (FSMA) in 2011 (Figure 3.1).

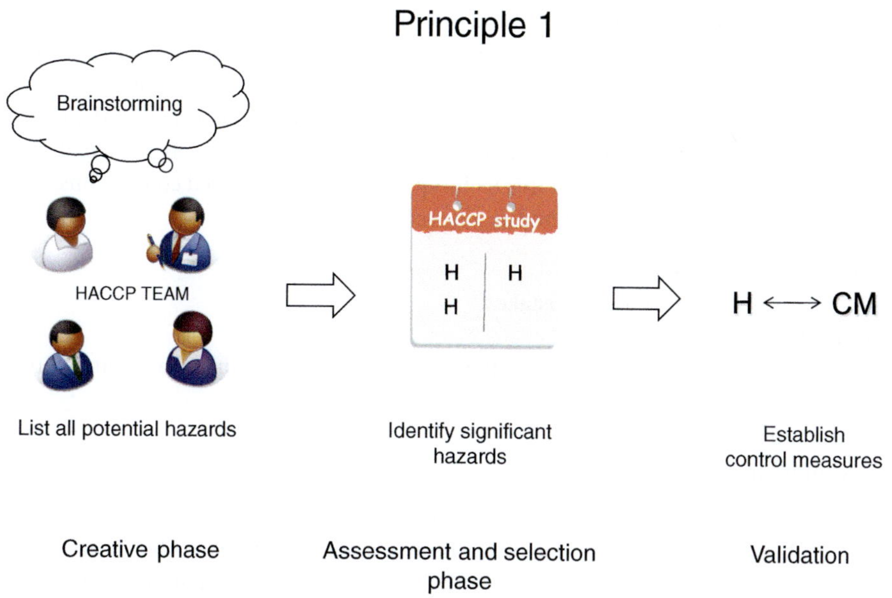

Figure 3.1 Principle 1 phases.

3.1.1 Brainstorming

In the first phase of the hazard analysis, the HACCP team must be able to list all potential hazards that may arise from the production process examined. Brainstorming is a task in which all team members must be able to contribute without feeling censorship, impediment, or limitation. We are facing the most creative phase of hazard analysis, where what matters is to give free rein to the imagination and try to think about any fact or circumstance that may lead to a failure or a problem capable of generating a hazard as strange as it may seem at first sight. The objective of this phase is to maximize the ability to anticipate problems and reduce the occurrence of unforeseen events as much as possible.

To organize the work of the HACCP team, it is necessary that someone from the team, either the leader or the secretary, has the necessary capacity to plan, prepare, and convene the working group sessions. This is an expense, consumption of the company's resources, which management, as a sign of its commitment to food safety, must assume from the beginning, authorizing team members to neglect their obligations temporarily so that they can have enough time to participate in these work sessions, without pressure and without the need to be aware at all times of their usual tasks.

Group dynamics techniques can foster the development of HACCP team meetings. There are several tools available for problem analysis and consensus decision-making, such as the nominal group or the Delphi technique; however, in the case of hazard analysis, the most appropriate tool and also the one most commonly cited is brainstorming, although it is not usual to find records or evidence that can prove its use by the HACCP team. Brainstorming is a group dynamics technique specially designed to stimulate the creativity of participants in order to generate new ideas and seek different approaches to known problems, avoiding the tendency to follow a natural or habitual pattern of thinking. The basis of brainstorming is to encourage the participation of all group members and to do so in the most uninhibited way they can.

The first thing needed to prepare the brainstorming is to summon the team members to reserve enough time in their agendas to participate in the sessions without haste or stress. One of the people in the group must be trained to be the coordinator or facilitator of the work session. The role of the leader or facilitator consists of raising the issue or problem to be analyzed, ensuring that the rules of brainstorming are respected, maintaining a pleasant environment conducive to participation, and guiding the process through the phases of this technique to reach the desired conclusions.

The rules of brainstorming can be summarized as follows:

- No one criticizes or judges anyone.
- Eliminate self-criticism; we can all make mistakes.
- Better to be wrong than not to intervene.
- The more ideas emerge, the better; it does not matter how crazy or strange they may seem at first sight.
- All team members must participate equally.
- Be brief in the exhibitions.

The development of a brainstorm goes through several specific phases. However, some of them, such as the initial explanation of the technique or warm-up exercises, can be avoided or abbreviated when the participants already have experience with this tool or when it is a more mature and cohesive group for a long time.

If participants need to learn this tool, it is necessary to explain the objective, the steps followed, and the rules of brainstorming. In the next phase, called warm-up, the aim is for the participants to relax and be uninhibited and to create the most appropriate environment for them to intervene in the most productive way possible. For this purpose, team members can use exercises such as riddles, logic games, or brainstorming on a fictitious topic unrelated to the subsequent work.

Then, the central part of the brainstorming begins, which is composed of these four phases:

- definition of the situation;
- creative phase;
- categorization; and
- critical phase.

The definition of the situation in the case of the HACCP study consists of listing the potential hazards of a particular activity. For this task, participants must previously have all the available information about the process analyzed, such as the product description and the flow diagram.

The most characteristic part of brainstorming is the creative phase; its objective is that the participants contribute as many ideas as possible about the hazards that can be caused in the different operations of the elaboration process, taking into account the causes and/or the circumstances that originate them. The role of the facilitator in this phase is vital to stimulate the participation of all HACCP team members equally, without any fear of expressing their ideas. All contributions are necessary, both those of the members with better training in food hygiene and those of the team members who work directly on the production line. The hierarchy or positions must be left out of the session.

As ideas emerge, a team member should write them down on a whiteboard or flipchart. Brainstorming can also be done in writing; after raising the situation to be analyzed, participants are given time to write their ideas and then share them with the team. Another possibility that can also be considered is to stop the session for a few hours or even continue it the next day or later. This pause is known as the incubation phase, and there are times when it may be convenient to allow participants to get away from the problem for a moment and leave time to mature ideas or for

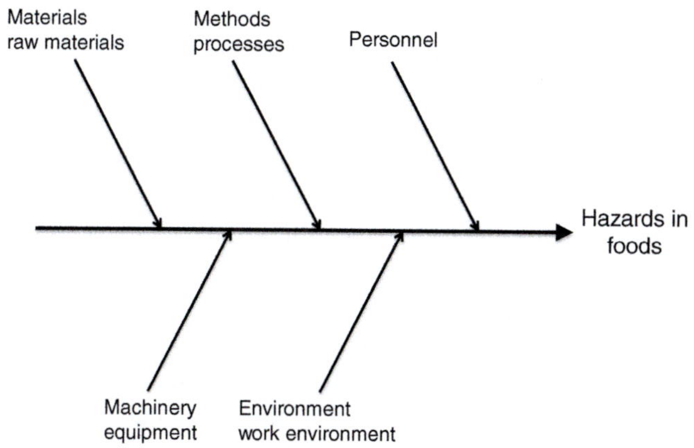

Figure 3.2 Five causes of hazards.

new ones to emerge. Brainstorming sessions usually last one to three hours; when the scheduled time for the meeting is running out, it is better to postpone the task for another day than to try to accelerate the debate and have to make decisions in a hurry.

In the categorization phase, the team examines the ideas and may categorize or classify them according to different concepts, such as the origin or nature of the identified hazards. For example, one way to classify hazards is according to whether they originate from the environment, equipment, personnel, process, or raw materials.

The critical phase represents a U-turn in the session's focus and the working group's objective. After the creative part, we now have to evaluate and select the circumstances and hazards that, from a practical and realistic sense, should be considered for examination in the next step of the HACCP study. Companies' resources are limited, and the success of management systems depends largely on knowing how to determine risks judiciously and on focusing available resources on aspects that affect production. With this last brainstorming phase, the HACCP team prepares the work for the next step of principle 1 (Figure 3.2).

3.1.2 Hazard Identification

In the process of identifying the hazards associated with the manufacture of a given food, the HACCP team should use two types of information sources:

- Internal or specific to the elaboration process.
- External or general.

The internal source is the one that derives from the analysis of the phases of the specific manufacturing process in the plant of each food operator, i.e. as described in the HACCP systematics, analyzing each of the steps and operations that make up the flow diagram to see which hazards may originate or the conditions for their appearance may occur. That is the core of HACCP.

In the case of external sources, what the HACCP team does is leave the process for a moment and look outside the company for all the information that may exist about the hazards associated with the type of food they produce, the raw materials used, the technology of the process or the food sector to which the food operator belongs.

Appendix C of the NACMCF HACCP Application Guide includes several factors that can assist the HACCP team in carrying out hazard analysis in a more structured manner. Among the items included in this appendix are the following:

- **Ingredients**: In addition to allergens, in which the ingredient itself is the hazard when it belongs to one of the identified groups of allergens, raw materials and ingredients are often associated with specific chemical, physical, or microbiological hazards, e.g. *Salmonella* and *Campylobacter* in chicken meat, aflatoxins in cereals, or mercury in some fish species.
- **Intrinsic factors**: Physical characteristics and composition (e.g. pH, type of acidulants, fermentable carbohydrate, water activity, preservatives) of the food during and after processing. As we saw when we referred to the product description phase, the composition and physicochemical characteristics of the food are decisive to knowing what type of microorganisms may be present and which can survive or multiply in them, and therefore establish the safety of the food.
- **Processing procedures**: Manufacturing processes can influence food safety in several ways; some processes are designed or serve to eliminate or reduce the presence of hazards to an acceptable level, e.g. heat treatment or pickling of canned vegetables, while in other cases, there are food processing operations in which, incidentally or unwantedly, other types of hazards arise, such as, for example, with toxic compounds produced in the frying process, such as acrylamide formed by a heat-induced reaction between asparagine and glucose and fructose, naturally present in some foods, or 3-monochloropropane-1,2-diol, a contaminant of the chloropropanol group, generated during the manufacture of soy sauce and hydrolyzed vegetable proteins produced by acid hydrolysis.
- **Microbiological content of the food**: The HACCP team must take into account what the expected microbial content is in a food or raw material when obtention and processing hygiene practices are correctly applied and from which values the content of microorganisms can be considered excessive, indicating a lack of hygiene or an increased risk to consumers; as well as the reasons why this microbial growth occurs. There are several sources to consult the reference values regarding the type and quantity of microorganisms considered acceptable in different foods.
- **Facility design**: The design and distribution of the facilities are essential to achieve a hygienic manufacturing environment and facilitate good handling practices. The HACCP team must analyze the flows of raw materials, people, or air to assess at which points contaminations may originate, how to differentiate dirty and clean areas, or the need for clean rooms with a controlled environment.
- **Equipment design and use**: In many cases, hygiene and the effectiveness of operations carried out with food depend on the right equipment. The HACCP team must consider the capacity, reliability, maintenance, and accuracy of equipment and measurement systems to evaluate the control of processes. It is also necessary to consider the equipment's most appropriate cleaning procedure and if they have systems for detecting or eliminating hazards such as magnets, filters, metal detectors, sifters, screens, bone removal devices, dud detectors, or X-rays.
- **Packaging**: Most of the time, the packaging is inseparably linked to the food. In many cases, food marketing cannot be conceived without the packaging containing it. Packaging plays diverse critical roles in food safety. Depending on the characteristics and properties it presents, the choice of the type of packaging is an essential element of the hazard analysis. In addition to the physical protection of the contents, issues such as hermeticity, selective permeability to CO_2, water vapor, or oxygen must be analyzed, as well as the use of modified atmospheres, or the increasing use of active packaging, which can absorb or release certain substances as appropriate. The packaging also supports the information transmitted to the consumer, such as the expiration or best-before date, storage conditions, instructions for use, or the list of allergens it contains. Last but not least, the packaging must be considered a source of hazards because of the harmful substances they can transfer to food, such as bisphenol A, melamine, or inorganic tin. The HACCP team must

consider the type of packaging used to ensure compliance with the principle of inertia of food contact materials, a legal requirement established in European Regulation 1935/2004.

- **Disinfection**: The cleaning and disinfection program is a prerequisite of the HACCP system; depending on the risks and the type of contamination occurring in the plant, the HACCP team and those responsible for cleaning programs must decide on the procedures and frequency of cleaning in the different manufacturing areas and equipment, to avoid the presence of physical, chemical, microbiological, and allergenic hazards. This analysis includes considering the hazards of improper use of cleaning and disinfection products.
- **Employee health, hygiene, and training**: Food handlers can affect food safety in several ways: due to their hygienic or manufacturing practices or because they can themselves be a source of contamination of microbiological and physical hazards, by their clothing and personal effects. When they have foodborne illnesses, it is necessary to separate them from food premises and monitor them medically. The training and awareness of handling personnel are essential to achieve the hygiene required in the food industry. In the past, handlers' bodies as the cause of food contamination attracted much attention. However, today, we know that what is crucial is the handlers' minds because it will condition their attitude and behavior and their adherence to HACCP hygiene procedures.
- **Storage conditions between packaging and end user**: Once the food has been processed, the HACCP team also needs to consider the next steps in the food chain, such as transport, distribution, and storage, to analyze the circumstances in which there may be contamination of the food or the microbial growth that occurs when there is abuse at the indicated storage temperatures.
- **Intended use**: The HACCP team must consider the product's intended use by the consumer, for example, if the product is ready for consumption or if it requires some treatment by the consumer. In those products that the end user does not consume in their entirety but can remain in the containers once opened, the manufacturer has to consider the new recommended expiration and if it requires different storage conditions to those before opening it.
- **End consumer**: This is another aspect already analyzed in the previous steps of the HACCP study. The hazard analysis varies depending on the target population of the food, mainly when it includes the most vulnerable groups, such as children, the elderly, or the immunocompromised. For example, the pathogen *Cronobacter sakazakii* poses a severe threat to children below six months of age and also to some older people or those with a weakened state of health, while for the rest of the adult population, it is not considered relevant. For this reason, European Regulation 2073/2005 establishes that *C. sakazakii* is a food safety criterion for dried infant formulae and dried dietary foods for special medical purposes intended for infants below six months of age but not for other products.

The HACCP team should also consult external sources to identify the hazards associated with each food category. These sources of information include:

- Food legislation.
- Food alerts; the Rapid Alert System for Food and Feed (RASFF) system.
- Generic or sectoral HACCP guides.
- Scientific literature.
- Epidemiological data.
- News about incidents in food companies.
- Web pages.

Some European regulations and rules contain provisions concerning the different chemical, physical, microbiological, and allergenic hazards that may be present along the food chain. European

Regulation 2073/2005 on microbiological criteria for foodstuffs identifies microbiological hazards that may be found in certain types of food. Therefore, it is necessary to consider the microorganisms, toxins, and metabolites associated with specific food categories by this regulation when carrying out the hazard analysis. The microbiological criteria laid down in Regulation 2073/2005 are based on scientific knowledge, empirical data, and risk assessment, so we must accept that there is sufficient epidemiological evidence to assert that these hazards may pose a risk to the consumer.

The application of food law as an external source to identify hazards that should be relevant in food is specifically recognized in point 7.4.2 of the guidance for the application of ISO 22000:2005 in the following terms:

> ""In cases where legal and regulatory authorities have established maximum limits, objectives, targets or criteria for finished product and/or processes for a specific hazard/product combination, the hazard in question automatically becomes relevant to that product.""

The reason for acting in this way should not come as a surprise since, as we said, these food standards rely on the risk analysis carried out by a panel of experts, such as the European Food Safety Authority (EFSA) in Europe, based on the available information and epidemiological data for each hazard, and the conclusion they reach is that the agent or hazard in question is significant enough to adopt management measures, such as the establishment of legal limits or microbiological criteria. In this case, risk analysis makes the job easier for hazard analysis.

The most important – but not the only – reference standard for chemical hazards is Regulation 2023/915 on maximum levels for certain contaminants in food. Annex I to this Regulation sets out maximum limits for six groups of contaminants in food: mycotoxins, plant toxins, metals and other elements, halogenated persistent organic pollutants, processing contaminants, and other contaminants. So, we can know the contaminants and limits we must meet for each food.

Another cause of hazards is food contact materials. European Regulation 1935/2004 on materials and articles intended to come into contact with food, known as the "framework regulation," establishes the principle of inertia whereby materials and articles, including active and intelligent materials and articles, shall be manufactured in compliance with good manufacturing practice (GMP) so that, they do not transfer their constituents to food in quantities which could endanger human health, bring about an unacceptable change in the composition of the food or a deterioration in the organoleptic characteristics thereof. In addition to this regulation, the HACCP team can take into account legislation on specific materials such as plastics, regenerated cellulose, or active materials, among others, or the rules on substances such as nitrosamines, BADGE (bisphenol A diglycidyl ether), BFDGE (bisphenol F diglycidyl ether), and NOGE (Novolac glycidyl ethers). Food businesses are users of these materials and articles. Hence, the control of these risks relies on the guarantees given by the supplier that the migration limits established in the conditions and the intended use for each of the foods are not exceeded, which justifies the requirement by the food operator of the "Declaration of Conformity" to the manufacturers of various types of materials.

There is no legal standard for physical hazards in food, so the sources of information for the HACCP team should be based on their experience in the production process, on the registration of incidents or consumer complaints, depending on the product type.

Concerning allergens, as mentioned in the product description step, European Regulation 1169/2011 on the provision of food information to consumers includes in its Annex II the fourteen groups of substances or products causing allergies or intolerances, whose declaration is mandatory on the labeling of food.

The FBO must identify which of these allergens are part of the composition of their food and when and where cross contamination may occur between their processes to prevent it as much as possible. Once the FBO has assessed the risks of the presence of these substances due to various causes, it must declare them on the label.

Another important external source of hazard information for the HACCP team is the RASFF. Created in 1979, the RASFF is a robust and efficient rapid information exchange system involving the 28 member states of the European Union, the European Commission, EFSA, EFTA Surveillance Authority (ESA), Norway, Liechtenstein, Iceland, and Switzerland. The RASFF allows food safety authorities to rapidly exchange information on health risks derived from food or feed so that they can take immediate action to avert the risk. Through round-the-clock service, the RASFF ensures that urgent notifications are sent, received, and responded to collectively and efficiently. Thanks to this rapid exchange of information between countries, some products can be withdrawn from the market before they pose a risk to consumers.

Depending on the seriousness of the risks associated with food products and their distribution on the market, the RASFF classifies notifications into alerts, information, border rejection, and news. With all this information collected, the RASFF has a vast database on the risks and hazards associated with different types of food, which can be very useful to know what hazards should be considered in each food group and to assess what their significance may be based on the incidence data, and the trends recorded by the system. An interactive, searchable online database, called RASFF Window, offers public access to summary information about the most recently transmitted RASFF notifications and allows searching for information on any notification issued in the past. This information is available to consumers, business operators, and authorities worldwide, who can search the RASFF Window by type of notification, type of product (food, feed, or food contact material), hazard detected, and category of food product, making it an effortless way to obtain helpful information for hazard analysis.

Generic, specific, or sectoral HACCP guides are also helpful and reliable sources of hazard information for HACCP teams. Taking into account the significant similarity that usually exists between companies in a particular food sector, these guides, provided that they are developed by professional organizations or teams of people with sufficient expertise in food hygiene and the processes characteristic of each sector, provide an excellent basis to start developing the HACCP system.

The practical guidance contained in these guides can help the HACCP team to know which hazards are considered significant in each food product and what control measures can be applied effectively in each case and also to document or justify the decisions they make for the application of the principles of the system, since they are considered as a helpful means for HACCP system validation.

The HACCP team must analyze which parts or elements of the guide can be applied in their establishment and thus adopt the guide's content partially or in its entirety if they deem it appropriate. They should also study where there are substantial differences between their production processes and those described in the HACCP guide and, in that case, carry out a specific hazard analysis to complement the parts not included in the reference guide.

In addition to those mentioned above, there are other sources that the HACCP team can use to obtain information on the hazards associated with food products, as well as to seek information regarding the identification of emerging risks in food, such as the discovery of new toxic substances in food, the appearance of outbreaks by exotic species so far, upward trends in the cases of certain diseases, and new methods of detection of pathogenic microorganisms or those related to changes in production methods and the environment. Among these sources of consultation are scientific publications, specialized journals, epidemiological data collected by various agencies, websites of research centers, professional groups, or official inspection bodies, where the HACCP team can find much information related to hazards and the identification of risks in food (Figure 3.3).

The hazards identified must be described in a precise manner, indicating the nature of the hazard and its cause. Only in this way is it possible to set up appropriate control measures that act directly on the origin of the hazard to try to avoid it.

Depending on the type of product and the operations that make up the process, microbiological hazards, in some cases, can be described generically as "microbial contamination" or "microbial proliferation" without implying any inconvenience in establishing valid control measures. However, in other cases, it is necessary to define the microbiological hazards in a more specific way, indicating the genera and/or species of pathogenic microorganisms that may be involved in each food: *Salmonella* spp., *Listeria monocytogenes*, *Clostridium perfringens*, *Staphylococcus aureus*, etc. The reason for doing so lies in the need to consider the microbial ecology that characterizes each of these pathogens:

- their way of behaving in specific environments,
- the mechanism by which they develop their pathogenesis, and
- the factors that can affect their growth or survival.

For this reason, it is important to differentiate which microorganisms are spore forming, which produce toxins, or if the toxin is thermostable, as happens with *S. aureus*, because in this specific case, a subsequent phase of cooking or regeneration in a prepared meal would not be an effective control measure.

Another aspect that should also be included in the hazard analysis is the description of the nature of the hazard, distinguishing, in particular, between these four cases:

Presence: when the hazard or causative agent is already present in the raw material, as received from the supplier.

Contamination or introduction: when there is a contamination or introduction of the hazard during any of the manufacturing process phases.

Growth: in the case of microbiological hazards, when there are favorable conditions for the growth of certain microorganisms in raw materials or processed products.

Survival: when a microbiological hazard survives a specific treatment designed to eliminate it due to a failure during treatment or because it is inefficient to eliminate the pathogen.

These four elements, necessary in many cases to correctly describe the nature of the hazard, have become popular among professionals working in food safety under the acronym 'PIGS': Presence–Introduction–Growth–Survival (Figure 3.4).

The importance of including the nature of the hazard in its description is plain to see when it is necessary to establish control measures; as we can see in the example proposed in the following table, the control measures differ in each case and are not interchangeable (Figure 3.5).

The correct description of a hazard must include the following elements: the hazard itself, i.e. the physical, chemical, or microbiological agent that is capable of causing harm to the consumer; the nature of the hazard, i.e. whether it is presence, contamination, growth or survival, and finally, the cause or source that causes it. The application of this guideline can be illustrated with the following examples:

- Microbiological growth due to an incorrect temperature of conservation of the products in the cold refrigeration equipment.
- Survival of *Salmonella* due to insufficient cooking in chicken meat.
- Presence of inhibitor residues in milk due to contamination in farms of origin.

Examples of microbiological hazards	
Gram-negative bacteria	
Salmonella spp.	*Cronobacter* spp.
Campylobacter jejuni	*Aeromonas hydrophila*
Yersinia enterocolitica	*Plesiomonas shigelloides*
Shigella spp.	*Francisella tularensis*
Coxiella burnetii	***Escherichia coli* group**
Brucella spp.	Enterotoxigenic *E. coli* (ETEC)
Vibrio cholerae	Enterohemorrhagic *E. coli* (EHEC)
Vibrio vulnificus	Enteropathogenic *E. coli* (EPEC)
Pseudomonas aeruginosa	Enteroinvasive *E. coli* (EIEC)
Gram-positive bacteria	
Clostridium perfringens	*Listeria monocytogenes*
Staphylococcus aureus	*Mycobacterium bovis*
Bacillus cereus	*Clostridium botulinum*
Streptococcus spp.	*Enterococcus* spp.
Clostridium difficile	
Protozoa and helminths	
Toxoplasma gondii	*Anisakis simplex*
Giardia lamblia	*Diphyllobothrium* spp.
Entamoeba histolytica	*Nanophyteus* spp.
Crystosporidium parvum	*Eustrongylides* spp.
Cyclospora cayetanensis	Amoeba associated with gastrointestinal disorders
Trichinella spp.	*Ascaris lumbricoides*
Taenia spp.	*Trichuris trichiura*
Virus	
Norovirus	Rotavirus
Hepatitis A virus	Other viral agents
Hepatitis E virus	Astroviruses Enteric adenovirus Parvoviruses Picornaviruses Sapoviruses
Other pathogens	
Transmissible spongiform encephalopathies (TSEs) agents: prions	
Examples of chemical hazards	
Natural toxins	
Allergens	Biogenic amines: histamine
Ciguatoxin	Pyrrolizidine alkaloids
Scombrotoxin	Poisonous fish
Tetrodotoxin	Grayanotoxin
Mushroom toxins	Phytohaemagglutinin
Mycotoxins	Erucic acid
Ergot sclerotia and ergot alkaloids	
Shellfish toxins Paralytic shellfish poisoning (PSP) Diarrhetic shellfish poisoning (DSP) Neurotoxic shellfish poisoning (NSP) Amnesic shellfish poisoning (ASP) Azaspiracid shellfish poisoning (AZP)	

Figure 3.3 Examples of hazards.

Environmental contaminants	
Arsenic	Mercury
Polychlorinated biphenyls (PCBs)	Lead
Cadmium	Hexachlorobenzene
Perfluorinated compounds (PFCs)	Nitrate
Dioxins and furans	Tin (inorganic)
Polycyclic aromatic hydrocarbons (PAHs)	
Residues from plant protection products	
Pesticides	Fertilizers
Residues of veterinary medicinal products	
Chemotherapeutic agents	Corticosteroids
Antibiotics	Agents acting on the reproductive system
Antiparasitics	Growth hormones
Anti-inflammatories	Substances acting on the nervous system
Processing contaminants	
3-Monochloro-1,2-propanediol (3-MCPD)	BADGE, BFDGE, NOGE
Isopropyl thioxanthone (ITX)	Semicarbazide
4-Methylbenzophenone	Ethanol
Acrylamide	Polar compounds from frying
Other contaminants	
Industrial lubricants	Paint products
Cleaning products	Refrigerant products
Disinfectants	Water treatment chemicals
Coverings and coatings	Pest control chemicals
Packaging chemicals	
Plasticizers	Adhesives
Vinyl chloride	Bisphenol A (BPA)
Printing inks	Perfluorochemicals
Prohibited or unauthorized substances	
Sudan dyes	Melamine

Examples of physical hazards		
Material	Possible damages	Origin
Glass	Cuts in mouth, throat, stomach; possible need for surgery to remove it	Bottles, windows, lights, utensils, etc.
Wood	Cuts, infections, possible need for surgery to remove it	Pallets, wooden crates, building material, installations, and surroundings
Stones	Broken teeth, mouth injuries, choking	Raw materials, buildings, land surrounding the factory
Metals	Cuts, infection, possible need for surgery	Machinery, tools, wires, personnel
Bones	Cuts, choking	Raw materials
Plastics	Choking, cut, possible need for surgery	Bottles, envelopes, packaging, equipment
Insulating materials	Choking, long-term health effects (asbestos)	Building materials
Personal items	Cuts, choking, broken teeth, possible need for surgery	Employees

Figure 3.3 (*Continued*)

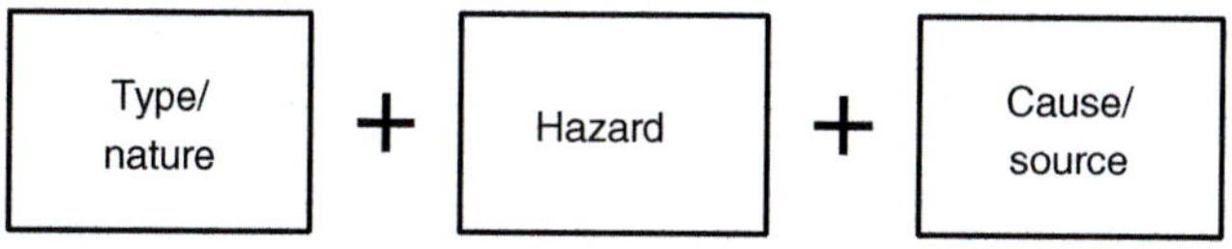

Figure 3.4 Hazard description elements.

Type of hazard/nature	Control measure
Presence	Supplier control; controls at receiving
Introduction (contamination)	GHP, GMP
Growth	Storage temperature control
Survival	Control of treatment parameters, equipment

Figure 3.5 Types of control measures according to the type and nature of the hazard.

As can be seen, when the proposed model is followed, the control measure to be applied is easy to determine. Thus, in the first example, it would be to monitor that the storage temperatures are adequate, and in the second, to ensure that sufficient culinary treatment is applied. In the third one, the control measures could be several: in the reception of the milk in the dairy center, the rapid detection test would be made to the tank or container; in case of a positive result, in addition to deciding the destination of the tank, it would be necessary to ensure that the farmer responsible for the contamination stops delivering milk while he cannot ensure that it is free of the contaminant or inhibitor detected and investigate in the farm of origin the ultimate cause, or root cause, of the presence of the inhibitor in the milk (poorly identified treated animals, not following the procedure of the milk withdrawal period, etc.) (Figure 3.5).

The description of hazards is one of the aspects that, without a doubt, the auditor will consider in the document review of the system to evaluate the application of principle 1. Many manuals still contain hazards described as follows: "loss of traceability," "improper handling," "lack of documentation," "product out of specification," "product contaminated at source," or "manufacturing defects." As these examples show, although they refer to real problems that influence the manufacturing processes or the control of products, as described, it is challenging to establish what hazard it is, what its nature is, and, therefore, what would be the most effective control measures. These generic or imprecise definitions end up conditioning the proper development of the HACCP study.

Once the HACCP team has worked out the list of potential hazards, the next step in implementing HACCP principle 1 is to identify the significant hazards and establish the relevant control measures.

As in the previous stage, what the HACCP team tried to do was to avoid setting limits in order to be able to think about the maximum number of possible hazards and the circumstances that originate them. Now, there is a radical change of direction because the aim is to assess, from a practical and realistic point of view, which hazards must be prioritized and how to allocate the necessary resources for their control.

This phase is crucial within the HACCP because it will condition the study's development and the entire system's effectiveness. The key to overcoming it is that the HACCP team knows how to determine which hazards have a reasonable probability of appearing and which can most affect production. Risk assessment consists of comparing probability with possibility. A potential hazard is possible, but is it likely? How many times has it happened among all the units produced by a company? This question is one of many that the HACCP team must answer; an FBO cannot organize the factory's activity nor establish control procedures as if a highly improbable event could occur at any time. "HACCP is not about that. He does not try to be perfect; seeks to demonstrate due diligence" (ASQ 2002). In any productive activity, it is impossible to control all possible hazards, no matter how rare or unlikely they may be. This effort to control everything would mean an unaffordable cost and an excessive expenditure of resources, which would end up conditioning the company's economic viability, which has to obtain its products by adjusting costs to market them at a competitive price.

For HACCP to be perceived by food operators as a valuable and credible tool for food safety management, it is necessary to take a pragmatic approach and develop systems designed to be applied by companies operating in the real world where zero risk does not exist. Only in HACCP applied to NASA's spaceflight programs could it make sense to set such an objective. However, those foods were not comparable to commercial production – where production costs are decisive – and economic constraints were not a problem.

The evaluation of hazards is not an easy task; it requires making decisions that, over time, can be verified that they were not the most accurate, so the HACCP team must always be attentive and willing to review and modify the system when experience shows that the expected results are not obtained.

According to the International Life Sciences Institute (ILSI), a significant hazard can be defined as a hazard of such a nature that its elimination or reduction to an acceptable level is essential for producing safe food, which brings us to another crucial concept in the HACCP: the acceptable level. For many agents of a biological or chemical nature, the potential hazard is not always a significant hazard. Still, it depends on these agents' ingestion dose or exposure level. For this reason, there are "maximum residue limits" (MRLs) and "acceptable daily intake" (ADI) established for these hazards, as well as the "as low as reasonably achievable" (ALARA) concept used in practice when no limits have been set.

The HACCP team must consider legal and regulatory requirements, the customer's intended use, and other relevant data, such as those related to the risk assessment for that hazard, to determine the acceptable level to be achieved.

The criteria commonly used to establish the significance of a hazard are the severity of adverse health effects in relation to the likelihood of their occurrence. This assessment can be carried out in a more structured way, using various tools, such as risk matrices.

The risk matrix is an essential component of the failure mode and effects analysis (FMEA), the predecessor of HACCP. During the first years of HACCP implementation, risk matrices were not very common but are now increasingly used to carry out and document hazard assessments. In the FMEA, the level of risk is calculated by combining the severity, probability, and detection – the latter estimates how well the controls can detect a failure and operates in the opposite way to the other two factors; combining these three factors, the FMEA team can obtain the risk priority number, which determines the measures to be applied.

In HACCP hazard assessment, the risk level is defined at each process step by the hazard's severity or effect in relation to the likelihood in which the hazard can occur: risk = probability × severity. Depending on the number of categories considered for each factor, we will obtain the total number of risk levels used to classify the hazards. For example, if we use the levels high, medium, and low for each factor, we will obtain a 3×3 table with five levels of risk.

In Appendix 2 of the 2022/C 355/01 Commission Notice on the implementation of food safety management systems (FSMSs) covering good hygiene practices (GHPs) and procedures based on

Likelihood						
	High	**4**	4	5	6	7
	Medium	**3**	3	4	5	6
	Small	**2**	2	3	4	5
	Very small	**1**	1	2	3	4
			1	**2**	**3**	**4**
			Limited	**Moderate**	**Serious**	**Very serious**

Severity

Figure 3.6 Risk matrix.

the HACCP principles, we can see an example of hazard analysis using a semiquantitative risk assessment procedure that uses a 4×4 table, with seven levels of risk (Figure 3.6).

The four classification categories for severity and probability proposed in this European Commission document are described as follows:

LIKELIHOOD

1 = very small

- theoretical chance – the hazard never occurred before;
- the control measure or the hazard is of such a nature that when the control measure is failing, no production is possible anymore or no useful end products are produced (e.g. too high a concentration of colorants as additives);
- it is a very limited and/or local contamination.

2 = small

- the control measures for the hazard are of a general nature like good hygiene practices (GHP) and these are well implemented in practice;

3 = medium

- failing or lacking of the (specific) control measure does not result in the systematic presence of the hazard at this step but the hazard can be present in a certain percentage of the product in the associated batch.

4 = high

- failure or absence of the (specific) control measure will result in a systematic error; there is a high likelihood that the hazard is present at this step.

SEVERITY

1 = limited

- there is no problem for the consumer related to food safety (nature of hazard, e.g. paper, soft plastic, large-sized foreign materials);
- the hazard can never reach a dangerous concentration (e.g. colorants, *S. aureus* in a frozen food where multiplication to higher counts is highly unlikely or cannot happen because of storage conditions and cooking).

2 = moderate

- no serious injuries and/or symptoms or only when exposed to an extremely high concentration during a long period of time;
- a temporary but clear effect on health (e.g. small pieces).

3 = serious

- a clear effect on health with short-term or long-term symptoms which results rarely in mortality (e.g. gastroenteritis, microbiological hazards such as *Campylobacter* or *Bacillus cereus*);
- the hazard has a long-term effect; the maximal dose is not known (e.g. residues of pesticides, etc.).

4 = very serious

- the consumer group belongs to a risk category and the hazard can result in mortality;
- the hazard results in serious symptoms from which mortality may result, including in the long term (e.g. *Salmonella*, *L. monocytogenes*, dioxins, aflatoxins, etc.);
- permanent injuries.

The 2022/C 355/01 Commission Communication also indicates what the control measures can be for each of the risk levels of the matrix: GHPs, operational prerequisites (OPRPs), or critical control points (CCPs), providing the following example of the determination of GHPs, OPRPs, and CCPs:

Risk levels 1 and 2: no specific actions, control covered by "routine" GHP.

Risk levels 3 and 4: possible OPRP. Additional question to be answered by the HACCP team: Are the general preventive measures as described in the GHP enough to control the identified hazard?

- If YES: GHP
- If NO: OPRP.

Risk levels 5, 6, and 7: look into CCP determination. When making a final decision on a CCP/OPRP at a certain step, consideration should be given to:

- the presence of a following step that will eliminate the risk or reduce its occurrence to an acceptable level: see decision trees in Appendix 4;
- the severity and likelihood of deviation and capacity to detect deviations.

The use of risk matrices provides the HACCP team with a tool with which the hazard analysis is better structured, the development of the sessions and the participation of the members of the group is favored, and also, by indicating a concrete value foreseen in a scale or a graduation, it makes it easier to reach consensus on the significance of each hazard.

Another advantage of using risk matrices is that they provide a simple recording tool to document the performance of hazard analysis. In addition to being a requirement of principle 7 of the HACCP system, the hazard analysis record is also a requirement of several food safety management standards with more extensive diffusion among food operators, which may explain their increasing use in recent years.

Despite the usefulness and benefits mentioned, the auditor must assess risk matrices cautiously because when he reviews the documentation to evaluate the application of principle 1, he can find risk matrices in many manuals. However, their mere existence does not ensure adequate hazard analysis in all cases. It is not the tool itself that ensures that the HACCP team makes the best decisions for the development of HACCP; it is the expertise of the team members, their professional judgment, and the information available to them that will enable the hazard assessment to be accurate and that the reasoning and deliberations of the group rely on appropriate science, the experience, and specific data of the process evaluated in each case (Figure 3.7).

Stage	Potential hazard	Justification	Should it be included in the plan? yes/no	Control measure
Fast cooling of meals	Germination and growth of *C. botulinum* and other spoilage bacteria	There is epidemiological evidence that *C. botulinum* spores can germinate after cooking some meats, and that growth can occur if cooling is slow	Yes	Rapid cooling of meals in a blast chiller. Ensuring a temperature drop from 60 to 10 °C in less than 2 h

Figure 3.7 Example of hazard control chart.

The hazard analysis record can be helpful for an FBO who wants to get the most out of HACCP as a genuinely beneficial tool because it is possible that over time, and based on the results obtained, the FBO can verify that specific hazards present a more significant risk than expected and then the control measures would have to be modified or, on the contrary, that there are hazards whose real risk, observed in the production plant, is lower than that initially considered in the hazard analysis, and this allows control measures to be less strict. Thus, hazard analysis documentation and the notes of deliberations and consensus reached by the team are very convenient as a reference when it is necessary to review and/or update the HACCP system.

The practical principles of risk analysis define it as a systematic and disciplined process carried out by governments to ensure public health protection. That is a powerful tool for conducting science-based analysis, which serves for decision-making and the search for solutions to problems related to food safety in the population. Risk analysis consists of three components: risk assessment, risk management, and risk communication. In turn, the risk assessment is composed of the following phases:

- **Hazard identification**: identification of physical, chemical, and biological agents that may cause harmful health effects and which may be present in food.
- **Hazard characterization (or dose-response assessment)**: quantitative and/or qualitative assessment of the nature of adverse health effects related to hazards present in food or determination of the relationship between the magnitude of exposure (dose) to a chemical, physical, or biological agent and the severity/frequency of associated adverse health effects (response).
- **Exposure assessment**: qualitative and/or quantitative assessment of the likely ingestion of biological, chemical, and physical agents through food and, if necessary, exposures from other sources.
- **Risk characterization**: estimating the probability of a harmful effect and its severity by an identified hazard in a particular population. This phase integrates the evaluation carried out in the previous three and is the most decisive element in deciding risk management (Figure 3.8).

As we have seen, hazard identification and risk assessment are terms used both in HACCP and in risk analysis; this is something that might create confusion because, although they have the same name, they refer to different activities carried out in different contexts and by agents that vary in each case. The following table shows some elements that may help to understand both concepts (Figure 3.9).

Governments are responsible for promoting the research necessary for risk analysis and assessment based on observation of disease incidence, laboratory information, field research, and other

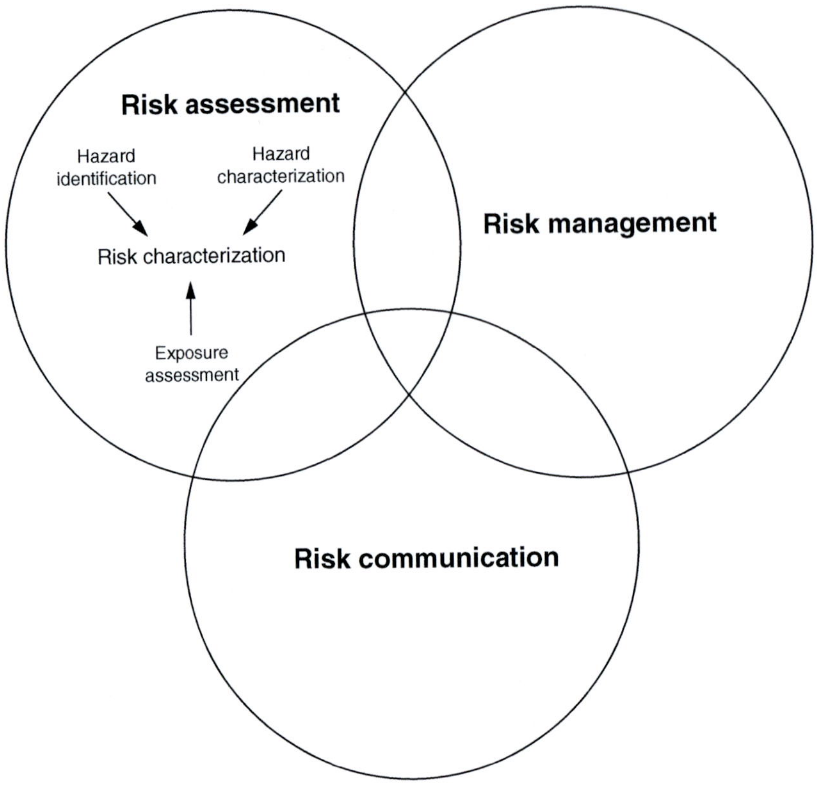

Figure 3.8 Risk analysis components.

Risk analysis	Hazard analysis
Government	Company
Expert panel	HACCP team
Global	Local
Months to years	Weaks to months
Food supply chain	Flow diagram
Population	Product
Exposure assessment	Hazard assessment
Quantitative	Qualitative
Risk management	Control measures
R° (EC) 178/2002	R° (EC) 852/2004

Figure 3.9 Comparative risk analysis and hazard analysis.

epidemiological data through the participation and collaboration of independent experts, scientists from research centers, and universities. At the next stage, taking into account the social and economic factors related to risk management and the cost of implementing control measures, governments can make an informed decision on what should be the acceptable level of risk for a disease or the appropriate level of consumer protection, which is termed "appropriate level of protection" (ALOP). An ALOP for an agent or a disease is the number of cases of disease per million inhabitants per year in a population, for example, three cases of listeriosis per million inhabitants per year. However, this public health goal, related to the incidence of a disease, is not valid for food operators because they need an intelligible value to know what they must do to reach the established level. For this reason, years ago, the concept of food safety objective (FSO) was created as a way to translate public health goals into parameters that can be understood and used, both by food companies and by official control agents, in order to evaluate whether companies are operating at the appropriate level of performance to achieve the established FSO.

An FSO is "the maximum frequency and/or maximum concentration of a hazard in a food at the time of consumption, which provides or contributes to the Appropriate Level of Protection (ALOP)." An FSO transforms a public health goal into a number that expresses the concentration and/or frequency (level) of a hazard in a food (ICMSF 2006). In addition to FSOs, other parameters, such as the performance objective (PO), microbiological criteria, and legal limits for contaminants, provide operators at the previous steps of the food supply chain with reference values to determine the level they must meet to achieve the FSO (Figure 3.10).

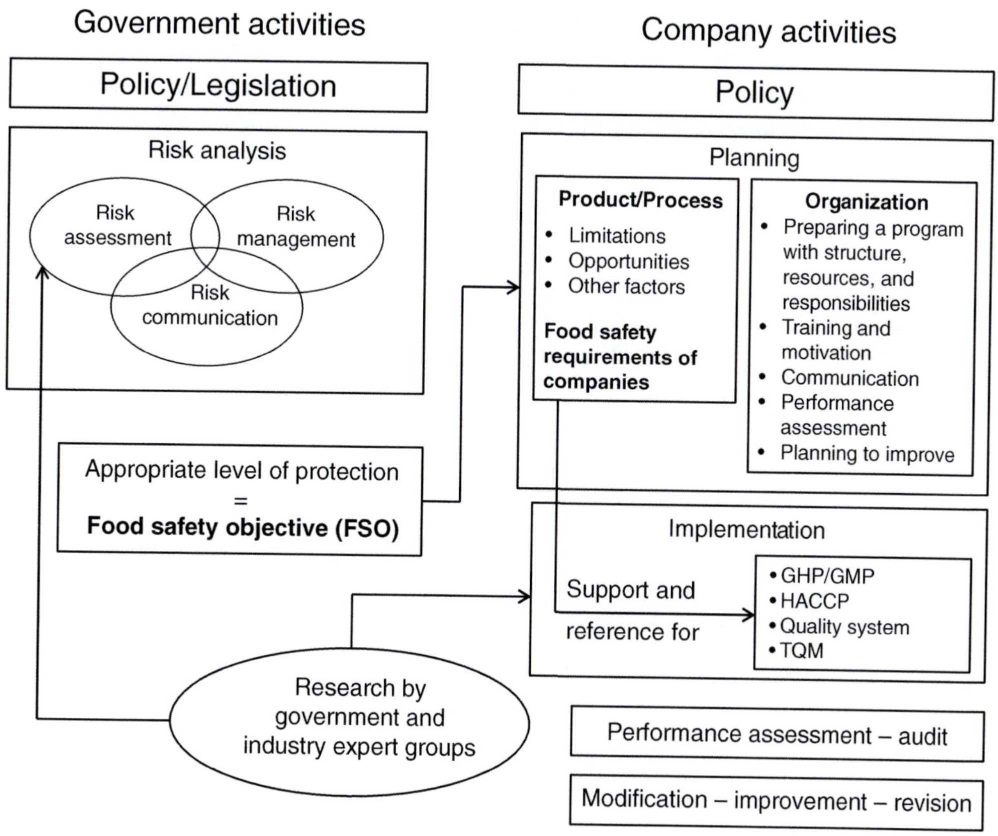

Figure 3.10 Comparison between government activities and companies.

3.1.3 Establish Control Measures

The third and final stage in applying HACCP principle 1 is establishing control measures capable of preventing, avoiding, or minimizing the hazards identified in the previous step.

According to the Codex Alimentarius, a control measure is defined as follows:

> **Control measure**: any action or activity that can be used to prevent or eliminate a hazard or reduce it to an acceptable level.

For each hazard identified as significant, it is necessary to determine what control measures need to be applied to eliminate or reduce them to an acceptable level. Sometimes, several control measures are necessary to eliminate a hazard, as we saw with the use of hurdle technology and the employment of the intrinsic factors of the food; on the contrary, other times, a single measure serves to control several different hazards, such as a heat treatment applied in a food processing (Figure 3.11).

When carrying out the document review, it is crucial to recognize the control measures corresponding to each identified hazard, so that everything is not a miscellany of different hazards located in the same phase, linked to a set of varied control measures, where we cannot know what control measures act on each hazard. Moreover, a hazard that originates in a specific phase does not have to be controlled in that same phase since it can be controlled in a later one.

When establishing measures for the control of hazards, the HACCP team should consider which measures are already in place by the FBO and in which cases it is necessary to establish new measures or modify existing procedures to ensure adequate control of hazards.

Control measures include the various hygiene programs and operations designed to prevent contamination of the production environment and to reduce or eliminate hazards identified at different stages of the production process. Among the control measures can be distinguished the following:

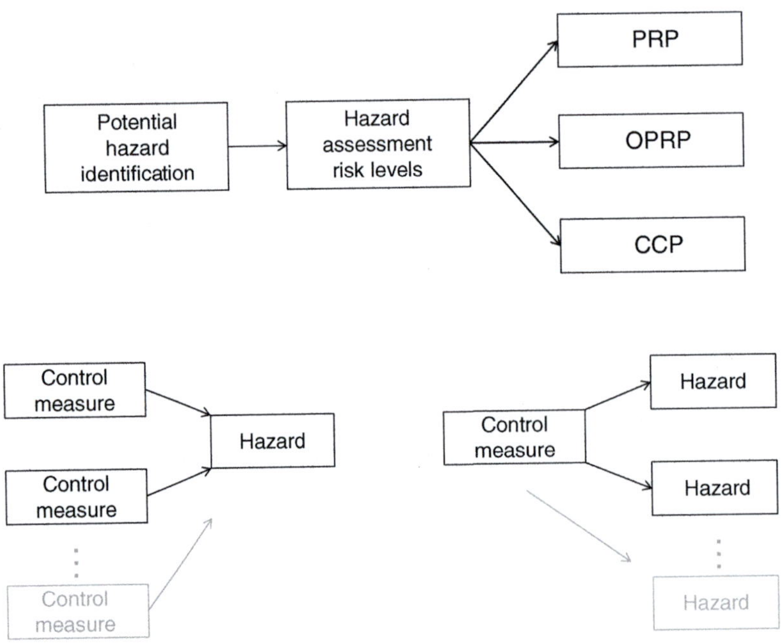

Figure 3.11 Hazard and control measures.

- PRPs aimed at controlling the working environment: pests, water, maintenance, temperature control, etc.
- PRPs related to GHPs and GMPs.
- Supplier control.
- Control of specific operations of the production process.
- Control of the intrinsic factors that determine the food product's safety.

Control measures in the FSMS may be documented as programs, plans, general procedures, specific procedures, or work instructions.
In any case, control measures must meet the following requirements:

- Act on the cause of the appearance of a hazard.
- Effective for the control of all significant hazards.
- **Feasible**: Are they easy to apply? Are they realistic?
- **Validated**: Will they be able to control the hazards?

A preventive measure is an action taken to eliminate the cause of the occurrence of a hazard or other potentially undesirable situation. Preventive measures are control measures, but not all control measures are preventive. For this reason, although the generic use of the term "preventive measure" can be seen in the literature and some manuals, "control measure" is the most appropriate denomination when we refer to the set of measures taken to prevent, reduce, or eliminate a hazard. Another failure that can occur in some cases is the confusion between control measures and surveillance activities; for example, if we establish that the control measure against the survival of bacteria in a product is frying at a specific temperature and time, checking the times and temperature of the oil are surveillance activities to verify that the control measure (frying) is done within the established parameters.

Regarding the validation of control measures, some do not allow quantifying their effect on specific hazards due to their nature. Therefore, prerequisites acting on nonspecific hazards or controlling working environment conditions will only sometimes be considered a priority for validation. In any case, there are other factors, such as the level of risk associated with the hazard, historical data, and the scientific and technical feasibility of the validation, that the HACCP team must also consider to determine the need to validate control measures. We will address this issue more in depth later when we discuss the application of HACCP principle 6.

As we saw when considering the risk assessment, the different levels of risk associated with each of the identified hazards also help the HACCP team select the most appropriate control measures for each type of hazard. Thus, CCPs would be associated with the hazards with the highest level of risk, according to the formula: risk = probability × severity; hazards with a lower level of risk would correspond to PRPs, and those at intermediate levels would include hazards that can be controlled through operational prerequisites (Figure 3.12).

Key elements of principle 1

Among the elements to which the auditor will pay attention in assessing the application of principle 1 are the following:

- List of potential hazards.
- List of significant hazards.
- Definition of hazards.
- Reference documentation.
- Justification of decisions.
- Use of risk matrices.

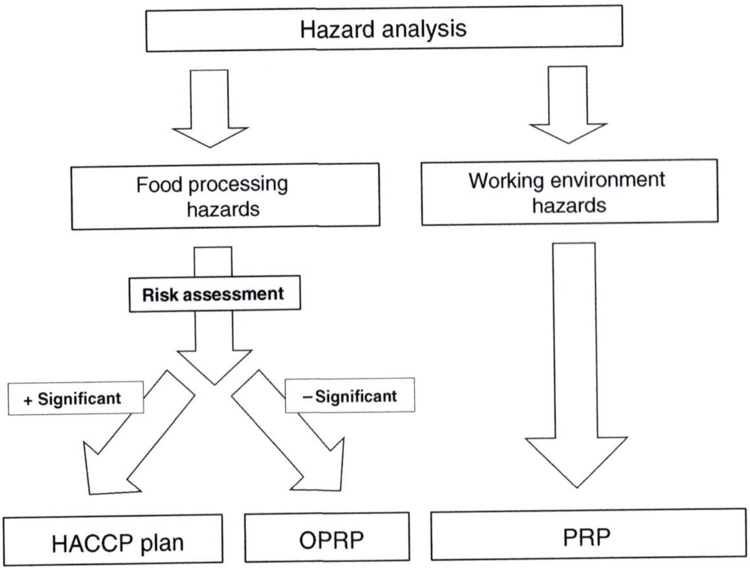

Figure 3.12 Apply different controls depending on the risk assessment associated with each hazard.

- Selection of control measures.
- Validation of control measures.
- Sessions records of the HACCP team.

3.2 Principle 2

HACCP principle 2 is to identify CCPs. The application of this principle is decisive because it conditions the development of the following phases of the HACCP study and the system's final structure, depending on the resulting number of CCPs.
According to Codex, a CCP is defined as follows:

> **CCP**: A step at which a control measure or control measures, essential to control a significant hazard, is/are applied in a HACCP system.

Along with the type of products produced and flow diagrams, the number and location of CCPs are one of the first aspects that the food safety auditor will surely pay attention to when preparing the audit.

The determination of CCPs is something that the HACCP team must address from a practical and realistic approach. A system HACCP is not safer because it has more CCPs than others, but because it has selected the CCPs that must be reasonably established in each specific situation and for managing them correctly. On the contrary, having many poorly chosen CCPs entails a larger expenditure of resources and can mislead the FBO about the system's usefulness.

As defined by the Codex, HACCP principles 2, 3, 4, and 5 are linked together, not only sequentially, so that one cannot move on to the next without having applied the previous principle, but also with one-to-one correspondence, that is, in which each corresponds to the others reciprocally. Thus, when a CCP is determined, a critical limit must be defined for that CCP. Subsequently, it is necessary to establish the monitoring of the critical limit established by principle 3. Finally, according to principle 4, it is necessary to foresee the corrective actions to apply when the monitoring indicates that the critical limit of a CCP has been exceeded.

The rest of the HACCP principles are not closely connected; not all hazards identified in principle 1 will be CCP. Nor does applying these principles have to be done sequentially; for example, principle 7 of documentation and registration must begin to be applied at the beginning of the HACCP study. Likewise, the validation, included in principle 6, can be applied to the system's different elements as the HACCP study progresses.

For each CCP identified, the HACCP system requires a monitoring system effective enough to detect deviations as quickly as possible or to take action in time. When a critical limit is exceeded, it is mandatory to take appropriate corrective action to bring the situation back under control and keep records to provide evidence that the FBO has taken the necessary measures to ensure the safety of processed foods. CCP management entails a significant increase in the workload in implementing HACCP in the company's day-to-day activities. Therefore, due to the consequences that result, applying principle 2 is a task that will test the expertise and professional competence of the HACCP team to select CCPs in the most effective way to manage the system.

From another angle, it is not wrong to have CCPs despite the work involved. On the contrary, when well identified, a CCP is a stage or an operation that successfully eliminates a hazard or reduces it to acceptable levels, which is a great leap to achieve food safety. The bad thing is not selecting them well and having superfluous CCPs, which are not essential to product safety, or too many, which only drain resources and distract attention from the crucial points of the process, which contradicts precisely the spirit and purpose of the HACCP system.

Following a logical and reasoned approach, the HACCP team must identify the CCPs, taking into account the characteristics of the production plant, the production process, the availability of human resources, and the company's organizational structure. There is no one-size-fits-all recipe. Companies in the same food sector or producing the same type of products do not have to coincide in the number or location of CCPs. What matters is that once a CCP is selected, those responsible for implementing HACCP must know what it means within the system and how to manage it.

When CCPs are selected arbitrarily, misapplying criteria, or starting from wrong assumptions, it is easy to lose their true meaning and role within the system. This poor application of principle 2 leads to trivializing CCPs, and the staff no longer perceive them as essential for food safety. If in colloquial or common language, any person understands the meaning of "critical" as something decisive, serious, or that cannot be overlooked, in the HACCP system, it may happen that due to a high number of CCPs or because they are not in the right places, those responsible for HACCP management no longer see them as actual "critical" control points, but as something that is not so decisive for food safety; so, even if they are out of control, that does not imply that any action has to be taken for the production process or the product; on the contrary, it is something that can be relativized or negotiated in some way according to each case. In this way, the message that permeates those responsible for HACCP and the company's staff is that CCPs are identified in the manual even though this is virtually irrelevant to their daily work or the production line. When this happens, it is clear that something is wrong: CCPs need to be better chosen and more credible to staff, or it is necessary to train and raise awareness among operators about the actual risks associated with their manufacturing process and the need to control specific points: the CCPs.

HACCP principle 2 states that CCPs should be identified following a logical or reasoned approach and through effective teamwork. Both the Codex guidelines for applying HACCP and NACMCF indicate that one of the ways to identify the CCPs is through decision trees. Nevertheless, in both cases, it is clear that their use is only an option. The use of decision trees is not mandatory.

Since its publication by the Codex, the decision tree has been commonly used in training courses to exercise teamwork and explain the logic and level of understanding needed to determine CCPs. However, despite their usefulness for training purposes, decision trees cannot be considered an essential element of HACCP but an accessory one.

The Codex document also contains some recommendations or clarifications on the use of decision trees, which the HACCP team should take into account:

- They should be applied flexibly.
- Its use is indicative, not binding.
- May not apply to all situations.
- In some cases, they can be modified.
- Its use requires training.
- There are other equally valid strategies.

Despite these recommendations, it is still easy to find HACCP manuals in which we can observe that the application of decision trees is too mechanical without considering the consequences of identifying CCPs by people who lack proper training or in whom adequate teamwork is not evident. The key issue is that we must understand that the decision tree is simply a tool to identify CCPs and that in no way replaces the knowledge, experience, and professional judgment of the HACCP team.

As with risk matrices, it is not the decision tree that ensures the correct application of principle 2 but the professional competence of those who use it that provides the best decisions in each case. In any case, when it is in good hands, the decision tree brings positive aspects such as the following:

- Facilitates teamwork.
- Provides a structured approach.
- Helps to focus the team's discussion on the premises necessary to decide.
- Gives consistency to the HACCP study by analyzing the hazards phase by phase.
- Forces team members to think about the result of the whole process.
- Serves to document decisions and as a record of the HACCP study.

Several models of decision trees are available for their use in HACCP. They all follow a similar scheme and contain the same type of questions. Although there are some differences in terms of:

- **Number of questions**: 3, 4, 5, or 6.
- **Utility**: for raw materials or processes.
- Some include prerequisite questions.
- Some are valid to differentiate CCP and OPRP.

Before applying the decision tree, the HACCP team must consider the result of the previous phases and the application of principle 1; otherwise, they cannot answer the questions with sufficient judgment. In particular, the HACCP team should consider the following:

- The phases of the diagram, or flow diagrams, must be clearly defined.
- The tree is used after hazard analysis.
- Each hazard must have control measures in place.
- References of what are considered acceptable levels for hazards are needed.
- Prerequisites must be previously developed and implemented.
- Some hazards do not have to be controlled at the stage where they occur but at some later stage, where the CCP may be (Figure 3.13).

To use the decision tree, the team must have all the information related to the process and the product they analyze; its members must know how to correctly interpret the meaning of the questions that compose it, which is not easy in all cases.

When using a decision tree, it is necessary to keep a record where the team can write down the outcome of its application so that it serves to document the justifications that support the decisions

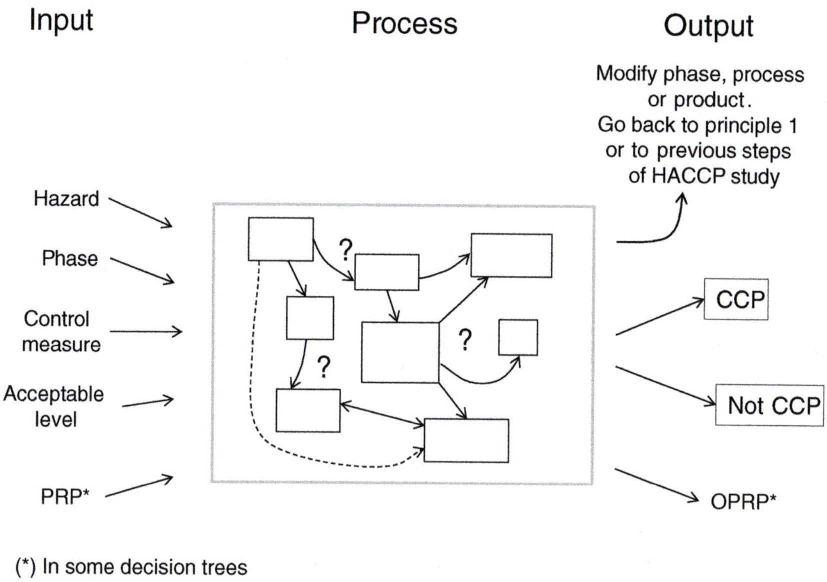

Figure 3.13 Decision tree application as a process.

Process stage	Hazard	Control measure	Q1	Q1a	Q2	Q3	Q4	CCP yes/no	HACCP team notes

Figure 3.14 Record model answers of decision tree.

taken, as well as other notes that the team considers appropriate to keep, for example, for future revisions of HACCP. Those are some of the records that the auditor will surely want to review to see how the HACCP team has identified the CCPs and how the application of principle 2 is documented (Figure 3.14).

Since its first edition, the Codex has included a decision tree model in its guidelines for implementing HACCP. This decision tree consists of four questions and a supplementary one and has been one of the most used reference models for teaching HACCP and, of course, for its application by food operators. The latest revision of the Codex guidelines from 2022 provides a different decision tree model than the one included in the past revisions. The changes are that this model contains four questions without the supplementary one, and the formulation of the four questions differs. However, the logic of the questions is practically equivalent to determining the CCPs, just like the other models we will see (Figure 3.15).

For didactic purposes and because we can find that many FBOs still use the first Codex model, we will take it as a starting model to analyze how to apply the decision trees. On the other hand, this remains an equally valid model among the many available for application in HACCP.

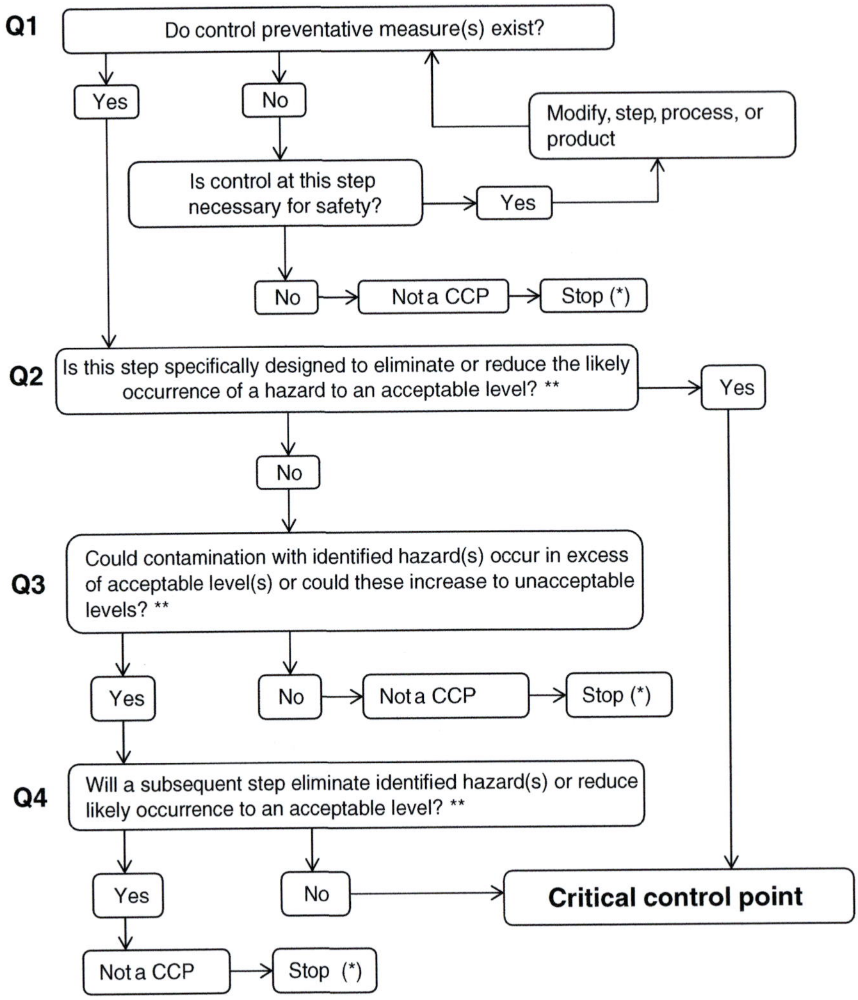

(*) Proceed to the next identified hazard in the described process

(**) Acceptable and unacceptable levels need to be defined within the overall objectives in identifying the CCPs of HACCP plan.

Figure 3.15 Original Codex decision tree with 4Q.

Its sequence of questions is as follows:

Q1 Do preventive measure(s) exist?

With this first question, we already see the need to have all the information obtained through applying principle 1. It is necessary to check whether there are preventive measures for each hazard. If the answer is "yes," we move on to the next question.

If the answer is "no," then the decision tree leads us to the supplementary question (Q1a): Is control necessary at this stage for safety?

At this point, several alternatives arise; the team may consider that control at this stage of the identified hazard is not necessary either because it considers its risk or likelihood to be low or because there are subsequent operations in the process that can control the hazard, including those that depend on the preparation of the food by the consumer. In that case, the decision tree indicates that this step is not a CCP for that hazard and to continue with the following hazard identified in the process.

If the answer to the supplementary question 1a is "yes," then the decision tree concludes that modifying the stage, the process, or the product is necessary. That answer means returning to the HACCP study's initial steps and rethinking the hazard analysis from the beginning. Question 1 may seem obvious because it deals with something that the application of principle 1 must have resolved. Whether there are control measures for each hazard, and if there are none, consider the need to modify the stage or process, so much so that a simple instruction could replace it: ensure that all necessary control measures have been taken (Besseling and Kalyankar 2014). The exhaustive nature inherent to the HACCP system could justify its inclusion in this Codex decision tree. Instead, the new Codex decision tree model has replaced this question with a different one.

> Q2 Is the step specifically designed to eliminate or reduce the likely occurrence of a hazard to an acceptable level?

Question 2 seems straightforward to answer; however, in practice, it may raise some doubts, and reaching a consensus on an adequate answer can be challenging.

The key to answering it is that Q2 refers to the process step, not the control measure that is being questioned. If team members answer thinking about control measures, then the answer will be "yes" in many cases, and that will result in more CCPs than necessary. The point here is that there are steps of the process that are also control measures; those are the ones that this question seeks to identify. What the question is actually asking is whether the step itself controls the hazard (Wallace et al. 2011).

"Specifically designed" means that the step has been added to the process as a necessary control measure to eliminate an identified hazard. Therefore, the step would not be necessary if there were no such hazard. If we understand it this way, it is easier to answer this question correctly.

When we apply this reasoning to each stage of the production process, there are times when it is easier to decide when a step meets this condition, such as pasteurization, acidification, or aseptic packaging, and others when there is room for different interpretations. Thus, in the literature, we can see how sometimes a specific phase is identified as CCP and others in which that same phase is not considered CCP. As we anticipated, it is the judgment and professional experience of the team that must dictate for each specific situation when to establish a CCP and what this entails, and not the mere use of the decision tree.

Some examples can better illustrate this; let us think of the phase of "cold storage of raw materials"; if the reader takes time to do a search in the literature or on the Internet, he will be able to see how there are cases in which this phase is considered to be specifically designed to eliminate or reduce a hazard to an acceptable level; therefore, this means answering "yes" to question 2 and consequently identifying a CCP; and we can also find other guides or publications in which it does not consider it so and, therefore, do not identify this cold storage as a CCP, but as a phase where a control measure is applied to prevent microbial growth. Particularly, if I had to choose, I adhere to this second option, but as we said, there are interpretations in both directions. Moreover, it is not easy to decide if one is better than another because it depends on the context and each organization.

On the other hand, if we think now of a production of fish that is consumed raw, such as anchovies in vinegar, a typical Spanish meal, in this case, in the elaboration process, we must add a freezing step, which is the specific control measure to eliminate or reduce to an acceptable level the presence of *Anisakis*. Therefore, since this step is specifically designed to eliminate that hazard, question 2 must be answered in the affirmative, which is why we are dealing with a CCP. If there were no risk of *Anisakis*, freezing the fish would not be necessary. The culinary preparation does not require it. This example admits less doubt than the previous one.

With some food processing, something similar happens; for example, in a catering establishment, the cooking involves a heat treatment that will eliminate most microorganisms, and the HACCP team must consider it in the risk assessment. Still, these are not steps that have been specifically designed to eliminate microbiological hazards but to achieve the degree of transformation of raw materials and changes in physical and organoleptic properties that the culinary process requires.

And finally, let's analyze the case of bakery CCPs. Is baking bread a CCP? How do we answer question 2 in this case? Baking is a necessary step to make bread. Its temperature and duration are not established according to the microbiological hazards that, without doubt, will also be eliminated during it, but to the baking conditions that are required for the dough to undergo the necessary transformations and acquire the organoleptic characteristics of bread. Therefore, the answer to question 2 could be "no." However, finding guides where the opposite is stated is relatively easy. Let's accept that question 2 of the decision tree can be answered affirmatively and consider the baking step a CCP. The next question is: What is the point of identifying bread baking as CCP? From a practical point of view, there is little sense. Because once the bread comes out of the oven, the microbiological hazards have been eliminated much earlier, and it does not make much sense to establish critical limits nor a monitoring system for a process that is implicitly checked simply by seeing that the bread is already made. An argument that matches with one of the tips of the New Zealand Food Safety Authority in its "FAQ" about problems determining CCPs: "if the operational characteristics of a reliable process well exceed what would be required in a critical limit then it should not be a CCP. For example, if frying bacon until crisp." Do we still have doubts? Think then of the number of times there have been incidents or safety problems due to insufficient bread baking. There we have the answer: the probability is very low, and so is the risk; therefore, it does not seem that we are dealing with something we must categorize as critical.

When the answer to question 2 is "no," we move on to question 3.

> Q3 Could contamination with identified hazard(s) occur in excess of acceptable level(s) or could these increase to unacceptable levels?

With this question, the HACCP team faces the dilemma of possibility versus probability again. If we think only about the possibility, the answer is obvious, and it will be "yes" in most cases. Therefore, before answering, team members must assess the probability of contamination or that the hazards may increase to unacceptable levels in the specific context of the process studied and based on the experience and the historical production data. To this effect, the team must analyze the possible sources of the identified hazard and assess the likelihood of cross contamination due to the work environment, personnel, ingredients, or equipment or that the hazard may increase to unacceptable levels due to an abuse of time and temperature conditions, among other factors, without forgetting the cumulative effect that can occur in the whole process. In addition, the team must have some reference of the acceptable level for the identified hazard in order to be able to give an adequate answer to question 3. Besides, this is stated in this Codex decision tree note as follows: "Acceptable and unacceptable levels need to be defined within the overall objectives in identifying the CCPs of HACCP plan."

In some publications and reference texts, such as those of the European Commission's Better Training for Safer Food, question 3 includes the note: "If control measures fail." Several authors prefer to interpret this question in this sense: If control measures fail, can there be contamination with the hazards or increase to unacceptable levels? In other words, how much is the risk involved if control measures fail? If the risk is high, then we must conclude that the effect of control measures is high, increasing the need for a stricter monitoring system, which leads to answering question 3 in the affirmative (Besseling and Kalyankar 2014).

> Q4 Will a subsequent step eliminate identified hazard(s) or reduce likely occurrence to an acceptable level?

Question 4 is another that forces the HACCP team to think about the whole process and not in an isolated step. Thus, the team can accept the presence of hazards in the analyzed step, provided that there is a subsequent one in which the hazard will be eliminated or reduced to an acceptable level, including the last step, prior to consumption, which is the one that depends on the consumer when preparing or cooking the food.

This question helps to identify control measures located at different points in the process that can act on the same hazard, which avoids identifying too high or unnecessary CCPs since if there is a step that fully controls the hazard, the others do not have to be classified as CCPs allowing monitoring resources to focus on the phase that is effective for the control of that hazard.

When a later step controls the hazard, the answer is "yes," so the current one is not a CCP (Figure 3.16).

Example I of the NACMCF CCP decision tree consists of three questions. The second and third questions are similar to the Codex decision tree. However, the first operates differently, forcing the team to consider the degree of significance associated with each identified hazard. Question 1 acts as a filter that prevents hazards with low risk levels from being identified as CCPs, which would probably increase their number unnecessarily. Thus, when the team determines that the risk level of a hazard is not high enough and answers "no" to this question, the decision tree indicates that it no longer makes sense to continue with the rest of the questions. This result agrees with the one

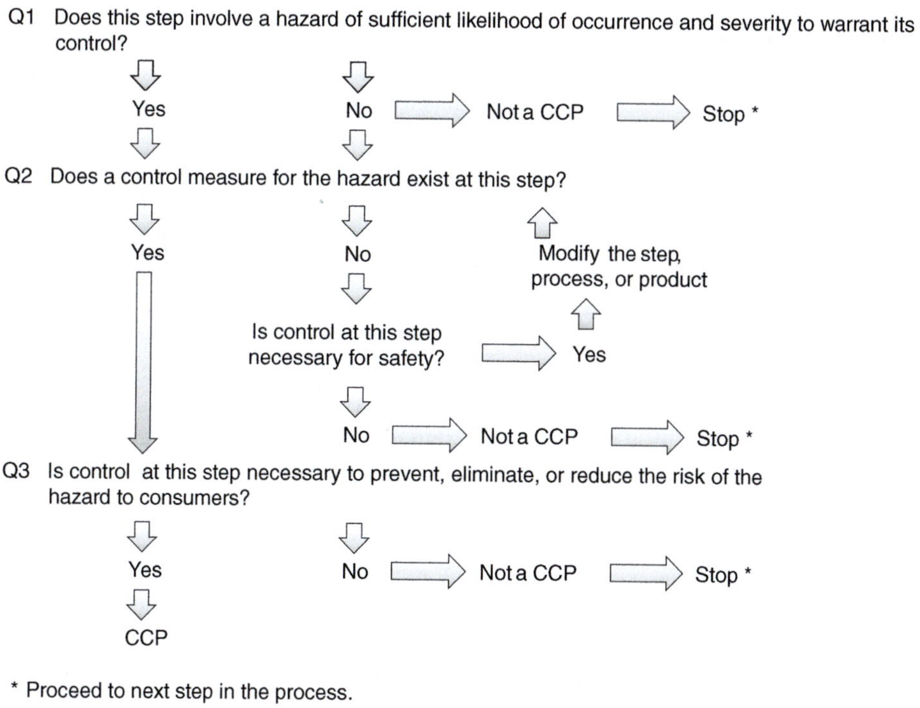

Figure 3.16 Example I – NACMCF decision tree.

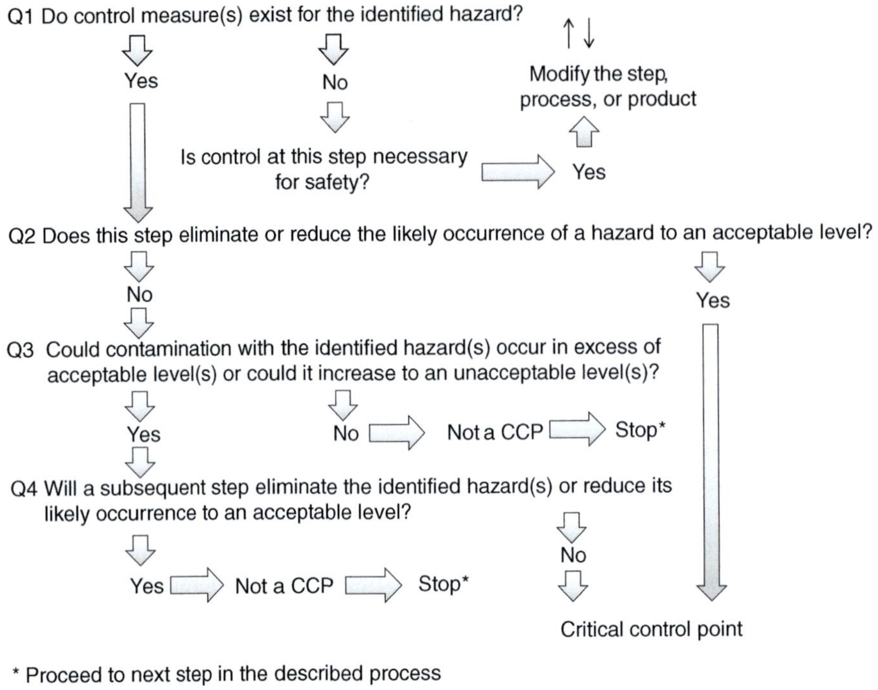

Figure 3.17 Example II – NACMCF decision tree.

obtained using the risk matrices, in which, as we saw, the hazard's medium and low risk levels are not associated with CCPs, but with PRPs or OPRPs (Figure 3.17).

Example II of the NACMCF CCP decision tree contains four questions and one supplementary question, which makes it practically comparable to the Codex tree. The slight difference is that the wording of question 2 does not include the phrase "specifically designed." So, the team that uses this decision tree must assess the sense in which it wants to interpret this question – and taking into account that it refers to the step of the process rather than to the control measures that exist for the identified hazards, in order to avoid unnecessarily increasing the number of CCPs.

For team members, it must be clear that there are no preset numbers or limits to CCPs that can be identified in an HACCP manual. However, experience shows that the ideal for good system management is that only a small number of CCPs should be monitored. In contrast, the existence of many CCPs is often not due to the correspondence between the identified CCPs and the process steps that should be considered critical to product safety but because the role of PRPs in controlling many types of hazards has not been taken into account by the HACCP team.

This lack of recognition of PRP's role in controlling hazards motivated the emergence of other decision tree models, in which, in addition to introducing hazards, control measures, and steps in the sequence of decisions, some questions are added regarding the control that the PRPs can have on the hazards. As happens with Campden's modified decision tree, in which an initial question is added to determine whether a PRP completely controls the analyzed hazard, if the answer is "yes," the program – or the programs – that control it is identified and we move on to the next hazard. When the answer is "no," proceed with the rest of the questions we already discussed (Figure 3.18).

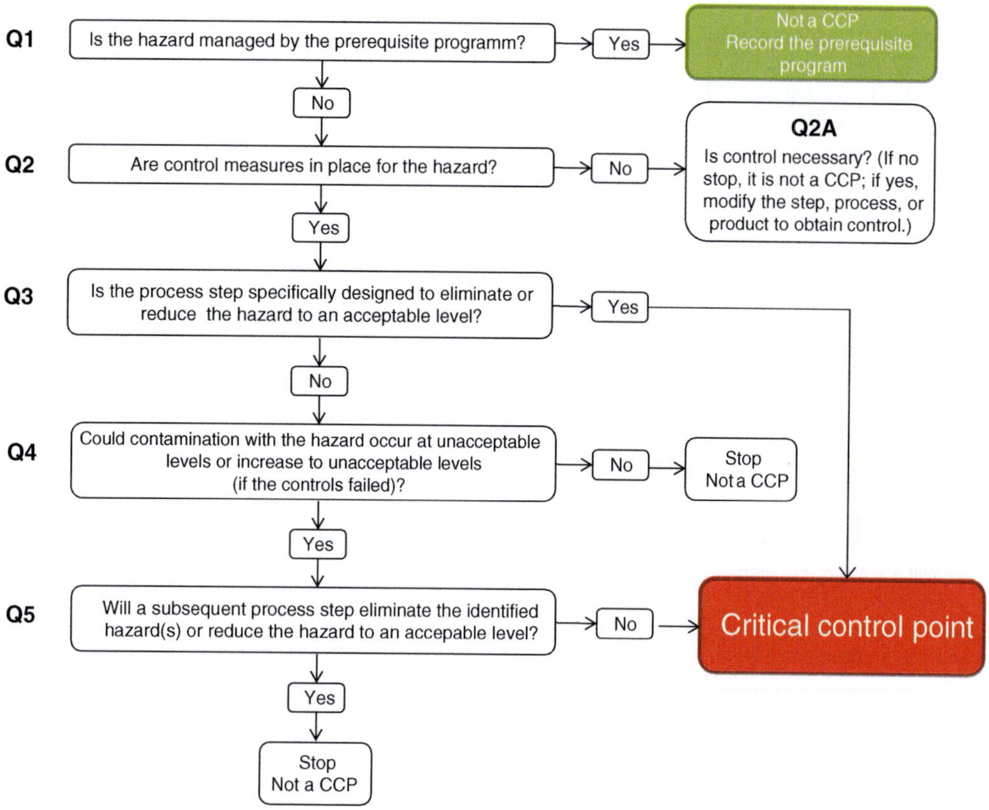

Figure 3.18 Modified Campden decision tree.

Another decision tree model, also available in the literature, introduces two initial questions, usually called A and B, to the questions of the Codex model. Those two initial questions act as a filter and must be answered before continuing with the rest of the sequence of questions:

> **A**: Does this hazard have a sufficient probability of occurrence and severity that requires its control?

If the answer is "no," it is not a CCP, and proceed with the next hazard. If the answer is "yes," move on to question B.

> **B**: Is this hazard completely controlled with a prerequisite?

If the answer is "yes," it is not a CCP.

If the answer is "no," move on to question 1 of the decision tree, which matches the Codex decision tree questions.

Now that we have seen several decision tree models, we can better understand the changes introduced in the Codex model of the 2022 revision. This new model differs quite a bit from the previous one, both in the wording of the questions and in their order, but with a perfectly valid logical sequence to decide whether a step should be a CCP (Figure 3.19).

The first question of this model is already familiar to us:

> Q1 Can the significant hazard be controlled to an acceptable level at this step by PRPs (e.g. GHPs)?

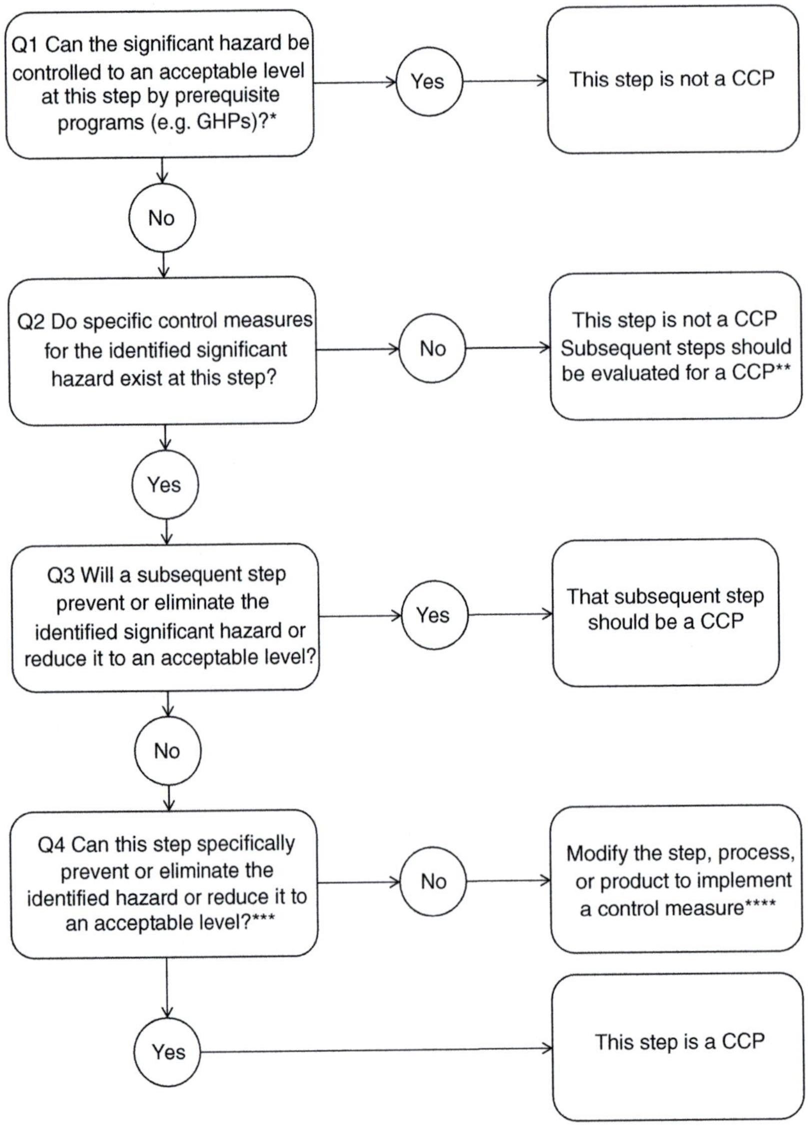

* Consider the significance of the hazard (i.e. the likelihood of occurrence in the absence of control and the severity of impact of the hazard) and whether it could be sufficiently controlled by prerequisite programs such as GHPs. GHPs could be routine GHPs or GHPs that require greater attention to control the hazard (e.g. monitoring and recording).
** If a CCP is not identified at questions 2–4, the process or product should be modified to implement a control measure and a new hazard analysis should be conducted.
*** Consider whether the control measure at this step works in combination with a control measure at another step to control the same hazard, in which case both steps should be considered as CCPs.
**** Return to the beginning of the decision tree after a new hazard analysis.

Figure 3.19 Codex decision tree 2022.

If the answer is "yes," this stage is not a CCP; if the answer is "no," we proceed with the second question.

Q2 Do specific control measures for the identified significant hazard exist at this step?

If the answer is "no," then this stage is not a CCP because a CCP must be controlled by specific control measures, not general control measures such as PRP. On the other hand, if there are no specific or general control measures to control this hazard at this stage, there must be a subsequent step, which leads us to question 3.

Q3 Will a subsequent step prevent or eliminate the identified significant hazard or reduce it to an acceptable level?

If the answer to question 3 is "yes," then that subsequent step should be selected as CCP. If the answer is "no," we move on to the last question.

Q4 Can this step specifically prevent or eliminate the identified hazard or reduce it to an acceptable level?

If the answer to question 4 is "yes," then this step should be selected as CCP. If the answer is "no," we must modify the stage, the process, or the product to implement a control measure, because we are dealing with an identified significant hazard that is not controlled either by the prerequisites or by any specific measure of the stages of the process. There is no choice but to modify the process.

Besides, Q4 includes an interesting footnote: "Consider whether the control measure at this step works in combination with a control measure at another step to control the same hazard, in which case both steps should be considered as CCPs." As mentioned, sometimes, more than one control measure is needed to eliminate or prevent a hazard.

In the end, we can see how this decision tree forces the HACCP team to focus the problem in terms very similar to those we had already analyzed in the other models. And in this way, a CCP can be understood as a specific control measure that can eliminate or reduce to an acceptable level an identified significant hazard, which has not been controlled by general control measures or by other specific control measures located at a subsequent step of the process (Figure 3.20).

Other types of decision trees were developed to apply to raw materials and not to process steps, such as the raw materials decision tree designed by Mortimore and Wallace or that of the ILSI, in which the first two questions refer to raw materials and the remaining four to the steps of the process being studied (Figure 3.21).

The decision tree model designed by Mortimore and Wallace to apply to raw materials is composed of three questions, trying to discriminate those raw materials that require a higher level of control. The result of applying the sequence of questions is that raw materials that will not undergo further processing or treatment by the operator or the consumer and may cause a risk of cross contamination in the operator's facilities must be considered sensitive raw materials or CCPs. These authors point out that in some cases, the management of this risk should then be done as a CCP by the supplier of that raw material.

In addition to the decision tree models mentioned, we can use other models to determine which hazards must be controlled through OPRPs; among the models developed are we can say Robert Gaze's prerequisite decision tree (Gaze 2009), Mortimore and Wallace's OPRP decision tree (2013) and the combined decision trees, in which the outputs of the sequence can be CCP or OPRP.

Some of the specific questions that are included in these decision trees to help determine when it is an OPRP are as follows:

- PRPs provide the basic conditions for maintaining a hygienic working environment (PRP).
- The hazard must be controlled by a specific process step (OPRP).
- There is a significant risk of cross contamination at the step (OPRP).
- There is a risk of contamination of the product by hazards introduced from the source of the hazard by a direct or indirect vector (OPRP).

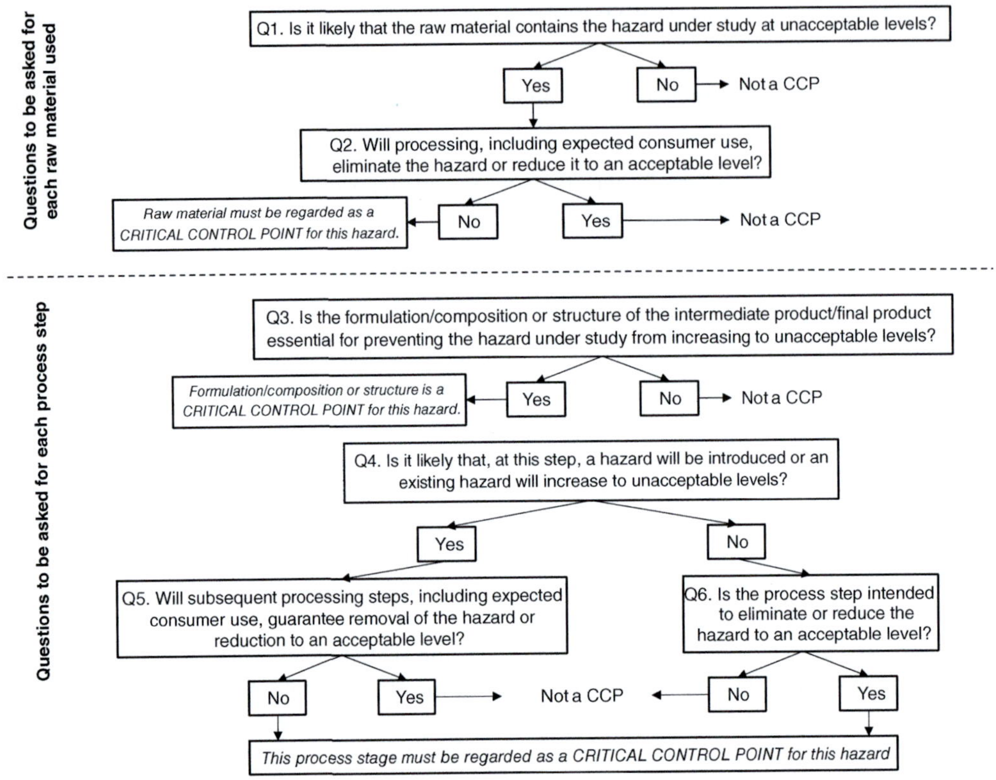

Figure 3.20 ILSI decision tree.

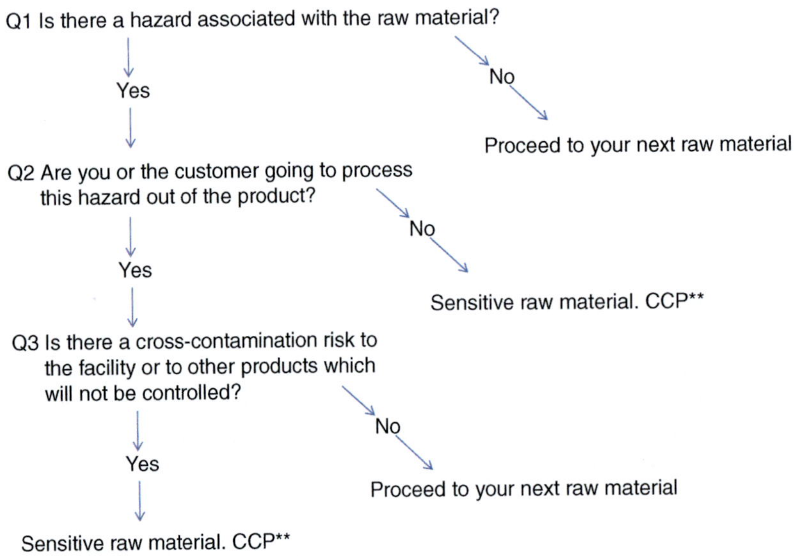

** High level of control required at the supplier, and therefore, this is a priority for supplier-approved activities. Following the hazard analysis, you are likely to find that this raw material must be managed as a CCP at the supplier.

Figure 3.21 Raw material decision tree.

- Control measures established at one step of the process can be monitored by critical limits (CCPs).
- Control measures established at one process step can be monitored by acceptance validity criteria but not critical limits (OPRP).

In parentheses, we can see the preventive measure associated with each question when answered in the affirmative.

In addition to decision trees, alternative methods are also valid for identifying CCPs, provided that the method used meets the Codex requirement of being systematic and reasoned. Thus, ISO 22000:2018 does not mention the use of decision trees to establish CCPs but indicates that the selection of measures for the control of identified hazards must be made through the use, by the HACCP team, of a logical approach based on the application of eight criteria included in its point 8.5.2.4: Selection and categorization of control measures.

Point 8.5.2.3 of ISO 22000:2018 Hazard Assessment states that for each identified food safety hazard, the organization should conduct a hazard assessment to determine whether its prevention or reduction to an acceptable level is essential. So, the organization shall evaluate this hazard with regard to:

> The likelihood of its occurrence in the end product before applying control measures.
> The severity of its adverse health effects in relation to the intended use.

These two factors are the same that make up the risk matrices, whose use, as we saw, helps to select the hazards with the highest degree of criticality to be controlled preferably by a CCP or an OPRP, according to the level of risk resulting from the formula risk = probability × severity.

ISO 22000:2018 states that based on the hazard assessment, the organization must select an appropriate control measure or combination of control measures that will be capable of preventing or reducing the identified hazards to defined acceptable levels. Then, the organization must categorize the selected control measures to decide whether they are managed as an OPRP or at a CCP. The selection and classification of control measures should be conducted using a systematic approach that considers the following criteria, included in point 8.5.2.4.1:

(a) the likelihood of failure of its functioning;
(b) the severity of the consequence in the case of failure of its functioning; this assessment shall include:
    1. the effect on the identified significant food safety hazards;
    2. the location in relation to other control measure(s);
    3. whether it is specifically established and applied to reduce the hazards to an acceptable level;
    4. whether it is a single measure or is part of the combination of control measure(s).

This evaluation must use three other criteria included in its point 8.5.2.4.2, which assesses the feasibility of:

(a) establishing measurable critical limits and/or measurable/observable action criteria;
(b) monitoring to detect any failure to remain within critical limits and/or measurable/observable action criteria; and
(c) applying timely corrections in case of failure.

We can comment on how we should interpret these criteria for categorizing control measures. It is clear that the greater the severity and probability of the hazard, the greater the risk, and therefore, as we saw when using the risk matrices, it indicates that we must select a CCP.

The same goes for the effect on hazards; the more significant the impact of a control measure on the level of hazards or their frequency of occurrence, the more likely we should manage it as a CCP. Control measures that eliminate or reduce a hazard to an acceptable level are more frequently associated with CCPs than control measures intended only to limit contamination or increased hazards, which often act in combination.

Regarding the location of the control measure in relation to other control measures, this criterion may mean that some measures are more effective when applied after others, for example, when they act after control measures that stress microorganisms. Other authors prefer to interpret this criterion with the existence of a subsequent control measure that can eliminate or reduce the hazard to an acceptable level. In such a case, the following control measure takes care of the failure of the first control measure, and thus, the risk of failure of the control measure is low. This line of thinking is the same that applies in Q4 of the first Codex decision tree that we saw: Will a subsequent step eliminate identified hazard(s) or reduce likely occurrence to an acceptable level (Besseling and Kalyankar 2014)

Criterion b.3 must already be familiar to us because it coincides with the question asked in some decision trees, and therefore, when the answer is "yes," we probably have a CCP.

The three criteria included in point 8.5.2.4.2 of ISO 22000:2018 are not related to principle 2 of HACCP but to principles 3, 4, and 5, all of which are associated by definition with the existence of a CCP. Therefore, an affirmative answer to each indicates this step should be a CCP.

3.2.1 Operational Prerequisites

The concept of operational prerequisite was introduced in 2005 by the ISO 22000 standard; since then, its use has been extended and assumed by an increasing number of bodies – as is the case of the European Commission – as well as by several certification standards, to the point that it has ended becoming normal as one of the usual reference elements in FSMSs.

The reason for the popularity achieved by OPRPs is probably due to the role they play in controlling certain types of hazards, identified in the hazard analysis as essential to food product safety but which cannot be categorized as CCP or do not meet all the conditions required by a CCP.

The best way to understand the concept of OPRP is by using the definitions included in ISO 22000:2018:

> **OPRP**: control measure or combination of control measures applied to prevent or reduce a significant food safety hazard to an acceptable level, and where action criterion and measurement or observation enable effective control of the process and/or product.

For this, we must also know the definition of action criterion:

> **Action criterion**: measurable or observable specification for the monitoring of an OPRP.
> **Note 1 to this entry**: An action criterion is established to determine whether an OPRP remains in control, and distinguishes what is acceptable (criterion met or achieved means the OPRP is operating as intended) and unacceptable (criterion not met nor achieved means the OPRP is not operating as intended).

And compare it with the definition of PRPs:

> **PRP**: basic conditions and activities that are necessary within the organization and throughout the food chain to maintain food safety.

The introduction of OPRPs changed the traditional division of control measures between PRPs and CCPs. It allowed the FBO to diversify strategies to improve hazard control according to the levels of risk they present. Thus, control measures can be categorized into PRPs, OPRPs, and CCPs. OPRPs occupy an intermediate position between PRPs and CCPs, trying to cover all the space, too broad, between the measures that provide the basic environmental conditions for the hygienic production of food and the steps of the process that are considered critical for the safety of food products. In a nutshell, an OPRP is an intermediate control measure between a PRP and a CCP, sharing characteristics of one and the other.

In addition to the different levels of risk that can be associated with each type of measure, for example, through the use of risk matrices, the main differences between PRPs and OPRPs are that PRPs are established prior to the hazard analysis and refer to the basic conditions necessary to achieve a hygienic working environment. In contrast, OPRPs are measures established in one or more production process steps to control hazards identified after the hazard analysis.

CCPs and OPRPs have in common that they follow the hazard analysis and refer to process steps; however, they differ in that OPRPs are control measures usually aimed at avoiding cross contamination at one or more points in the production process or at limiting the growth of a hazard to unacceptable levels, not to eliminate or reduce a hazard, as happens with the CCPs.

Another main difference between OPRPs and CCPs is related to their feasibility for monitoring; CCPs can be monitored at a point in the process using observable or measurable critical limits (HACCP principle 3); instead, OPRPs are programs that generally intervene in several steps of the process and that cannot be associated with a critical limit so that it is not possible to monitor as conclusively as that carried out on a CCP.

On the other hand, the HACCP system indicates that when a critical limit is exceeded, the manufactured product is considered nonconforming, and that requires corrective action to be taken on the product (HACCP principle 5); in the case of a failure in the OPRP, the action on the product is not mandatory although, if we take as a reference the requirements of ISO 22000:2018, it must be evaluated by those responsible for the management of HACCP, concerning the causes of nonconformity, and the consequences derived in terms of food safety and depending on the result of this evaluation can be released.

The idea of OPRPs as control measures that do not provide the same assurance as CCPs and do not allow such conclusive monitoring was already present in other concepts that appeared years ago but have yet to achieve the same acceptance among institutions and food operators. Such was the differentiation proposed by the International Commission on Microbiological Specifications for Foods (ICMSF) in 1988 between CCP1 and CCP2; CCP1 would be the current CCPs, and CCP2 would have the same function as OPRPs. Besides, references to "control points" and "attention points" were introduced in the past, which can be considered very similar to OPRPs in both cases (Figure 3.22).

In summary, the objective of HACCP principle 2 is to identify CCPs; regardless of the method or strategy followed by the team, the important thing is that the CCPs are chosen in a reasoned way and with a practical approach, thinking about the consequences that the management of each CCP implies in the day-to-day activities of the company. Principle 2 represents a fork in the path of the HACCP study. The decisions made at this stage will affect the application of the following principles and the rest of the study and, therefore, the final structure of the FSMS.

	PRP	OPRP	CCP
Working environment	Yes	No	No
Production process	No	Yes	Yes
Stages at which it acts	Several	1 or more	1
Significant hazard	+	++	+++
Monitoring	+	++	+++
Critical limit	No	No	Yes
Action on product	No	Assess	Yes
Validation	Not priority	Priority	Yes

Figure 3.22 PRP, OPRP, and CCP.

If we have doubts that we have identified the CCPs correctly, then we should check if they have the following characteristics:

- They are control measures that are part of the process.
- They are associated with high-risk hazards.
- They eliminate hazards or reduce them to an acceptable level.
- They can be monitored at one point.
- A critical limit must apply to them.
- They allow us to take quick action.

If one or more of these characteristics are unmet, we are probably dealing with an unnecessary CCP whose management is not essential for food safety. The application of principle 2 is linked to principles 3, 4, and 5 of HACCP; therefore, if we identify a CCP, but it is impossible to establish a critical limit that allows its effective monitoring, then it implies a failure in the system's coherence because we have a critical step but cannot ensure its vigilance. The same happens if we establish a CCP that does not allow us to take quick action to correct a deviation that we consider critical, which prevents further production of nonconforming products.

Key elements of principle 2

Among the elements that the auditor will review to evaluate the application of principle 2 are the following:

- The number of CCPs identified.

Consider situations where the HACCP team does not identify any CCP.

- Systematics used:
  - Decision trees:
    - Decision tree model used.
    - Registration with the answers to the tree.
  - Other methods.
- Characteristics of the CCPs identified.
- Justification of the decisions made by the team.
- Competence and experience of the HACCP team.

3.3 Principle 3

The Codex guidelines state that critical limits should be specified and validated for each CCP.

The definition of critical limit in the Codex document is as follows:

> **Critical limit**: A criterion, observable or measurable, relating to a control measure at a CCP which separates acceptability from unacceptability of the food.
>
> **The NACMCF defines it as follows**: A maximum and/or minimum value to which a biological, chemical or physical parameter must be controlled at a CCP to prevent, eliminate or reduce to an acceptable level the occurrence of a food safety hazard.

As we saw earlier, principles 2 and 3 are inextricably related; each CCP must have a critical limit, and each critical limit must correspond to a CCP. However, in some cases, it may be necessary to set more than one critical limit for a specific step. What does not make sense within the HACCP system is associating critical limits with system elements or other control measures that are not considered CCPs. A critical limit is a barrier that separates safe food from unsafe food; for this reason, it does not seem consistent to apply a critical limit to a control measure that, at the previous step of the HACCP study, was not considered critical or essential for the safety of the food.

The primary condition that a critical limit must meet is that it allows practical, effective, and rapid, or real-time, monitoring of the CCP so that any properly trained operator can clearly determine when the deviation occurs in a CCP and thus be able to take appropriate corrective action. Critical limits must be objective, measurable values, or easily observable characteristics. The critical limits are typically temperature, time, humidity level, pH, a_w, free chlorine level measurements, and sensory parameters such as appearance and texture.

When using critical limits that are not measurable values and that may involve some subjectivity, those responsible for monitoring must have clear instructions on how to differentiate acceptable from unacceptable, and where appropriate, they must receive the necessary training to know how to operate with this type of critical limits effectively and safely. In some cases, monitoring instructions or procedures may include photos or diagrams that help determine the limit of what is considered acceptable. During the on-site audit, the auditor may show interest in checking whether those responsible for monitoring can act with such critical limits without difficulty and demonstrate sufficient competence to determine when the critical limit is exceeded, which implies that the processed product becomes nonconforming.

Critical limits should be set according to the parameters that determine the effectiveness of the control measure, which are those that can be monitored, not on the hazard, as happens in some HACCP manuals, where we can see critical limits like "absence of pathogens," "particles greater than 0.5 mm," "absence of residues," "maximum histamine levels greater than 25 ppm," or "recount of mesophilic aerobic microorganisms less than 10 cfu/ml." No one denies that these values have to be met. However, these examples indicate the objective, the acceptable hazard level to achieve with the control measure, not the critical limits that the operator must monitor to determine whether the CCP is under control.

A mistake still found in some self-control manuals is the presence of critical limits defined ambiguously, imprecisely, or so subjectively that they cannot be considered valid for monitoring CCPs. Examples include:

- Correct handling.
- Compliance with the cleaning and disinfection plan.
- Satisfactory hygienic conditions.
- Compliance with legal regulations.

- Absence of dirt.
- GHP compliance.

The problem with these types of limits is that they do not allow the operator in charge to establish when the barrier of the unacceptable is exceeded. When does a manipulation stop being correct? What is meant by satisfactory hygienic conditions? It does not seem possible to operate with critical limits defined in this way, also taking into account the responsibility it entails for the operator who has to decide immediately if these limits are exceeded, which implies applying immediate corrective actions, stopping production in many cases and managing the product processed in those conditions as nonconforming, i.e. a decision that affects the entire production plant.

Besides, some of these examples refer to compliance with PRPs such as cleaning and disinfection or GHP, which are programs that affect different parts and processes of the establishment and, therefore, cannot be monitored at one point. Critical limits must be associated with a CCP, which is a step or operation of the process, not a program. When we set a critical limit as compliance with a PRP or an OPRP, what we say is that the control of a CCP ultimately lies with the application of a PRP or an OPRP, which does not seem to be very consistent with the application of principle 2, nor with the logic of the HACCP system. For this reason, several decision trees begin with the question of the effectiveness of PRPs in controlling hazards.

Each CCP may have one or more components that must be monitored to ensure that the control measure is effective. For example, in the case of pasteurization, the established treatment is 72 °C for 15 seconds or other combinations of time and temperature that achieve an equivalent effect. However, when monitoring this CCP, only the temperature of the treatment is usually monitored; the reason is that the time was already previously adjusted for each equipment with the length of the holding tube. On the other hand, in different types of heat treatment, such as canned food, there may be more factors to consider: the size and shape of the can, viscosity, temperature or initial weight, equipment used, etc.

Reference sources that can be used to establish critical limits include the following:

- Scientific publications, sectoral guides, food safety standards, codes of practice, and articles in scientific journals.
- Regulations and legal standards.
- Experimental data, tests, and trials carried out by operators.
- Private or university experts, official and private laboratories, and professional associations.
- Historical production data.

Microbiological testing can rarely be accepted as a critical limit for CCP monitoring due to several reasons: the time it takes to obtain the result does not allow for real-time or sufficiently fast monitoring of production, the sample may not be representative of the batch, and in general, there are limitations of analyses to detect production failures or product contamination when the percentage of defective products is sufficiently low.

Some critical limits may be defined by legislation, for example, the time and temperature of a pasteurization treatment, based on internationally recognized standards where the combinations of time and temperature that are required to achieve the desired lethal effect are established: elimination of pathogenic bacteria, such as *Mycobacterium tuberculosis*, and the almost total reduction of bacterial flora. However, this is different from suggesting that legal limits should be managed as critical limits when not associated with a CCP. This is a mistake that can be seen in some manuals where, for example, legally established temperatures, such as refrigeration storage or the ambient temperature of certain premises, are applied as if they were critical limits in steps where the

HACCP team had not previously identified a CCP, which is contrary to the application of HACCP principles.

Regardless of how HACCP is developed, the legal limits must be respected. Its infringement can entail sanctions imposed by the official control agents, depending on the circumstances of a specific case. However, this does not mean the processed product exceeding a legal limit should always be considered nonconforming, as it happens when a critical limit is exceeded under the HACCP systematics.

In short, a critical limit may be based on a legal limit, but a legal limit need not be a critical limit; a legal limit should not be a critical limit if it is not associated with a CCP.

Another difference between legal limits and critical limits is that legislation accepts that, in certain circumstances, certain specified limits may be exceeded, provided that this does not imply a health risk. For example, Chapter IX of Annex II to Regulation 852/2004 on food hygiene states that: "limited periods outside temperature control are permitted, to accommodate the practicalities of handling during preparation, transport, storage, display and service of food, provided that it does not result in a risk to health."

A practical application of this provision is the "2-hour/4-hour rule" found in the Food Standards Australia New Zealand and similar rules of bodies such as the FDA. The rationale behind these rules is that bacteria can take time to grow to high enough numbers to cause food poisoning, allowing food businesses to hold food between 5 and 60 °C, but only for short and measured periods of time.

According to the 2-hour/4-hour rule, food can be held between 5 and 60 °C for controlled periods according to the following scheme (Figure 3.23).

Critical limits must be validated prior to the implementation of the HACCP plan. Validation is one of the activities included within HACCP principle 6. As will be seen when discussing this principle, validation consists of checking that the elements of the HACCP plan effectively control the identified hazards. In the case of critical limits, the HACCP team must obtain evidence that the

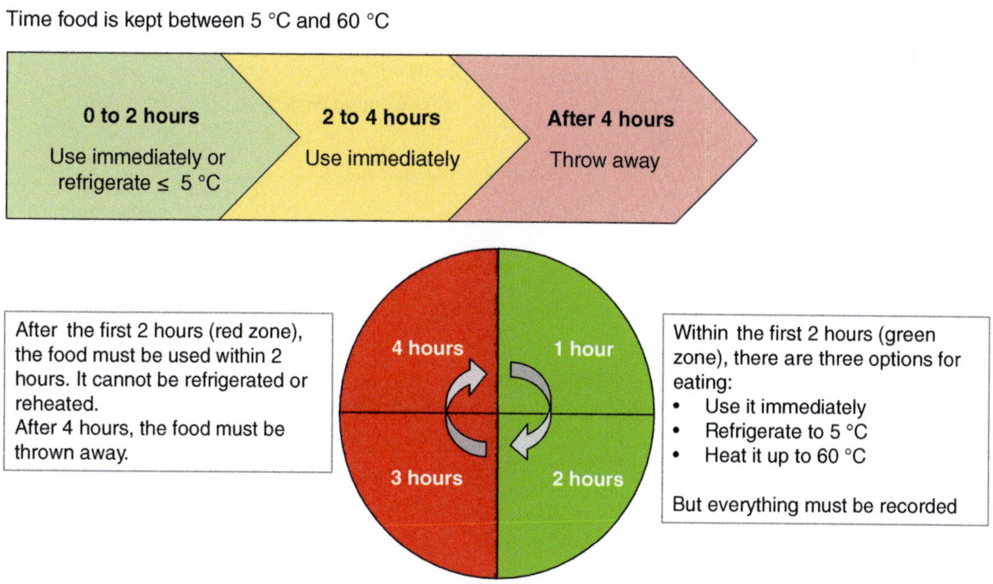

Figure 3.23 The 2-hour/4-hour rule.

established critical limits serve to ensure that the CCP can eliminate or reduce, to an acceptable level, the hazards identified at that stage of the process.

The validation of critical limits can be conducted in different ways, such as scientific literature, publications, legislation, sectoral or generic HACCP guides, codes of practice, international standards for food, specific tests, history of production, and the use of mathematical models.

3.3.1 Operational Limits

Depending on the characteristics of their production system, some operators find it helpful to set operational limits for monitoring CCPs.

The observation of trends in a production process can warn of the possibility of a loss of control and that critical limits are exceeded; for this reason, it seems reasonable to add operational limits, also called alarm or action limits, which must be stricter than the critical limits, so that CCP monitors can make adjustments before the observed parameter exceeds the critical limit (Figures 3.24 and 3.25).

The usefulness of setting operational limits lies in avoiding all the actions the FBO must take when a critical limit is exceeded: application of corrective measures, identification and segregation of the nonconforming product, deciding its final destination, and finding the root cause of the deviation. Depending on the process's variability and the measuring equipment's accuracy, the distance between the critical limit and the operational limit provides the food operator with a margin of safety to adjust the process before the observed parameter exceeds the critical limit.

Statistical process control (SPC) is one of the tools that can help food companies monitor that processes remain stabilized within certain limits and prevent the occurrence of safety or quality problems in the product. SPC assumes that the measurable variables in a process, such as temperature, time, pressure, weight, pH, show an inherent variability produced by the so-called common causes. Common causes include different factors (raw material, machinery, environment, handlers, etc.) whose combination generates a limited but predictable variation maintained within a known range over time. The amplitude of the range can be reduced by adjusting the process to a specific

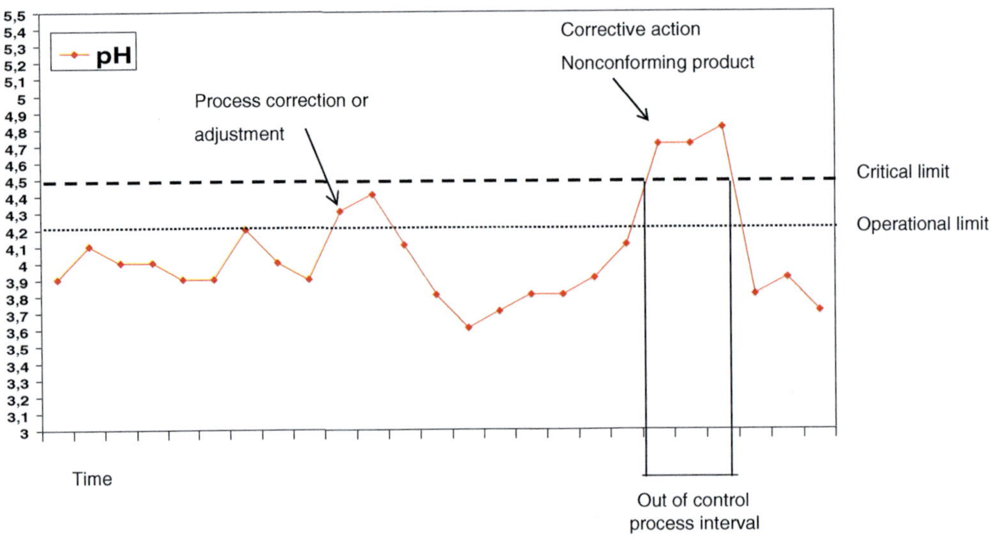

Figure 3.24 pH monitoring corrective action.

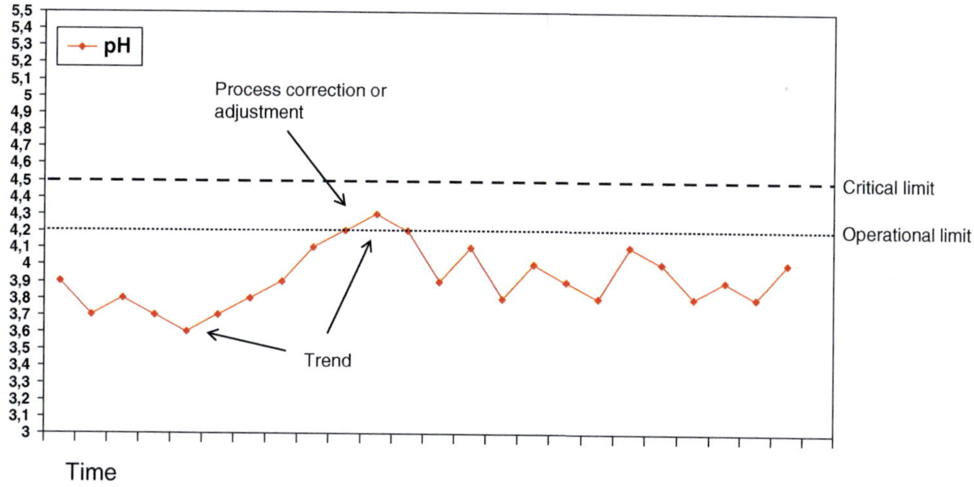

Figure 3.25 pH monitoring trend.

point that is considered optimal; however, it is complicated and expensive to try to suppress or limit the effect of common causes. In contrast, special causes of variation are specific or individual causes; only one causes the control parameters to exceed the specified limits, which requires the corrective actions necessary to bring the process back under control and the need to investigate the individual cause.

From the process variables and statistical values such as mean, range, or standard deviation, different types of control charts are helpful to monitor that the process remains within the established limits and to detect trends that make it possible to anticipate a loss of control in the process; this allows to adjust process before the control limits are breached. For example, when a certain number of values fall below the average in a row, this can be a warning sign that the process is no longer within its inherent variability (Figure 3.26).

In other control chart models, the zone between the upper and lower limits is divided into six zones, located on each side of the mean, or control line, at a distance of 1б (standard deviation), 2б, and 3б. Depending on the number of values in each zone, different types of warnings can be established that the process is affected by a special cause of variation. For example, a single value that falls beyond zone A, eight consecutive increasing or decreasing values, two out of three in zone A, or a combination of four out of five values falling in zone B or outermost means according to pattern analyses that the process is no longer stable (Figure 3.27).

When the food operator uses this tool, the auditor may be interested in checking whether the control charts show that the process remains stable and whether there is evidence that FBO takes corrective action when a deviation is detected.

Figure 3.26 Control chart model.

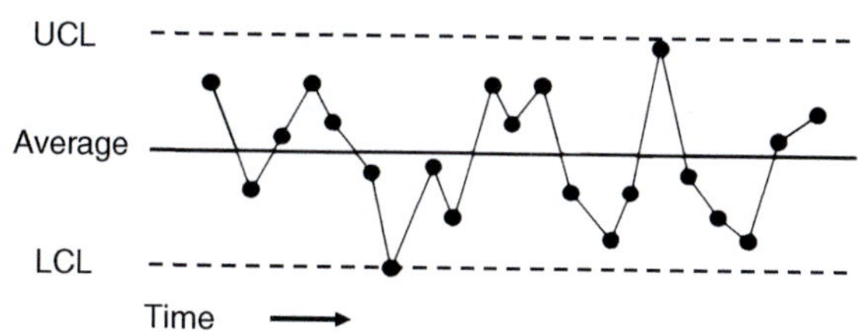

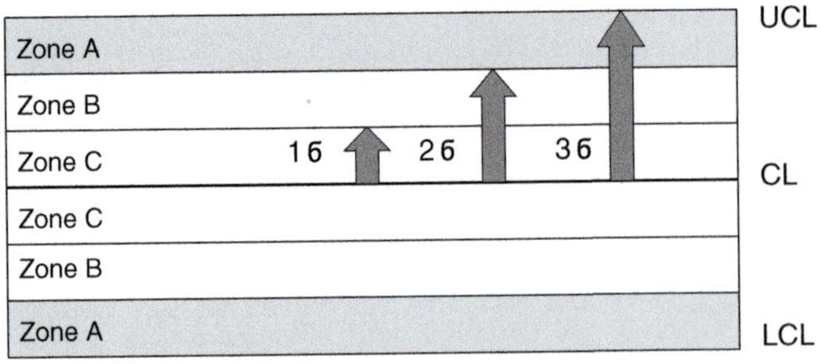

Figure 3.27 Six sigma.

Key elements of principle 3

Among the elements that the auditor will examine to evaluate the application of principle 3, we can mention the following:

- Critical limits are associated with a CCP.
- Each CCP must have one or more critical limits.
- Critical limits must be observable or measurable values.
- They should be established for the control measure, not the hazard.
- Where critical limits are based on observations, precise instructions for their application should be provided.
- Critical limits must allow for rapid action.

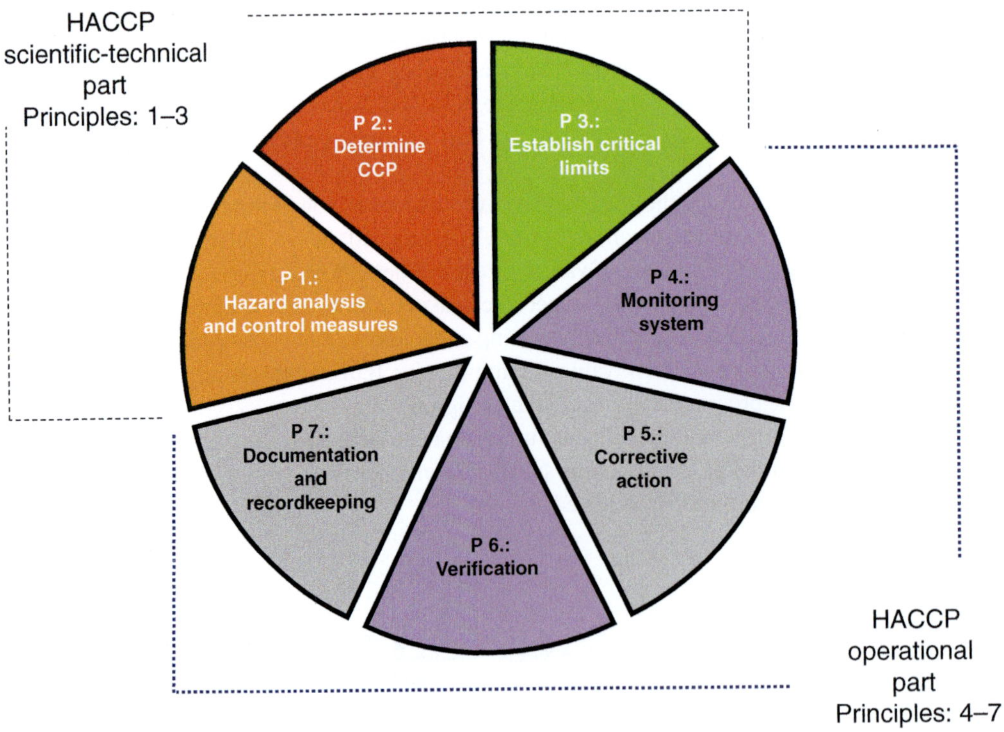

Figure 3.28 2 Part HACCP.

- It must be documented how they were established or their basis.
- Critical limits must be validated.
- Does the FBO use operational limits?

Having concluded principle 3, we have practically reached the halfway point of the HACCP study. We can split the seven principles into two parts for purely didactic purposes: the first includes the three principles, and the second is composed of the following four. Thus, principles 1–3 represent what we call the scientific-technical part of HACCP, in which the application of knowledge in different matters of hygiene and food technology is fundamental, just like using reliable scientific and bibliographic sources. The first part is also where the HACCP team makes the main decisions affecting the design and structure of HACCP, where control measures are selected, and the effect of each measure on food safety. On the other hand, the following four principles, except the validation included in principle 6, involve a more operational part, common in many aspects to other management systems, which seeks to develop the necessary procedures to specify how, when, and who have to carry out CCP monitoring, take corrective actions, and conduct verification activities, as well as those related to the registration and documentation of the HACCP system (Figure 3.28).

3.4 Principle 4

The objective of principle 4 is to establish a monitoring system for each CCP. Once the CCPs have been selected and the critical limits defined, after applying the two previous principles, the HACCP system requires the food operator to conduct regular monitoring to verify that the critical limits set are not exceeded.

The 2022 version of the Codex guidelines defines "monitor" as follows:

> **Monitor**: The act of conducting a planned sequence of observations or measurements of control parameters to assess whether a control measure is under control.

The previous edition of Codex defined "monitor" as conducting a planned sequence of observations or measurements of control parameters to assess whether a CCP is under control. This former definition could lead us to think that only the CCPs should be monitored, which implies a terrible misunderstanding since the operator must monitor the implementation and effectiveness of all control measures and other data and parameters to ensure control of all identified hazards.

The difference between CCP monitoring and other control measures is that PRP monitoring is not as strict as CCP monitoring; the frequency and consequences of detected deviations differ. In addition, PRP monitoring is not done on critical limits, defined only for CCPs, but on compliance with procedures and practices included in a particular program.

Effective CCP monitoring requires that those responsible for conducting it can accurately determine when a value or measure exceeds the set critical limits and then take rapid action to correct the process and bring the CCP back under control with the least possible delay. The fulfillment of these conditions is directly related to the correct application of principle 3; when critical limits are appropriately defined, using observable or measurable values that do not cause imprecision or ambiguity, CCPs can be reliably monitored.

Depending on several factors, such as the nature of the process, the characteristics of the production line, the equipment, or the technology used, monitoring systems can be of several types:

- online;
- offline,

- continuous; and
- discontinuous.

Online monitoring is possible when the FBO has automated equipment or devices capable of measuring the values generated during the production process, usually physical or chemical parameters, continuously or at specified time intervals.

Offline monitoring is carried out manually by the operator in charge, using appropriate equipment to measure the process control parameters or values measured on the product at different stages of manufacture. When critical limits are based on observable characteristics, monitoring is carried out visually and following adequate instructions that allow the operator to determine if the critical limit is exceeded.

Continuous monitoring is possible when the FBO has automated equipment capable of continuously measuring and often recording process control parameters.

In the case of discontinuous monitoring, a fundamental issue is to define the most appropriate frequency to ensure that the process is kept under control in order to act in time to correct the process and avoid prolonged periods during which the processed product would be considered nonconforming.

From the point of view of the reliability of the measures and the reduction of the burden of operational management, whenever feasible, continuous and online monitoring is the best for several reasons:

It provides complete information on the status of the process at all times.
It allows quick action.
There is no risk of human error due to carelessness or lack of diligence.
It can sometimes include an automated registration system, making it easier to comply with this requirement without increasing the management burden for operators.

In addition, with the current technological development, other functionalities can be added, such as the sending of alarm signals whenever the critical limits are exceeded, by SMS or email to the selected people, or the possibility of remotely consulting the surveillance data at any time and from anywhere, thanks to the Internet and the use of Wi-Fi data loggers.

The establishment of an effective monitoring system requires the food operator to develop a procedure determining how monitoring activities must be conducted, specifying the following:

- Who is the person or persons responsible for monitoring each of the CCPs. These people must demonstrate the appropriate training to use the measuring equipment correctly, understand the critical limits established, and know how to operate with them. In the case of critical limits based on subjective criteria or on the observation of certain characteristics, they must have received specific training to carry out such monitoring. The monitoring procedure shall also provide for the designation of the necessary substitutes to replace the usual monitors in case of absence.

Some procedures also include supervision, but we must understand that this activity is carried out by those responsible for the management system at a higher level to verify that the monitors conduct their tasks in the specified manner.

- What needs to be monitored, and how it should be done. This task requires operators in charge to know what parameters or values to monitor, what equipment or instruments to use, and where to monitor. These aspects may be set in a documented procedure or work instruction, which must be provided to those responsible for monitoring. In addition, the auditor will probably check whether the measuring equipment is calibrated and correctly identified.

- It is also crucial to determine the frequency of monitoring. When continuous monitoring is impossible, it is necessary to specify how often CCPs should be checked. To make this decision, the HACCP team must consider the characteristics of the process and the speed of the line or the amount of product produced at any given time. Many times, this decision is based on reaching a compromise between the workload involved in increasing the frequency of monitoring and the amount of processed product that the FBO can accept for management as nonconforming, which happens whenever the FBO has stopped production and withheld the product due to deviation from the critical limit.
- Finally, the procedure should specify the monitoring registration system. The HACCP team must decide the data to be recorded and design the appropriate type of form, either on paper or in electronic format. In case there is no automatic registration of data, it is necessary to establish who is responsible for filling in and signing the records. The procedure should also define the identification of records, for example, using a code or a reference to the management system, and lay down the necessary arrangements to control access, protection, retention time, and disposal of records.

If we apply flexibility criteria for implementing HACCP, we can accept that monitoring procedures limit data recording to "record by exception"; that is, only values that exceed the set limits or observations considered incorrect are recorded, which allows for substantially reducing the workload involved with the recording of monitoring data.

Key elements of principle 4

The elements that the auditor may take into consideration in assessing principle 4 are as follows (Figures 3.29 and 3.30):

CCP	Critical limit	Monitoring			
		What	How	When	Who
Trimming or condemnation	Macroscopic contamination (milk, feces, or ingestion) not visible	Each carcass or piece of meat	Visual observation	Each carcass or piece of meat	Slaughter line operator
Organic acid spray	Organic acid concentration, at least 2%.	Acid concentration	Kit	Start and end of production day	Slaughter line operator/designee
	Each carcass or part	Application of acid to carcasses or pieces of meat	Visual observation	Each carcass and offal	Slaughter line operator/ designee
Carcass cooling	Carcass surface temperature < 40 °F in 24 h	Carcass surface temperature	Temperature monitoring device	25% of the carcasses, at least one head and offal per 20	Slaughter line operator

Figure 3.29 Monitoring chat example.

Stage number/ CCP	Hazard	Critical limit	Monitoring			
			What	How	When	Who
10. Syrup addition, CCP: 3	Growth of *Clostridium botulinum* due to insufficient acidic pH	pH < 4.2	pH	pH meter calibrated each day at the start of the day	At the start of the day and every hour during production	Responsible operator

Figure 3.30 Example of monitoring of CCP in canned vegetables.

- Based on the control of critical limits.
- Allow quick action.
- Types:
  - Continuous.
  - Discontinuous: frequency.
- Documented procedure:
  - Work instructions.
- Responsible:
  - Training.
  - Substitutes.
- Measuring equipment:
  - Calibration.
- Recording:
  - Model
    - Paper
    - Electronic.
  - Code and reference to the management system.
  - Automatic.
  - Access.
  - Retention time.
  - Disposition.

3.5 Principle 5

Principle 5 of HACCP states that the FBO must establish the corrective actions to be taken when monitoring indicates a deviation from a critical limit at a CCP has occurred.

According to the statement of this principle, corrective actions should be taken when there is a deviation in a CCP; that is, the monitoring system has detected that a critical limit is exceeded.

Knowing how the Codex defines these concepts is imperative to understand the HACCP systematics well.

Deviation: Failure to meet a critical limit or to follow a GHP procedure.

Corrective action: Any action taken when a deviation occurs in order to re-establish control, segregate and determine the disposition of the affected product, if any, and prevent or minimize reoccurrence of the deviation.

The HACCP Principles and Application and Guidelines of the NACMCF include two other definitions that are also of interest:

> **Deviation**: Failure to meet a critical limit.
> **Corrective action**: Procedures followed when a deviation occurs.

As we can see, the NACMCF only refers to failures in the CCPs, while according to the Codex definition, a deviation also happens when the FBO does not follow a GHP procedure.

In the HACCP system, corrective actions consist of a set of measures whose objectives are the following:

- Get the process back under control.
- Identify and segregate the nonconforming product.
- Decide the final destination of the nonconforming product.
- Analyze the cause of the failure or deviation.
- Establish the corrective action that avoids the repetition of the failure.
- Document the actions taken.

Corrective actions should be documented in a procedure describing the actions to be taken when a deviation occurs and who is responsible for carrying out each. In order to avoid the doubts or hesitation that may arise for the person responsible when he has to act – taking into account the economic or productive consequences arising from his decision – it is essential that the measures to be applied are clearly defined in advance and that the person responsible for carrying them out has the appropriate training and sufficient authority to take the measures with the speed that the situation requires.

When a deviation occurs, a group of measures must be applied immediately, as soon as the monitoring detects the failure in the process; instead, other actions may require more time for their study or to carry them out and are not so urgent.

The corrective actions that the FBO must carry out immediately are the following:

- **on the process**: it is necessary to stop the process or put it back under control to prevent further production of nonconforming products.
- **on the product**: identify and segregate the product produced when the process was outside the critical limits to prevent it from being mixed with the conforming product, continuing forward in the production process, or being released for consumption.

Depending on the nature of the manufacturing process, the type of equipment used, and the characteristics of the monitoring system, the possibilities for rapid intervention in the process and for segregating the product concerned vary widely.

For example, milk pasteurization equipment has probes to measure milk's temperature continuously during treatment. When the milk does not reach the set temperature, the diverter valve is automatically activated so that the milk that did not receive the programmed treatment returns to the buffer tank to be processed again. Similarly, egg grading machines now have devices with the latest technology in visual systems with artificial intelligence (AI) to detect leakers, dirt, cracks, and blood in eggs and automatically remove them from the processing line. However, there are many types of production processes in which, due to their lower degree of mechanization or the frequency of monitoring, this automation is not possible, and the action of an operator is required to be able to intervene in the process with the required speed, once the deviation is detected.

The time available to act on the product and prevent it from passing to the following processing stages or from being released for consumption also differs considerably from one process to another.

Thus, in some activities or food sectors, the duration of the processes or the time that the products remain in the operator's facilities, for example, when they require some ripening stage that can last days or months, allows sufficient time to act, without the urgency that, on the other hand, is required in other types of establishments, as can be the case of a catering company that distributes ready-to-eat hot meals, where the release toward consumption is immediate and there is very little time to act.

The need to intervene without delay when there is a deviation in the process requires that the responsibility for taking these measures is usually assigned to an operator close to the production process or someone who can act quickly. As we have said, this operator and those who have to replace him must be adequately trained and feel they have the authority to act without hesitation.

The ISO 22000 standard differentiates the different types of actions that FBO must carry out in the event of a deviation and defines them as follows:

- **Correction**: action to eliminate a detected nonconformity.
- **Corrective action**: action to eliminate the cause of a nonconformity.

In its notes, it adds:

> A correction includes the handling of potentially unsafe products and can therefore be made in conjunction with a corrective action.
> A correction may be, for example, reprocessing, further processing, and/or elimination of the adverse consequences of the nonconformity.
> There can be more than one cause for a nonconformity.
> Corrective action includes cause analysis.

According to ISO 22000, a correction has to eliminate the adverse circumstances of a nonconformity and the handling of potentially unsafe products, which are the actions that the FBO must do immediately. Instead, corrective action must act on the cause of the deviation. Except for some cases where the cause is quite obvious, determining the cause requires those responsible for managing the HACCP system to take the time to analyze what happened and identify the root cause that ultimately caused the deviation.

The root cause is the fundamental or underlying reason behind a failure or problem, so when this cause is corrected or eliminated, the problem can be prevented from happening again. Often, the root cause is hidden or masked by apparent causes or contributing factors, and its determination requires systematic investigation and the use of specific problem-solving techniques by people with the proper training or even with the help of experts.

For these reasons, the responsibility for determining corrective action should be assigned to higher-level management positions, such as the quality or production manager or a technical director of the company, who are usually the people in charge of carrying out the investigation of the causes of the deviation and proposing the corrective action that they consider effective in solving the problem. The company's management can be informed and sometimes intervene in decision-making or approve the proposed corrective action.

When a deviation occurs in a CCP, the person or persons assigned that responsibility must complete a record describing the incident and the actions taken to correct the process. Once the incident is recorded, the corrective action procedure should indicate who is responsible for carrying out the root cause analysis and proposing and approving the subsequent corrective action.

To investigate the root cause of a deviation or a problem, the food operator can use various methods designed to analyze the events and determine the cause or causes that originate them, such as

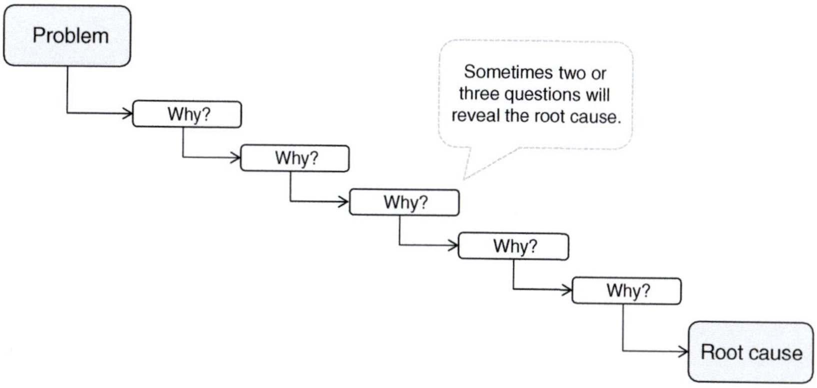

Figure 3.31 The 5 whys technique.

the Pareto chart, fishbone or Ishikawa diagram, the 5 whys technique, Is/Is Not Analysis, or the FMEA, among others (Figure 3.31).

The recording of an incident results in the initiation of a series of tasks by the operator. These activities follow each other as if they formed a cycle, which closes when those responsible for the HACCP system verify that the corrective action implemented has effectively eliminated the cause of the deviation and that the problem should not occur again. The phases that make up the cycle that follow the corrective actions are the following:

1. Record of the incident and opening of the corrective action.
2. Root cause analysis.
3. Approval and execution of corrective action.
4. Verification of its effectiveness and closure of the corrective action.

When a deviation occurs in a CCP, the HACCP system establishes that the processed product must be managed as a nonconforming product since the critical limit differentiating between acceptable and unacceptable has been exceeded. On the other hand, if the FBO fails to meet the PRP, it does not entail any action on the product because the associated risk does not justify it. OPRPs occupy an intermediate position, and although in principle, as the critical limits are not breached, the HACCP system does not require an action on the product, it may be appropriate for those responsible for HACCP management to evaluate the situation produced and, depending on the causes and the risk generated in each case, make a decision on the product produced when an OPRP is breached (Figure 3.32).

ISO 22000 also establishes different management for products affected by failure to remain within critical limits at CCPs, which are always considered nonconforming products, from those affected by a failure to meet action criterion for OPRPs, which could be released for consumption after the result of an evaluation, as indicated in Section 8.9.4 of this standard. In all cases, the FBO must document these actions.

The management of nonconforming products by the food operator comprises the following activities:

1. Identify the product concerned by the deviation.
2. Separate the product concerned from the rest of the production.
3. Remove the product from the market if it has come out of the operator's control.
4. Evaluate the risk associated with the product.
5. Decide the final destination of the product.

<table>
<tr><td rowspan="3">Company logo</td><td rowspan="3">Logging of incidents and corrective actions</td><td colspan="2">No incident, corrective action</td></tr>
<tr><td colspan="2">Date, hour</td></tr>
<tr><td colspan="2">Name of the person who detects</td></tr>
<tr><td colspan="4">Description of the incident or deviation</td></tr>
<tr><td colspan="4">Actions in place, corrections</td></tr>
<tr><td rowspan="4">Product concerned</td><td>No</td><td colspan="2"></td></tr>
<tr><td rowspan="3">Yes</td><td colspan="2">Identification of the product concerned: decription, lot, quantity</td></tr>
<tr><td colspan="2">Decision on the product/disposition</td></tr>
<tr><td>Responsible</td><td>Signature</td></tr>
<tr><td colspan="4">Identification of the root cause</td></tr>
<tr><td rowspan="2">Proposed corrective action</td><td colspan="3">Description</td></tr>
<tr><td colspan="2">Deadline estimate</td><td>Completion date</td></tr>
<tr><td colspan="4">Corrective action verification</td></tr>
<tr><td>Responsible

Signature</td><td colspan="2">Result:</td><td>Date:</td></tr>
</table>

Figure 3.32 Incident and corrective action log format.

The food operator's traceability system must make it possible to identify the product affected by the deviation correctly; it is necessary to describe the product involved: name, quantity, and the lot or lots involved; and separate it from the rest of the production in an appropriate place, or in an area delimited for that purpose.

While the product is being held, those responsible for HACCP management must assess the risk associated with the product produced during the deviation, depending on the causes, the conditions of its processing, and the consequences for its safety. To conduct this evaluation, the FBO must consider the following points:

- the combined effect of control measures;
- any evidence other than monitoring showing that control measures have been effective;
- the result of sampling and any analysis of the retained product;
- any other information indicating whether the product concerned is within the levels acceptable for the hazard in question.

Once sufficient information has been collected, the persons in charge should decide on the final destination of the product concerned. Here are the possible options:

- reprocess the product for its intended use;
- reprocess the product for a different use;
- destroy the product;
- eliminate the product and manage it as a by-product not suitable for human consumption; and
- release it for consumption if the FBO can prove that there is no risk to the consumer.

The food operator must keep records of all actions carried out on the nonconforming product and must be able to demonstrate the correlation between the amount of product retained and that of the product at the final destination (Figure 3.33).

Since its origin, and in essence, HACCP has always been defined as a preventive system whose purpose is to establish the necessary control measures to prevent potential hazards in food. However, it is practically impossible to prevent deviations arising from established plans or any failures that occur in production. When this happens, the food operator must react and take the necessary measures to limit the deviation's consequences as much as possible.

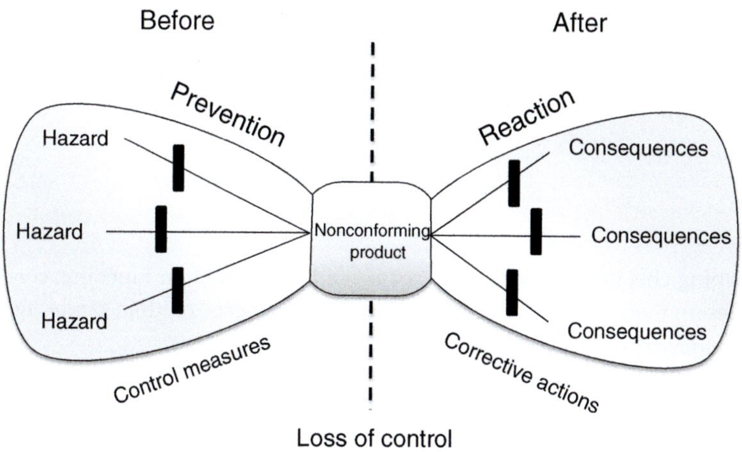

Figure 3.33 Bow tie diagram.

HACCP principle 5, since applied after a loss of control or deviation in a CCP, can be considered a reactive part of HACCP, and many of the tasks it comprises have to do with minimizing the negative consequences of the detected failure, as are all those related to managing the nonconforming product to prevent it from reaching the consumer.

Qualifying a measure as preventive or reactive is somewhat arbitrary and depends on the choice of the undesired event taken as a reference. For example, for the food operator, identifying and segregating the nonconforming product is a corrective action. However, from the consumer's point of view, it is a preventive measure because it prevents adverse health effects that could arise from consuming potentially unsafe food (Besseling and Kalyankar 2014).

Corrective actions represent a good indicator of the functioning of a management system. When the system is mature, its machinery greased, and each of those responsible at different levels of the organization fulfills its mission, it is expected that both the execution of the corrective action and its verification and closure will be carried out as prescribed in the FBO's procedure. The HACCP system auditor will indeed show interest in verifying compliance with the actions related to the resolution of deviations, as well as the documentation and records related to the management of the nonconforming product and its final destination.

The application of principle 5 can be assessed based on the following elements:

- Immediate actions
  - **Correction**: readjustment of the process.
    - Responsible.
  - **About the product**:
    - Identification.
    - Separation.
    - Retention.
- **Corrective action**:
  - Root cause analysis.
  - **Proposal and approval of corrective action**:
    - Term.
    - Responsible.
  - Execution of corrective action.
  - Verification and closure.
- **Product concerned**:
  - Decision on the final destination.
  - Traceability.
- Documentation and registration of the actions taken.

3.6 Principle 6

Principle 6 requires establishing checking procedures to see if the HACCP system functions correctly. The FBO can use different methods, procedures, trials, verification tests, random sampling, and analysis techniques to achieve this objective.

According to the statement of this principle, within its scope, we can include a set of diverse activities:

- review of the HACCP system and its records;
- examination of deviations and disposal systems for products;

- inspection of operations;
- confirmation that CCPs remain under control;
- validation of HACCP components;
- calibration of equipment;
- sampling;
- analysis of final product, intermediate product, or processing area surfaces;
- auditing (internal and external); and
- review of consumer and customer complaints.

The complexity and variety of activities encompassed by principle 6 may make us think that we are dealing with the most undefined principle of HACCP. However, despite the different activities and methods it comprises, the objective of principle 6 is very specific: to verify that HACCP works appropriately and that it is effective in achieving the objective of the FSMS. Therefore, principle 6 is called "the principle of truth": the one that tries to verify the system. As its etymology indicates, from the Latin *verus* + *facere*: to verify that it is true, that is to say, its provisions are fulfilled and effective.

Therefore, in accordance with the provisions of the Codex, the FBO should undertake verification activities as relevant to the business, to check that GHP procedures have been implemented effectively, monitoring is occurring, where planned, and appropriate corrective actions are taken when requirements are not met.

In the past, the wording of Codex principle 6 only referred to verification. However, the 2022 version of the Hazard Analysis and Critical Control Point and Guidelines for its Application amended it to read as follows: Validate the HACCP plan and then establish procedures for verification to confirm that the HACCP system is working as intended. In this way, the validation activities that remained hidden among so much verification become more evident to the FBO. Before this change in the Codex guidelines for HACCP, many authors and sources already considered principle 6 to comprise three activities:

- validation;
- verification; and
- revision.

Validation and verification are concepts defined in the Codex HACCP Application Guidelines and the NACMCF HACCP Application Guide.

Article 5 of Regulation 852/2004, the basis for making the system compulsory in the countries of the European Union, refers to the verification and updating of the system but does not expressly mention validation, although it is included in other official documents of the European Union such as 2016/C 278/01 Commission Notice (replaced from 2022) and 2022/C 355/01 Commission Notice.

The General Principles of Food Hygiene CXC 1-1969 defines validation and verification as follows:

> **Validation of control measures**: Obtaining evidence that a control measure or combination of control measures, if properly implemented, is capable of controlling the hazard to a specified outcome.
>
> **Verification**: The application of methods, procedures, tests, and other evaluations, in addition to monitoring, to determine whether a control measure is or has been operating as intended.

When people begin training in HACCP, many need help distinguishing these concepts well and how they work within the system. For this reason, in 2008, the Codex Alimentarius published the document CAC/GL 69-2008 Guidelines for the Validation of Food Safety Control Measures. That is quite a handy paper to understand the differences between validation, verification, and monitoring, and also as guidance to industry and governments on the validation of individual control measures.

Point 19.11.1 of the Codex guidelines helps us understand how the FBO should conduct the validation and integrate it into the HACCP system.

Validation consists of ensuring that all elements of the HACCP plan as a whole are capable of ensuring the control of significant hazards relevant to the business. Validation is done before the implementation of the system and only needs to be repeated if there are changes in the system that indicate the need for revalidation. Codex guidelines state that any changes that potentially impact food safety should require a review of the HACCP system and, where necessary, a revalidation of the HACCP plan.

The HACCP team must evaluate the changes produced in the system to determine in which cases they can modify the effectiveness of the control measures or the safety of the final product, which requires a new validation of the affected elements.

Validation could include a review of scientific literature, using mathematical models, conducting validation studies, experimental trials, surveys, and/or guidance developed by authoritative sources. Where HACCP guidance developed by external experts has been used, the HACCP team should ensure that this guidance fully applies to the specific operation and product(s) under consideration.

Validation should ensure that all elements of the HACCP system, such as identifying the hazards, CCPs, critical limits, control measures, frequency and type of monitoring of CCPs, corrective actions, frequency and type of verification, and the type of information to be recorded, are capable of ensuring the control of significant hazards relevant to the FBO.

Unlike validation, verification is a regular, periodic activity planned with the appropriate frequency to confirm that HACCP is working correctly.

The verification is done after the implementation of the system. For that purpose, the FBO must document the procedures indicating who is responsible for the verification tasks, the verified elements, and the frequency with which they must do it.

Verification includes observations, records review, calibration, sampling and testing, and auditing (internal and external) that should be performed on an ongoing basis to ensure the HACCP system functions as intended and continues to operate effectively. According to its definition, verification must verify how the elements of the HACCP system work; for so, it is an activity performed at a higher level. We can verify monitoring, but we cannot monitor verification. That is how HACCP works. Therefore, and for the same reason, verification should be carried out by someone other than the person who is responsible for performing the monitoring and corrective actions. No one can verify or audit their own work. This is an essential auditing principle. Where the FBO cannot carry out certain verification activities with its personnel, the verification must be performed on behalf of the company by external experts or qualified third parties.

The differences between validation and verification can be summarized in two questions.

> **Validation must answer**: Will the system work?
> **Verification must answer**: Are we doing the things as we have established?

In addition, due to its nature of searching for data and scientific, technical, or experimental information with which to base and justify the establishment of control measures, validation is an activity that represents the most scientific part of HACCP, where the need for scientific knowledge of the members of the HACCP team is perfectly evident. In contrast, verification represents a part clearly related to the system's operational aspects and the tasks involved in the implementation of HACCP in the day-to-day life of the food business.

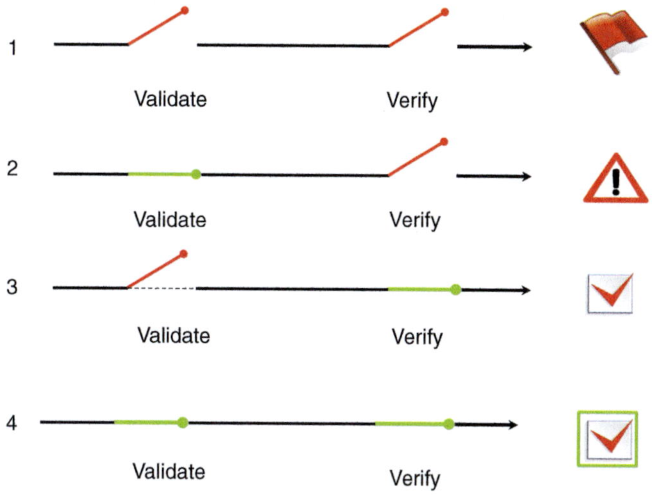

Figure 3.34 Validation and verification work in tandem.

According to HACCP principles, the FBO must first validate the control measure to know whether it will control the hazard and then verify that the measure is applied. If this is done, the hazard will be under control. Validation and verification must work in tandem to make HACCP effective. A validated control measure that is not applied is useless, and applying a measure that is not validated does not assure us of its effectiveness.

As we see in Figure 3.34, we can find four situations:

1. when there is neither validation nor verification, we are in a situation of total lack of control of the hazards, in which HACCP is neither validated nor applied;
2. if we validate but do not verify, we know what we have to do, but we do not know if we are doing it, and therefore, we cannot guarantee that the hazards will be under control.
3. the third situation deserves further attention because it can be relatively frequent; these HACCP plans are being applied as described in their system but do not have the validation carried out or documented. Control measures are, therefore, being implemented and have yet to be validated. In these cases, the operator may be taking for granted some measures based on a production history or because he has not detected problems with an incidence rate that makes them doubt their effectiveness, but that should not be the only basis for considering the measures valid;
4. this is the ideal HACCP situation: we have a validated measure; its effectiveness is demonstrated, and we verify that it is applied correctly (Figure 3.35).

3.6.1 Interrelationships among Validation, Monitoring, and Verification

In addition to knowing the differences between validation and verification, it is essential to understand how to differentiate them from monitoring and to analyze the interrelationship between the three concepts.

Monitoring and verification are two ongoing activities to check whether control measures are being adhered to and demonstrate that they work as intended. However, there are significant differences between the two activities. Thus, following the Codex Guidelines for the Validation of Food Safety Control Measures:

	Validation	Verification
When	Before implementation	After implementation
How many	1 time[a]	Regularly
Key question	Will it work?	What is established is done?
Nature	Scientific part of HACCP	Operational part of HACCP
How	Studies Tests Microbiological analysis Challenge test Scientific documentation	Audits Periodic analyses Record review
Who	HACCP Team	Designated person Quality manager
Documentation related	HACCP supporting documentation	Verification procedures

[a]And when there are changes that require it.

Figure 3.35 Validation vs. verification.

Monitoring of control measures is the ongoing collection of information at the stage when the control measure is applied. Monitoring activities generally focus on "real-time" measurements and the performance of a control measure. When monitoring a CCP, the information collected makes it possible to determine whether the measure operates within the established critical limits. Detecting a deviation of the control parameters above the critical limits involves corrective action, which results in the production of nonconforming products; this requires that the frequency of monitoring allows a rapid response.

Verification is an ongoing activity to determine whether control measures have been implemented as intended. Verification occurs during or after applying a control measure through various activities, including observation of monitoring activities and review of records, among others, to confirm that the FBO applies the control measures as stipulated in its self-control system.

The Codex guidelines for validating control measures provide the example of an uncooked fermented sausage to understand how these three concepts differ in practice. The first task that the FBO must perform is to validate the control measures. In doing so, it must consider the acceptable level established by the competent authority for the identified hazard, *Escherichia coli*, in this case.

The validation indicated that the production process must achieve a logarithmic reduction specified by ensuring a specific pH decrease during fermentation and a specific decrease in the activity of the water during maturation, along with ensuring that the raw materials have less than a specified level of Pathogenic *E. coli* based on statistically based microbiological testing.

The parameters to monitor are the drop in pH and the reduction of a_w (or weight) during maturation.

The verification should consist of periodic analysis to verify that incoming levels in the raw materials are within specification and that fermentation and maturation achieve the intended outcome in the semi-finished or finished product, and the examination of monitoring records (Figure 3.36).

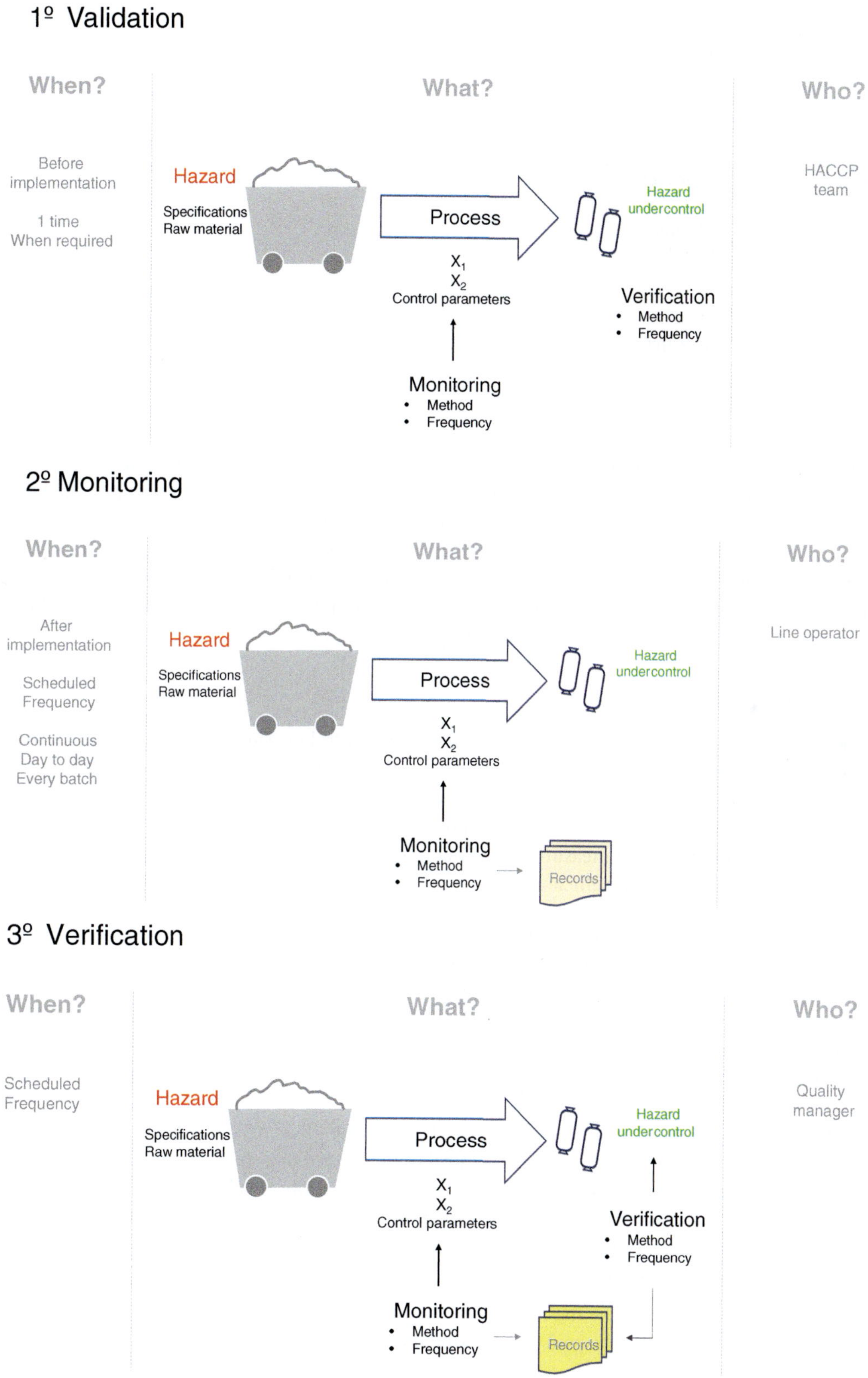

Figure 3.36 Interrelationship among validation, monitoring and verification.

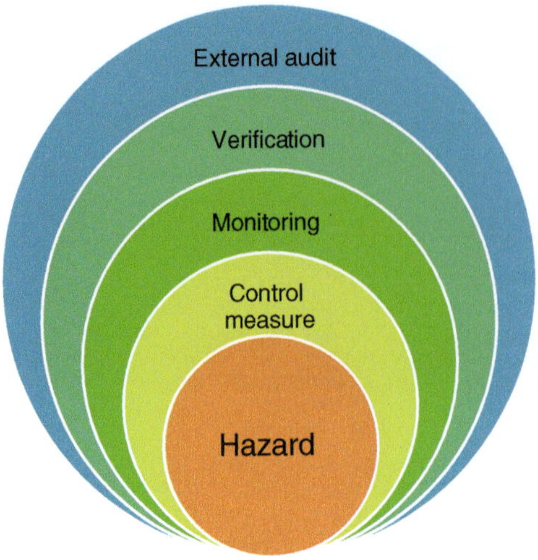

Figure 3.37 The four levels of control in HACCP.

As we just saw in the example above, verification includes monitoring; therefore, it is a control activity at a higher level, whose responsibility must be assigned to people who occupy a higher position within the company's management system. In this way, verification adds another level of safety for hazard control; the first level is control measures, the second is monitoring activities and corrective actions, and verification is at the third level (Besseling and Kalyankar 2014). Following this scheme, we can still add a fourth level: the audit, whose purpose also includes checking that the verification procedures are applied as intended (Figure 3.37).

3.6.2 Validation Methods

Depending on the product, the nature of the hazard, the control measure to be implemented, the technology used, and the safety management system, among other factors, various approaches are available to validate control measures, which, in turn, can differ considerably in terms of cost, difficulty, or the level of scientific and technical knowledge required to apply them.

Before validation of control measures, the FBO must complete certain tasks so that validation can be achieved effectively and efficiently.

3.6.2.1 Tasks Prior to Validation Include:

Identify the hazards that are intended to be controlled in the product and/or environment concerned.

Identify the food safety outcome required. The FBO must determine if FSOs exist for the hazards identified in a food. In the absence of FSOs established by the competent authority, the targets should be identified by the industry. FBO may also set stricter targets than those set by the competent authority.

Identify the measures that are to be validated. For this, the FBO must consider the importance of the control measure in achieving the control of the hazard to a specified outcome, whether the control measure has already been validated, and the priority of validation.

Concerning the validation priority, it must be taken into account that validation is a process that can be resource-intensive; thus, to decide which control measures have priority for validation, the FBO must consider the following criteria:

Adverse health effect: The higher the potential for an adverse health effect from a hazard, the more attention should be paid to validating the control measures selected.

Historical experience: Control measures have less priority for validation when there is long experience of their effectiveness for hazard control than when there is little or no experience in their application for the control of a specific hazard. These historical data may obviate the need to conduct validations in certain instances. However, it is important to avoid assuming that a food production or processing system is safe based solely on historical experience.

Ability to monitor and verify the control measure: Control measures that can be monitored and verified after implementation have higher priority for validation. Control measures of such a nature that it is not feasible to determine their quantitative effect on specific hazards cannot be considered a priority for validation. In prioritizing control measures for validation, consideration should be given to any scientific and/or technical challenges to validating the measure.

The availability of resources of the food operator: particular validation activities, such as experimental trials, process capability studies, surveys, mathematical modeling, product or environmental sampling, and analytical testing, mainly when applied with appropriate statistical treatment, require significant resources.

According to the Codex guidelines, the validation process of control measures should include the following steps:

Decide on the approach or combination of approaches.
Define the parameters and decision criteria that will demonstrate that a control measure or a combination of control measures, if applied correctly, can control the hazard consistently.
Gather relevant validation information and conduct studies when necessary.
Analyze the results.
Document and review the validation.

3.6.2.2 Approaches for Validating Control Measures

The following approaches to validation may be used individually or in combination as appropriate:

Reference to scientific or technical literature, previous validation studies, or historical knowledge of the performance of the control measure: Before considering the validation of a control measure, the food operator should research whether any previous scientific or technical publications or validation studies have demonstrated the effectiveness of the control measure they intend to apply. Undoubtedly, this is the most direct, fast, and economical way, due to the saving of time and resources that it entails, that the food operator has to validate, document, and justify a control measure. Sources of information that the FBO may use include scientific literature, government guidelines, GHP guidelines, and HACCP control measures with a known history of good functioning validated by the competent authorities or independent scientific authorities, international standards or guidelines (e.g. Codex Alimentarius), and validation studies of the industry and/or equipment manufacturers. When using this approach for validation, the food operator must check if the models or references it intends to apply fit with the particular conditions of its food safety control system and the characteristics of its production process and, if so, to analyze which parts can be applied and how to do it.

In other cases, the operator can resort to safe harbors, that is, control measures whose validation is recognized for their foundation and vast experience in producing safe products by the food industry.

For example, a safe harbor for milk pasteurization is to perform a minimum heat treatment of 72 °C for 15 seconds so the FBOs can implement it without revalidating the process.

Experimental tests and studies: One of the ways to validate control measures is by obtaining data and conducting experimental studies. There are several techniques that the operator can use for validation, such as the study of the physical and chemical parameters of the processes and the performance of microbiological analyses of the final or intermediate products, in order to check if they are effective in achieving the established objective, which can be defined as an FSO, a PO, a process criterion or a microbiological criterion.

These studies can be carried out through laboratory tests or, on a larger scale, in pilot plants where they try to reproduce the parameters and the real conditions of the process so that the extrapolation of the data and the results can still be considered valid when moving to the commercial scale of production.

Variability of microbial load at process stages and throughout the food chain may affect the estimation of the ability of control measures to achieve the FSO or the value that has been set as acceptable to validate the process. When calculations are made using only average values, an error is made because the risk associated with the variation in the values used is not taken into account, both in the initial levels of microbial contamination of the raw material and in the reduction effect achieved by the control measures, or estimating the microbial growth that can occur in the food. Therefore, when the variation in actual production is taken into account and the associated standard deviation is added to the different values used, this requires increasing the capacity of the process to ensure that the products meet the established objective with an acceptable margin of safety (ICMSF 2011).

Challenge test: A challenge test is a microbiological test in which microorganisms are inoculated to the food we want to validate the control measures and other purposes. Challenge tests are done for three reasons:

- Evaluate the effectiveness of control measures to eliminate microorganisms.
- Determine the capacity of microbial growth in a food.
- Establish the shelf life of food.

The use of this type of test by the food operator requires a high level of knowledge and skills in microbiology, physical chemistry, engineering, etc., so it may be necessary to resort to external experts in order to constitute a team capable of designing and executing this type of protocols with guarantees.

When designing a challenge study, it is necessary to take into account several factors that vary depending on the process, the type of product, or the purpose of the trial, among others:

- The selection of the microorganism or group of microorganisms to be inoculated, either pathogens or pathogen surrogates.
- The concentration of microorganisms in the inoculum. It is higher in cases where we want to validate the lethality of a process than in the other objectives.
- Preparation of the inoculum.
- Inoculation method.
- Duration of the study.

Laboratory challenge tests must mimic process conditions and industrial or pilot plant trials of particular aspects of food processing unit operations. Scale-up of laboratory-based experimental trials in a pilot plant helps ensure that the trials properly reflect actual processing parameters and conditions. However, this almost always entails the availability of appropriate non-pathogenic surrogate microorganisms since viable pathogenic microorganisms should never be introduced into a food production facility.

Mathematical modeling. Mathematical modeling is a means of mathematically integrating scientific data on how factors affecting the performance of a control measure or combination of control measures affect their effectiveness in achieving the intended FSO. The food industry can use these models to study and compare processes and validate control measures. Combining the intrinsic and extrinsic factors that influence microbial growth, such as a_w, pH, NaCl, initial load of microorganisms, and temperature, and selecting the type of microorganism these microbiological models enable predictions regarding:

- Possibility of microbial growth.
- Thermal inactivation.
- Survival in nonthermal processes

Mathematical modeling allows the food operator to know in advance the effect on microbial growth or survival of modifying one or more food composition factors. Existing programs display results in graphical form – like a curve – or numerically. Several models are available on the Internet, such as the US Department of Agriculture (USDA) Pathogen Modelling Program (USDA 2006) and ComBase (2010), which are free to use. When the food operator uses these models to check the effectiveness and document the validation of the control measures, he must consider adapting it for a particular production and the variability associated with these models so that additional testing may be necessary. Mathematical models, such as pathogen growth models to evaluate the impact of changes in pH and water activity on pathogen growth control or using *z*-value models to determine alternative heat treatments, are widely used in the food industry.

Collecting data during operating conditions: One approach to validate a process is using biological, chemical, or physical data obtained under normal production conditions. Data collection is scheduled for a specified time, for example, three to –six weeks, on a full production scale, including situations in which activity may be intensified or when a deviation from ideal operating conditions may occur. The use of this validation method by the food operator requires the appropriate design of the sampling plans and the statistical treatment of the data to achieve statistically significant and reliable results.

Historical record of production: In many situations of food production and processing, there is an enormous amount of accumulated experience in applying specific control systems or measures without having detected problems, complaints, or incidents with that production. This experience can be interpreted as an empirical – and implicit – demonstration of the effectiveness of control measures, which have been used for years in the food industry, and can therefore be accepted as a way to validate control measures; however, as already mentioned, the FBO should not use the historical experience as the sole way to validate control measures.

It is worth remembering that before the advent of HACCP, the food industry produced safe food for years. Although the development of HACCP is intended to replace the previous systems, it should not conflict with them but should take advantage of the practical experience accumulated over the years. Therefore, historical data obtained from GMPs, quality control, and final product analysis, provided they are objective and quantifiable, can also be used to validate HACCP plans (ILSI 1999).

However, it is necessary to check whether historical information can still be used today or whether changes in the process, technology used, or hazards indicate the need for further assessments.

Evaluation of heat treatments: A particular type of validation is performed to calculate the control parameters of heat treatments in foods that include some stage of microbial destruction in their preparation. The study of the values that define the thermal process is specific to each food, according to its formulation, method of preparation, size of the tin that is processed, and the equipment used for the treatment.

In order to establish the time and temperature of the treatment, it is necessary to conduct the study of temperature distribution in the equipment used and the heat penetration in the container to determine the temperature of the product at any moment of the treatment, at the slowest heating point of the container. The total lethality of the F0 treatment is calculated by integrating the cumulative lethal effect of all combinations of time and temperature recorded at the slowest heating point.

Shelf life studies: A significant part of the validation activities carried out by the food operator involves determining the shelf life of food. Shelf life is the length of time that the food stays fit for consumption from a health point of view, maintaining its sensory, physical–chemical, and nutritional characteristics within quality parameters defined as acceptable.

The current regulations oblige to include in the labeling the information on the shelf life of the food, which can appear as an expiration date or date of minimum durability. The expiration date is set for highly perishable foods and determines the point at which the food could be harmful to health. The best-before date indicates the last day the food is "good" to consume before it starts to break down. The best-before or best-by date denotes the optimal length of time during which the manufacturer feels their product will retain its original quality.

Shelf life studies allow the food operator to ensure that the food will meet the food safety criteria, or FSO, at the time of consumption. For example, in the case of ready-to-eat foods that support the growth of *L. monocytogenes*, the manufacturer must demonstrate, to the satisfaction of the competent authority, that the food will not exceed the limit of 100 cfu/g throughout its shelf life, or otherwise, apply the criterion of "Not detected in 25 g" before it has left the immediate control of the FBO who has produced it (European Commission 2005).

The operator can choose several types of shelf life studies according to their needs: direct studies, accelerated life studies, oxidative tests, etc. In direct studies in real time, the FBO must assess the changes that occur when the food is subject to the expected conditions during storage, transport, and distribution but taking into account that if it wants to apply sufficient safety margin, it is necessary to contemplate the worst conditions, or the worst case scenario, that a food may have to face during real-life conditions, and not ideal storage conditions.

On certain occasions, for example, when the operator wants to develop new products with a lengthy best-before date, it is not practical to wait until the end of their shelf life to obtain the study results. In these cases, the FBO can use accelerated shelf life studies to expose the food to more severe controlled conditions of temperature or other factors to shorten the study time, which can go from months to days. Although it is necessary to consider the study's margin of error, some deteriorations may appear that do not occur in normal circumstances.

3.6.3 Verification Methods

Once HACCP has been implemented, the objective of verification, as we saw, is to verify that the food operator is doing things as planned and that the HACCP system effectively ensures safe food production. For that purpose, the HACCP plan must describe the different types of verification activities and the tasks, responsibilities, frequencies, and verification methods by establishing documented procedures. One of the critical elements for the audit of principle 6 is to review how verification activities are documented in those procedures and whether they are implemented in the manner described.

Within the broad set of tasks that can be included in the verification, for didactic purposes, two types of activities can be distinguished, depending on their specific objectives (ILSI 1999):

Demonstrating conformity with the HACCP plan: A key objective of conformity is to gain confidence in the competence of the manufacturing plant. This is achieved by demonstrating that the established procedures are being followed effectively and that "we do what we say we do." This verification aspect has much in common with auditing in that documentation provides evidence for external (and internal) people that food operators are doing the right thing to ensure food safety. The above people must have a good scientific and technical knowledge background and familiarity with the product, process, and production.

Gathering of information: The objective of information gathering is to raise confidence in the HACCP plan. This is obtained by demonstrating that product safety requirements are being met and that the HACCP plan remains updated and adapted to the food operator's characteristics. Verification should ensure that the HACCP plan is still applicable for the equipment, processes, products, factory layout, distribution and storage, raw materials, and packaging and incorporates the latest information about hazards and consumer trends.

Methods for checking compliance with the HACCP system include:

- Review of CCP monitoring records.
- Review of deviations and corrective actions applied, including records related to the management and destination of nonconforming products.
- Inspection of the facilities and processing to verify the application of control measures.
- Review of PRP application records.
- Supplier audits.
- **Internal audits**: undoubtedly the best way to check compliance with the HACCP system.
- **Calibration of instruments and measuring equipment**: Some operators prefer to include it within the prerequisite of maintenance of facilities and equipment.

The second type of verification activities, which are carried out to obtain information on the effectiveness of the HACCP system, are the following:

- Analysis of monitoring data to observe if they remain within the established limits or if trends may indicate a possible loss of control.
- Physical and chemical analysis of processes: For example, the alkaline phosphatase test to check the effectiveness of milk pasteurization or the pH measurement of the rinse water to verify that the rinsing process included in the C & D plan is effective.
- Microbiological analysis:
  - Final product.
  - Intermediate product.
  - Raw materials.
  - Environmental:
    - Process line: contact surfaces with products.
    - Manufacturing environment.
  - Food handlers.

- Analysis of complaints and claims from customers and consumers: Having an effective claims and complaints management system is essential for any company, not only to improve its relationship with customers or consumers of the product but also to obtain precious information about problems or failures that are not being detected within the company and thus be able to initiate actions to solve the problem as soon as possible. Reviewing claims and complaints that have occurred during a given period can give an idea of the effectiveness of the system or the areas in which action needs to be taken.
- Review and analysis of incidents and/or management indicators.

In applying the principle of continuous improvement, the verification results should be evaluated systematically to modify or update those aspects of the HACCP system that are necessary to correct the faults detected, prevent the recurrence of the problems, or improve the performance of the processes. The NACMCF HACCP Application Guide states: "Maintaining an effective HACCP system depends largely on regularly scheduled verification activities. The HACCP plan should be updated and revised as needed."

3.6.4 HACCP Revision

Although the HACCP study can be completed at a given time, after the implementation of all the steps, in practice, HACCP should be understood as a living system, subject to changes and modifications that must be made to adapt it to the variations that can usually occur in the activity of food businesses over time. What we do not correct or do not update becomes obsolete and loses its usefulness for the organization.

The Codex guidelines mention the revision of the HACCP system as a verification activity and state that verification should include a comprehensive review (e.g. reanalysis or an audit) of the HACCP system periodically, as appropriate, or when changes occur, to confirm the efficacy of all elements of the HACCP system. This HACCP system review should confirm that all system elements remain valid to control the identified hazards, that monitoring and verification are occurring according to the plan and can detect deviations, and that appropriate corrective actions are taken. The HACCP system review can be carried out by the staff of the food operator or with the help of external experts.

Furthermore, the review is a common element of management systems and food safety standards, which include some provisions for management review, which should be carried out through meetings held at planned intervals, typically at least once a year or more frequently when there are significant changes in the system. In this way, we can differentiate two types of review:

- Planned at regular intervals.
- Unplanned, motivated by changes in the system.

In addition to the type of review, it is necessary to consider the factors that may convene the revision meetings or that should be considered as initial information for the revision of the system. A wide range of factors can motivate the review, which, according to their origin, are classified into internal factors, i.e. due to changes within the organization, and external factors, i.e. changes outside the organization.

Internal factors to consider for system review include:

- Changes in raw materials, ingredients, and product formulation.
- Changes in processes, equipment, or technology.
- New products.
- New suppliers.

- Changes in facilities or the work environment.
- Modifications to PRPs, e.g. cleaning and disinfection programs.
- Changes in organizational chart, responsibilities, or staffing.
- Results of verification activities.
- Analysis of complaints and claims.
- Results of internal or external audits.

External factors include aspects beyond the control of the food operator, which may affect the HACCP system, such as the following:

- Information on emerging pathogens.
- Changes in legislation.
- Advances in scientific knowledge or process technology.
- Alerts and information about issues that appear in other food industries.
- Changes in the environment.

Reviewing the system is one of the aspects to which the auditor will undoubtedly pay attention. Many of the nonconformities of the HACCP system audit have to do with its need for updating. For this reason, the HACCP team must always be aware of the changes that occur, especially those that can modify control measures' effectiveness or affect essential safety factors. Review activities, whether planned or motivated by any of the above reasons, are proof of management's commitment to the FSMS. In the meetings held for the review of the system, it is highly recommendable that, in addition to the members of the HACCP team, those responsible for the company's senior management also attend, as a proof of commitment and to arrange the resources required for the efficient maintenance of the HACCP system (Figure 3.38).

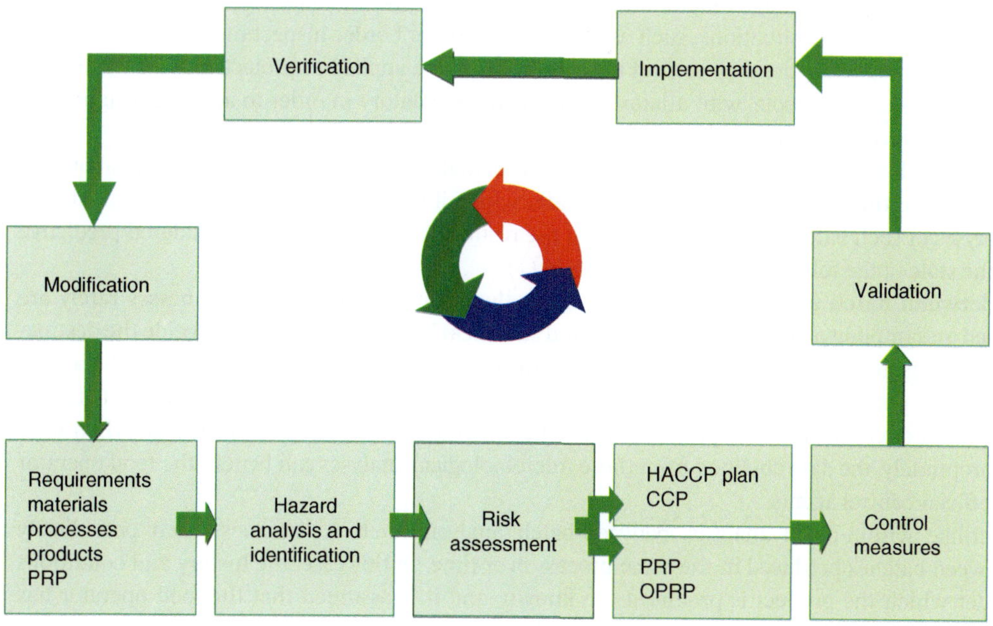

Figure 3.38 FSMS cycle.

3.7 Microbiological Analyses in the HACCP System

As we saw earlier, one of the reasons for using HACCP is precisely the limitation of final product analysis as the only way to ensure the safety of processed foods. In addition to being ineffective, the analysis alone causes a false sense of safety due to the low probability of contamination detection provided by the usual sampling plans when the concentration of microbiological hazards is below a certain level.

The low reliability of the sampling plans meant that the food safety model based on inspection and analysis of the final product, in force until the 1970s and 1980s, was gradually replaced by the application of hygiene programs and the HACCP system, recognized as the most effective way to manage food safety throughout the food chain. However, when used with an awareness of their limitations, in the right context, and in a well-planned way, microbiological analyses play an essential role in validating and verifying the HACCP system, among other issues.

ICMSF puts it this way: "When using microbiological tests to evaluate safety or quality of food it is important to select and apply these with knowledge of their limitations, their benefits and the purposes for which they are intended. In many instances, other assessment are faster and more effective than microbiological testing for food safety assurance."

Within-batch analysis: A significant part of food analysis involves the sampling plans that are carried out to decide about a product or a particular batch of foods. That is to say, the procedures by which a sufficient number of samples are selected from a batch or from the consignment inspected in order to have an adequate statistical basis to be able to determine, with a high degree of confidence, whether a product batch is acceptable since it meets the established criteria or whether it should be rejected for containing an unacceptable percentage of defective units.

Once the decision criteria are defined, as the percentage of defective units in a batch decreases, increasing the number or size of samples is necessary to maintain the same level of confidence. So, when the number of defective units is low enough, for example, when it is less than 5%, sampling may become economically or materially unfeasible for the food operator.

However, there are situations, such as at ports of entry or border inspection posts, where batch analysis is practically the only control measure that can be applied to protect consumers' health, as is the case, for example, with aflatoxin tests made mandatory in order to accept consignments of peanuts received from certain countries.

Another limitation of the within-batch analysis is that there is little or no knowledge about the product and the processes and conditions under which it was manufactured and distributed. The analyses of each batch are independent, and the results obtained cannot be considered predictive of the state of the following batches (ICMSF 2011).

Between-batch analysis: Microbiological analyses carried out in food businesses rarely are based on sampling plans with the intensity and statistical basis that is applied to decide the destination of consignments at ports of entry; to the contrary, periodic food sampling by the food operator often does not have the number and sample size that would be necessary to detect, with an acceptable level of confidence, the presence of batches with low contamination. However, when used appropriately, the data obtained from these microbiological analyses can benefit the food operator or official control agents.

Unlike within-batch analysis, between-batch analysis refers to samples taken periodically between batches produced in the same process over time. In this case, the history and conditions under which the product is produced are known, and it is assumed that the food operator has already established and validated in advance the processes and practices required for the control of the identified hazards; therefore, the purpose of the between-batch analysis is not to determine the acceptability of a batch but to verify that the processes are working as expected and that they are effective. In between-batch analysis, the number of samples taken from each batch is insufficient to determine its condition or quality; instead, the food operator increases the number of lots sampled, considering the process as if it were a continuous batch or a series of larger batches, defined by a period in which the individual batches overlap (ICMSF 2011). This is the basis of the "moving window" approach that is applied, for example, in the sampling plans included in the process hygiene criteria of the carcasses of the different species, established by European Regulation 2073/2005 on microbiological criteria in foodstuffs; in this way, each new result obtained in the sampling of the carcasses eliminates the oldest one, always maintaining the same number of sample units defined in the sampling plan.

As we have seen, the consequence of an unfavorable result in individual analysis or within-batch analysis is the rejection of the lot or consignment; however, in process control analyses or between batches, an unsatisfactory result indicates a loss of process control and a potential failure of the HACCP system. In this case, the food operator must take action to bring the process back under control and find the cause of the microbiological contamination to prevent it from happening again.

A precondition for between-batch analysis to fulfill its function of system verification is that the process is under control, i.e. that the inherent variability of the process has been reduced to the minimum possible, and this is the case when only common or unidentifiable causes of variation are involved, which is a large set of factors that would be very difficult or very expensive to eliminate. When a process is under control, a microbiological analysis above the ranges of variation that are considered normal, or a trend in the observed data, is a sign that the process may be out of control and that there is some assignable cause that is causing that variation, and then corrective actions must be taken. When testing programs are appropriately designed and the data obtained are reviewed and interpreted correctly, process control analyses are a powerful tool for evaluating and correcting the systems used to control processed products' safety and microbiological quality before established limits are exceeded and the product becomes unsuitable for commercialization (Figure 3.39).

Within-batch analysis	Between-batch analysis
Product control	Process control
Determine the destination of a batch or shipment	Determine that a process is controlled and functioning properly
Analysis at ports of entry or borders	Analysis in the food industry
Unknown product history and processing conditions	Known history and production conditions
Individual, non-predictive analysis	Periodic, predictive analysis (in controlled processes
It may be considered monitoring, in certain cases	Principle 6 of HACCP
Corrective action: batch or shipment rejection	Corrective action: correct the process

Figure 3.39 Within- and between-batch analysis.

For both individual and between-batch analyses, it is necessary to establish microbiological criteria that define the acceptability of a food product and that of the manufacturing, handling, and distribution processes.

Microbiological criteria are part of the risk management measures established to prevent diseases caused by foodborne pathogens. The development of microbiological criteria is a complex process that considers several factors, such as the risks associated with the food, the need to set criteria – when other more effective measures are not available – their practical application in food companies, and their effectiveness in protecting consumers' health.

Once established, the microbiological criteria apply to both health authorities and FBOs in order to be able to segregate acceptable and unacceptable raw materials, ingredients, products, or batches, as well as to determine whether the processes are being applied following general principles of hygiene and whether they are effective in controlling biological hazards.

According to CAC/GL 21 – 1997 Codex Guidelines for the Establishment and Application of Microbiological Criteria related to Food, a microbiological criterion consists of the following components:

- the purpose of the microbiological criterion;
- the food, process, or food safety control system to which the microbiological criterion applies;
- the specified point in the food chain where the microbiological criterion applies;
- the microorganism(s) and the reason for its selection;
- the microbiological limits (m, M) or other limits (e.g. a level of risk);
- a sampling plan defining the number of sample units to be taken (n), the size of the analytical unit, and where appropriate, the acceptance number (c);
- depending on its purpose, an indication of the statistical performance of the sampling plan; and
- analytical methods and their performance parameters.

Microbiological criteria can be applied, but are not limited to, to the following:

- bacteria, viruses, molds, yeasts, and algae;
- protozoa and helminths;
- their toxins/metabolites; and
- their markers associated with pathogenicity (e.g. virulence-related genes or plasmids) or other traits (e.g. antimicrobial resistance genes) where/when linked to the presence of viable cells where appropriate.

Microorganisms included in a microbiological criterion can be pathogenic, indicators, or spoilage microorganisms, although, whenever feasible, pathogen analyses take precedence over indicators. The microorganisms selected must be significant for the food and the technological process in which they are intended to be applied.

Ideally, microbiological limits should be based on FSOs: maximum frequency or concentration of a microbiological hazard at the time of consumption; however, having an FSO as a basis for setting microbiological limits is not always possible.

In any case, microbiological limits should be set realistically, considering the values that can reasonably be expected to be achieved under normal business conditions when a food operator is in acceptable compliance with GHP and HACCP.

The sampling plan specifies the number of sample units and the decision criteria to be applied in each case. Sampling plans should be established based on the health risks associated with the microorganisms analyzed and the likelihood of detecting the microorganisms in the batch. The lower the number of microorganisms in a food product, the greater the probability that the test will not detect and accept a batch that should have been rejected (consumer risk). The risk of making wrong decisions can be reduced by increasing the number and size of samples; however, there is no way to avoid a certain degree of error in the decisions made about the batch unless the whole lot is analyzed, which is obviously impossible. Therefore, in practice, the aim is to try to reach a compromise solution between a higher number of sample units and less decision error and a lower number of sample units and a higher probability of making a wrong decision so that sampling plans can be economically and administratively viable (Figure 3.40).

3.7.1 Types of Sampling Plans

- Attributes plans.
- Variables plans.

In attributes plans, microbial counts are sorted in categories or ranges, such as absence or presence, or values above a particular limit m, in two-class plans, or m and M in three-class attributes plans. In practice, most sampling plans used in the food industry are attribute based.

A variables plan is a sampling plan in which the criteria employ statistics of a sample, such as the mean and standard deviation, rather than the segregation of sample units into attribute classes. The mean and standard deviation of the sample log counts are combined with a constant or constants selected for the appropriate number of sample units for comparison with a limit value. Variable plans are used when the frequency distribution of microorganisms in a lot is known; if used appropriately, in certain circumstances, they can be helpful in the achievement of specific objectives.

Two-class attribute sampling plans: Categories can be defined as presence (positive result) or absence (negative result), or apply a limit m that discriminates the class with microbial rate above that limit (positive result) from the class whose rate is below (negative result).

Two-class attribute sampling plans are defined by the following parameters:

- *n*: number of sample units required for analysis;
- *c*: maximum permissible number of unsatisfactory results in the analysis;
- *m*: microbiological limit (e.g. cfu/g); A sample is considered positive when its microbial content exceeds this limit.

Two-class attribute plans provide a simple means of inspection where the sampling plan is defined by two values, *n* and *c*. When undertaking a microbiological analysis, any item contaminated at a concentration higher than *m* is considered nonconforming.

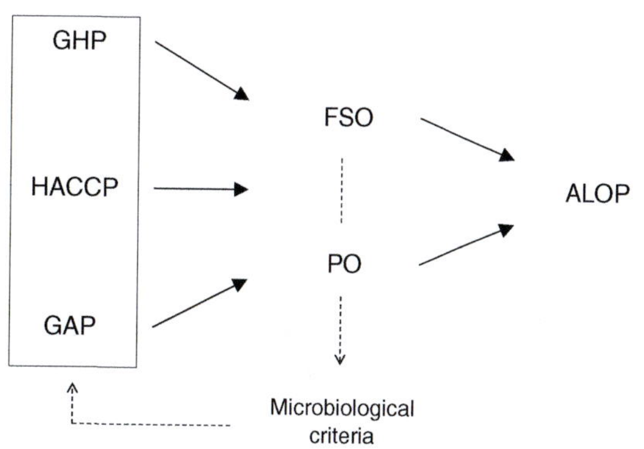

Figure 3.40 Interrelationship between HACCP, microbiological criteria, FSO and ALOP.

The stringency (probability of rejection) of two-class attribute sampling plans depends on the values of n and c. For a given value of c, the stringency of the plan will increase as n increases.

In the same way, for a given value of n, the stringency will increase as c decreases. Conversely, if c increases, the plan becomes less strict, so substandard batches (with some unsatisfactory results) can be considered conforming. Operating characteristic (OC) curves are used to visualize the probability to accept the lot of a two-class attribute plan according to the proportion of defective units in the lot and the values of n and c (Figure 3.41).

Three-class attribute sampling plans: These plans are applied to situations where the quality of a product can be divided into three categories or attribute classes: good quality, marginally acceptable quality, and unacceptable quality.

Three class attribute plans are defined by the values n, c, m, and M;

- n: the number of sample units required for analysis;

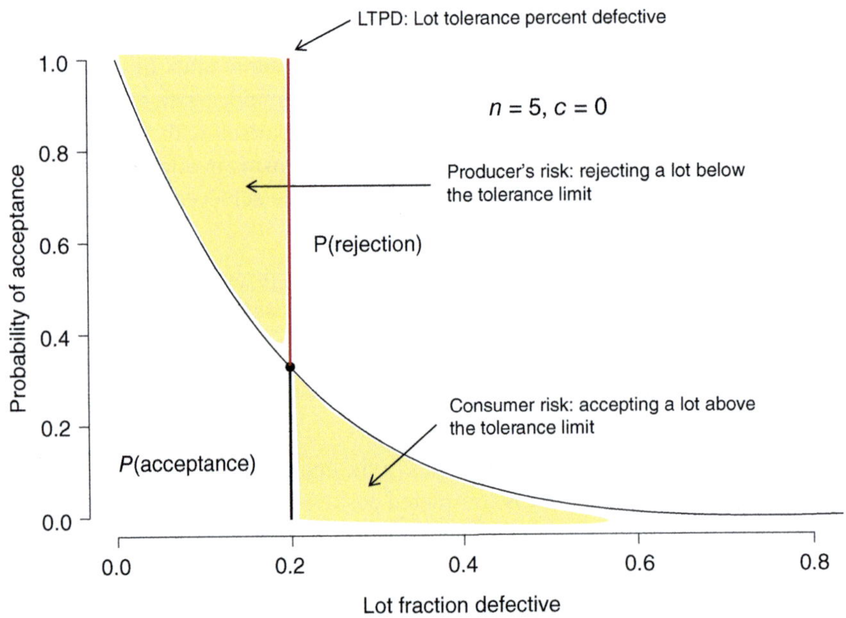

Figure 3.41 OC curve.

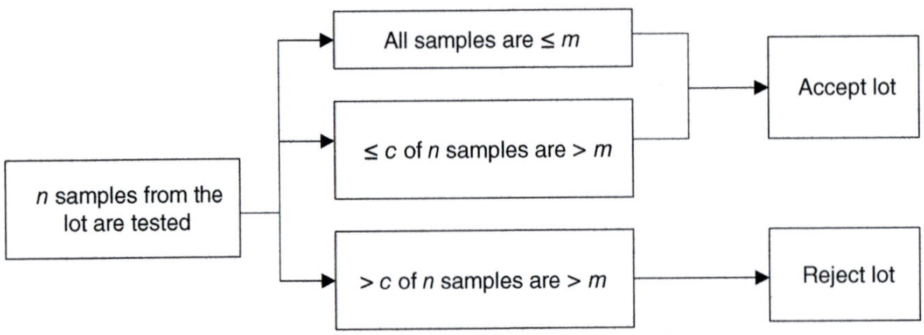

Figure 3.42 Diagram Two-class plans.

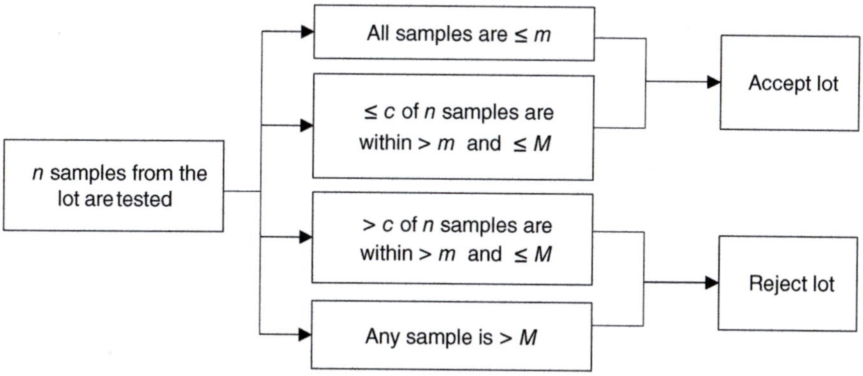

Figure 3.43 Diagram Three-class plans.

- *m*: microbiological limit separating the good quality class from the marginally acceptable quality class. The value of *m* is the concentration of the microorganisms that reflects the upper limit of a GMP;
- *M*: microbiological limit above which the sample is unacceptable and may pose a health risk. The value of *M* is a hazardous or unacceptable level of contamination caused by poor hygienic practice, including improper storage;
- *c*: the maximum number of permissible samples with a concentration between *m* and *M* limits, included in the marginally acceptable class. The lot is directly rejected when the concentration of microorganisms in any sample item exceeds *M*.

In general, three-class attribute sampling plans are less rigorous than two-class sampling plans and are used for microorganisms associated with a lower health risk.

The following diagrams show how the parameters *n* and *c* are used in two-class and three-class plans to decide the use of a batch based on defined limits (Figures 3.42 and 3.43).

In summary, the sampling plan's severity depends on *n*, *c*, *m*, and *M*; as *n* increases and the values of *c*, *m*, and *M* decrease, sampling plans become more stringent. Based on the health risk associated with the microorganisms analyzed and the conditions of use expected between the preparation of food (or the time of sampling) and its consumption, the ICMSF developed a scheme for the use of the sampling plans consisting of fifteen cases, which are the result of combining five levels of health risk associated with the microorganisms analyzed, and three different conditions of use: the one that decreases the risk between processing and consumption, the one that does not change the risk, and the one that increases it. The following table represents the fifteen cases of the ICMSF, with the values of n and c corresponding to each of them. In cases 1–3, the microorganisms included do not have health effects but affect economic and quality aspects, and that is why the plans used are more tolerant than the following ones. The health risk is significant in cases 10–15, so these plans are much stricter (Figure 3.44).

3.8 EU Regulation 2073/2005 and HACCP

This section discusses the relationship between EU Regulation 2073/2005 provisions and the HACCP system. Therefore, its content does not apply outside the European Union and is an expendable section for readers from other continents. However, it may be interesting to read to understand how microbiological criteria are related to the principles of the HACCP system.

Suggested sampling plans for various combinations of health risks and product use conditions "The 15 cases"			
Degree of concern relative to utility and health hazard	Conditions in which food is expected to be handled and consumed after sampling, in the usual course of events		
	Conditions reduce degree of concern	Conditions cause no change in concern	Conditions may increase concern
No direct health hazard **Utility:** e.g. shelf life and spoilage	Increase shelf life CASE 1 3-class, $n = 5$, $c = 3$	No change CASE 2 3-class, $n = 5$, $c = 2$	Reduced shelf life CASE 3 3-class , $n = 5$, $c = 1$
Health hazard: low, indirect **Indicator**	Reduce hazard CASE 4 3-class, $n = 5$, c =3	No change CASE 5 3-class, $n = 5$, $c = 2$	Increase hazard CASE 6 3-class, $n = 5$, $c = 1$
Health hazard: **moderate**, direct, limited spread	CASE 7 3-class, $n = 5$, $c = 2$	CASE 8 3-class, $n = 5$, $c = 1$	CASE 9 3-class, $n = 10$, $c = 1$
Health hazard **serious,** direct, potentially extensive spread	CASE 10 2-class, $n = 5$, $c = 0$	CASE 11 2-class, $n = 10$, $c = 0$	CASE 12 2-class, $n = 20$, $c = 0$
Health hazard: **severe,** direct	CASE 13 2-class, $n = 15$, $c = 0$	CASE 14 2-class, $n = 30$, $c = 0$	CASE 15 2-class, $n = 60$, $c = 0$

Figure 3.44 The 15 cases.

Regulation 2073/2005 on microbiological criteria for foodstuffs requires food operators to comply with the microbiological criteria set for different types of food. Although its provisions enforce FBOs in the context of their GHP and HACCP procedures, it also applies to health authorities who must know how to interpret the criteria and verify their compliance.

The publication of Regulation 2073/2005 brought about a radical change in the regulations on microbiological criteria in force up to that time. In addition to bringing together the microbiological criteria of products of animal origin, which were dispersed into several standards, and adding microbiological criteria for plant products that did not exist until then, Regulation 2073/2005 describes for the first time the use of microbiological analyses for the validation and verification of HACCP and other hygiene control measures.

For the development of this new approach, the CAC/GL 21-1997 "Principles for the Establishment and Application of Microbiological Criteria for Food" of the Codex Alimentarius Commission were taken into account, which considers microbiological criteria as legal tools for the management of microbiological risks within a certainly preventive system such as HACCP.

Another novelty introduced by Regulation 2073/2005 is the differentiation between two types of microbiological criteria, such as food safety criteria and process hygiene criteria, defined as follows:

Food safety criterion: a criterion defining the acceptability of a product or a batch of foodstuff applicable to products placed on the market.

Process hygiene criterion: a criterion indicating the acceptable functioning of the production process. Such a criterion is not applicable to products placed on the market. It sets an indicative contamination value above which corrective actions are required in order to maintain the hygiene of the process in compliance with food law.

By the provisions of Regulation 2073/2005, food operators must establish documented procedures that include:

- food product or element to be sampled (surfaces);
- the microorganisms and/or toxins to be analyzed;
- analytical method;
- sampling plans and methods;
- the number of sample units; and
- sampling frequency.

Regulation 2073/2005 lays down the microbiological criteria the food operator must meet for each food category, indicating the analytical method, the plan, and the sampling method for each case. Each microbiological criterion specifies the number of sample units required; however, it is possible to reduce the number of sample units provided that the food operator can demonstrate through historical documentation that HACCP-based procedures are working effectively.

The food operator decides the sampling frequency, except in cases where the regulation lays down specific frequencies, such as for sampling carcasses, minced meat, meat preparations, and mechanically separated meat. Once the HACCP system is in place, the operator may choose to reduce the sampling frequency based on trend analysis and where there is a history of favorable results, provided that reduction does not jeopardize the safety of the food products.

A key application of Regulation 2073/2005 for HACCP is its use as a reference for hazard analysis; since the microbiological criteria established for each food category are based on risk analysis, this indicates that competent bodies or agencies have already assessed microbiological hazards, and for that reason, the food operator must consider that these hazards are significant for the foods in which these microbiological criteria apply. In other words, when Regulation 2073/2005 sets a criterion for a microorganism in a food, that microorganism must be included as significant in the hazard analysis of the processing of that food.

In addition, Regulation 2073/2005 also establishes the corrective measures to be taken by the food operator when food safety criteria and process hygiene criteria are not met. When noncompliance occurs in a food safety criterion, the affected batch or product must be withdrawn or recalled from the market – by the provisions of European Regulation 178/2002 – and the other corrective measures established in the HACCP procedures. On the other hand, when process hygiene criteria are not met, the measures focus on improving the hygiene of the process or the selection of raw materials, and the operator is not required to take action on the product. These are aspects that the food safety auditor is likely to pay attention to in order to check how the operator acts when it obtains unfavorable analytical results and if it does what its procedures say it must do.

Finally, another application of Regulation 2073/2005 for the HACCP system is the use of microbiological criteria as a reference to establish the shelf life of food, for example, with the criteria and conditions set for foods that support the growth of *L. monocytogenes*. (Figure 3.45)

Key elements for the evaluation of principle 6

Validation

- Documentation of the validation of control measures and the system's elements in general.
- References to guides, documents, or publications used as a basis for validation.
- Justification of control measures.

HACCP	Regulation (EC) 2073/2005
Assemble HACCP team	
Describe product	**Shelf life studies**
Identify intended use	Intended use must be considered when establishing microbiological criteria
Construct flow diagram	
On-site confirmation of flow diagram	
Principle 1: Conduct a hazard analysis and identify significant hazards	**Microorganisms with microbiological criteria defined are significant hazards**
Principle 2: Determine CCPs	
Principle 3: Establish validated critical limits for each CCP	
Principle 4: Establish a monitoring system for each CCP	If there is no alternative, microbiological criteria can be used for monitoring
Principle 5: Establish corrective action	Unsatisfactory results require corrective action specified in the regulation
Principle 6: Validate the HACCP plan and establish verification procedures	**Microbiological criteria must be taken into account for the validation and verification of HACCP**
Principle 7: Establish documentation and recordkeeping	

Figure 3.45 Interrelation between HACCP and R° 2073/2005.

- Shelf life studies, where required.
- In establishments with thermal inactivation stages, documentation of the evaluation studies of the heat treatments.
- Need to revalidate parts of the system.

Verification

- Documented verification procedures:
  - Responsible.
  - Frequency.
  - Records.
- Results of the analysis.
- Claims and complaints management.
- Calibration records.
- Internal and external audit reports.
- Verification records.

System Review.
- Update of documentation: edition number and date of last revision.
- Changes in products, equipment, systems, or managers.
- Planning the system review at regular intervals.
  - Responsible.
- Causes for review, internal and external.

3.9 Principle 7

Besides being a HACCP requirement, a documentation and recording system is a sine qua non condition for the audit feasibility; the audit is only viable when the operator has an adequate documented system. Factual documentation becomes one of the main workhorses for the auditor, who will spend a significant share of his time reviewing the documents that make up the system and requesting records related to the practical application of HACCP and the hygiene prerequisites.

HACCP principle 7 states that efficient and accurate recordkeeping is essential to implementing a HACCP system. HACCP procedures should be documented. Documentation and recordkeeping should be appropriate to the nature and size of the operation and sufficient to assist the company in verifying that HACCP controls are in place and being maintained.

The documentation and recording system is, therefore, a requirement and, in turn, the result of the HACCP study, which for the auditor means that the documented system is part of the object of the audit. It is one more element to audit and the basis for checking the rest of the principles and the system as a whole. Thus, the auditor's task is to verify that the HACCP plan contains all the elements and parts resulting from the systematic application of the Codex principles and then check that the FBO implements its procedures as indicated.

For descriptive purposes, we can consider that the HACCP system includes three types of documents:

Supporting documentation to develop the HACCP plan.
Documentation describing the elements that make up the HACCP system and the resulting procedures for its implementation.
Records generated with the application of the HACCP system.

The supporting documents for the development of the HACCP system are all those that are often not explicitly included in the HACCP plan but which served as the basis for the HACCP team to document and justify the decisions they were making at the different stages of the study and which usually constitute the scientific or logical basis of the HACCP system.

These supporting documents include the following:

Data, publications, and information used for hazard analysis.
Risk matrices used in hazard assessment.
Data used to establish and validate control measures.
Information, references, and data used for product shelf life studies.
Documented application of decision trees or other approaches used to determine CCPs.
Data or benchmarks used to establish and validate the effectiveness of critical limits.
Experimental studies, publications, sectoral guides, or GHPs that support the team's decision-making.

Although principle 7 occupies the last place in the Codex numerical sequence for the application of HACCP, this does not mean that chronologically, it is the last to intervene; on the contrary, in practice, the HACCP team must start applying it from the very beginning, in order to document all the phases of the HACCP study (Figure 3.46).

The second type of document contains what we consider the central part of HACCP: the documentation that describes the different elements that make up the system. From the auditor's point of view, these documents constitute, on the one hand, the demonstration of the correct application of the HACCP phases and methodology and, on the other hand, the basis for the continuation of the on-site audit activities since they indicate what the FBO says it does. Within this type of document, we can relate the following:

Scope of the HACCP manual.
Composition of the HACCP team.
Description of the products produced.
Flow diagram(s).
Hazard analysis.
Description of preventive measures.
Identification of CCPs.
Critical limits.
Monitoring procedures.
Verification procedures.
Documentation system and formats of the records used.
It is very typical to use HACCP control charts or HACCP worksheets to integrate some of the above elements.

The development of the HACCP study determines the basic content of the HACCP manual, composed of the elements just mentioned; however, there is no indication that the HACCP manual must adopt a defined format. The Codex also does not specify the existence of mandatory procedures or records. It only suggests as an example a model worksheet or control chart, the use of which, moreover, has become widely accepted by HACCP practitioners. This lack of mandatory formats means companies have a broad range of possibilities to develop their self-control system most appropriately. The only requirement is that, whatever the structure or format chosen, it is possible to identify all the essential parts of the HACCP system.

When carrying out the audit document review, the auditor must be able to identify all the elements that make up the HACCP system, regardless of the multiple formats that the food operator can use to document its self-control system. This ability requires, on the one hand, a sound knowledge of the principles and systematics of HACCP and, on the other hand, the need to understand the general structure of the documented system in each case. For this purpose, it is necessary to take into account aspects such as the following:

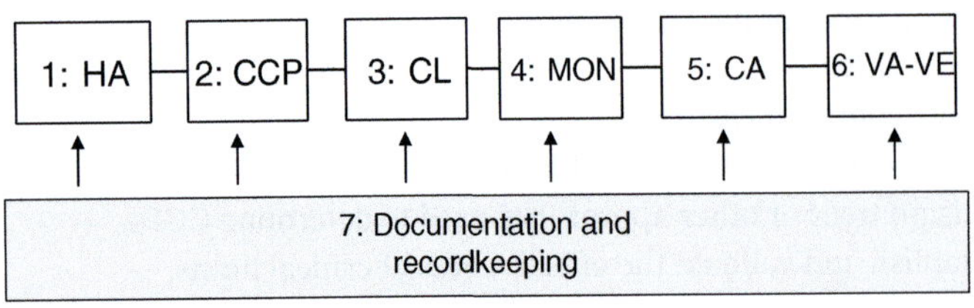

Figure 3.46 Documentation recordkeeping and other principles.

Scope of the manual.
HACCP Type:
- o Linear.
- o Modular.

Integration of HACCP into other management systems.

One of the first aspects that the auditor must review is whether the activities included within the scope of the HACCP manual correspond to those actually carried out by the food operator. It is also essential to know the type of HACCP used; when applying a linear approach, attention should be paid to the flow diagram(s) developed for each product or group of products, from the receipt of raw materials to their dispatch; instead, in a modular HACCP, the structure is composed of modules, each of which contains the operations to which different types of food pass through or according to the modules into which the overall production process is divided.

Another important question for the auditor, and one that happens relatively frequently, is the one that occurs when the food operator decides to integrate the HACCP plan into another quality or FSMS that he has implemented in his company. That is an option that many companies adopt in order to reduce the administrative burden involved in managing various quality or food safety systems, and the auditor must accept it. However, this should not undermine the HACCP auditor's capacity to access all the documents that make up the HACCP system or to review all the procedures and records related to its practical implementation. The operator must provide and make available to the auditor all the documentation necessary to conduct the audit effectively.

Finally, principle 7 calls for creating a registration system that proves that HACCP is being implemented appropriately. Depending on the company's characteristics and the nature of its production process, each operator must design the registration formats or models that best suit their needs and will better serve them to comply with this requirement satisfactorily.

Records are an essential part of the HACCP system; the completion and maintenance of records are closely linked to the day-to-day business activity and the implementation of preventive measures and controls established in the FSMS.

Although not specified in the HACCP guidelines, records control is a common requirement of management systems, which calls for companies to make arrangements for the identification, storage, protection, retrieval, definition of retention time, and disposition of records.

Some companies, especially those starting in HACCP, perceive data registration as a tedious task imposed by a legal requirement and not worth the work involved. However, when the records and the data to fill in are well selected, many companies end up recognizing the benefits of a proper registration system. First, the records provide evidence that the self-control system is operating correctly. This evidence works for the food operator, customers, auditors, and any party interested in verifying the guarantees of the safety of processed products. Internally, for the operator himself, the information contained in the records is helpful to establish internal traceability, know how processes are operating, detect trends, analyze failures or deviations, and often help determine the root cause of problems. Finally, maintaining an adequate registration system demonstrates due diligence, which, as we saw above, in litigation cases or in the event of the investigation of a file, can significantly help defend the actions of the food operator against a possible sanction.

In the HACCP system, we can distinguish several types of records, according to either their origin or their relationship with the different parts of the system.

According to their origin, we can distinguish the following:

- **Internal records**: created or generated by the owner of the management system.
  Created ad hoc: they derive from the application of the principles and systematics of HACCP.

Existing: these documents were already available and in use before the development of HACCP, such as commercial documents or documents derived from other uses.
- **External records** are documents or models not created by the company but by customers, raw materials suppliers, or service suppliers.

Depending on their relationship with the elements that make up the HACCP system, the records can be classified into those that derive from applying HACCP principles and those that have to do with implementing PRPs. Here are some examples of each of these two types:

Records related to HACCP principles:

- HACCP team composition.
- Product sheets.
  - Shelf life studies
- Supporting documents for hazard analysis.
- CCP monitoring records.
- Temperature data log sheet of a CCP.
- Records of corrective actions.
- Verification records:
  - Certificates of analysis.
  - Calibration certificates.
  - Audit reports.
  - Records of complaints and claims.
  - Records of system revisions

Records related to the implementation of PRPs: here are some examples:

- Training records:
  - Certificate issued by an external company.
  - Internal training registration.
  - Training program.
- Records of corrective maintenance operations:
  - Invoices from external companies.
  - Internal parts for carrying out repairs.
- Production records.
- Pest control plan review reports.
- Cleaning product sheets.
- Residual chlorine control records.
- List of suppliers.
- Supplier evaluation form.
- Waste and by-products disposal documents.
- Declarations of conformity of packaging materials.
- GHP monitoring checklist

An issue that the auditor must also consider is the increasing use of electronic record formats by food operators. In principle, there is no restriction on the use of electronic records in HACCP; however, the auditor must assess the guarantees offered by the operator regarding the authenticity and security of these records, depending on the measures that have been adopted in terms of access, use of passwords, and obtaining backups, as well as to prevent their alteration or unauthorized modification.

Key elements for the evaluation of principle 7

Scope of the manual: activities covered.
General structure.
Type of HACCP.
Integration with other systems.
Access to all parts of the system.
Supporting documents.
HACCP plan descriptive documents.
System formats and records.
Records control.

References

ASQ Food, Drug and Cosmetic Division (2002). *The Certified Quality Auditor's HACCP Handbook*. Wisconsin: ASQ Quality Press.

Besseling, P. and Kalyankar, C. (2014). *Hazard and Risk Analysis in Food Processing*. Précon Food Management.

Europe, European Parliament and Council (2002). Regulation (EC) 178/2002 of the European Parliament and Council of 28 January 2002 laying down the general principles and requirements of food law, establishing the European Food Safety Authority and laying down procedures in matters of food safety. *Official Journal of the European Union* 1–24.

Europe, European Parliament and Council (2022). Regulation (EC) No 178/2002 of the European Parliament and of the Council of 28 January 2002 laying down the general principles and requirements of food law, establishing the European Food Safety Authority and laying down procedures in matters of food safety. *Official Journal of the European Union* 1–24.

European Commission (2022). Commission Notice on the implementation of food safety management systems covering Good Hygiene Practices and procedures based on the HACCP principles, including the facilitation/flexibility of the implementation in certain food businesses (2022/C 355/01).

FAO and WHO (2008). *Guidelines for the Validation of Food Safety Control Measures. Codex Alimentarius Guideline, No. CXG 69-2008*. Rome: Codex Alimentarius Commission.

FAO and WHO (2022) General Principles of Food Hygiene. Codex Alimentarius Code of Practice, No. CXC 1-1969. Codex Alimentarius Commission. Rome. 2022.

Gaze, R. (2009). *HACCP – A practical Guide. Fourth Edition, Guideline No. 42*. Chipping Campden, UK: Campden BRI.

ICMSF (International Commission on Microbiological Specifications for Food) (2006). *A simplified guide to understanding and using Food Safety Objectives and Performance Objectives.*

ICMSF (International Commission on Microbiological Specifications for Food) (2011). *Microorganisms in Foods 8. Use of Data for Assessing Process Control and Product Acceptance*. Springer US.

International Life Sciences Institute (ILSI) (1999). *Validation and Verification of HACCP*. ILSI Europe.

International Organization for Standardization (ISO) (2018). *ISO 22000:2018. Food Safety Management Systems – Requirements for Any Organization in the Food Chain.*

Mortimore, S. and Wallace, C. (2013). *HACCP: A Practical Approach*, 3e. US: Springer.

Wallace, C.A., Sperber, W.H., and Mortimore, S.E. (2011). *Food Safety for the 21st Century*. Oxford, UK: Wiley-Blackwells.

4

The Prerequisites of the HACCP System

Food hygiene is the result of the implementation of hygiene prerequisites and procedures based on the hazard analysis and critical control point (HACCP) system. Before implementing HACCP, operators need to develop and implement hygiene programs, or codes of good hygiene practice (GHP), that are appropriate for each sector and each type of food establishment. One argument accepted by food safety practitioners is that HACCP cannot operate in a vacuum. A solid foundation must first be in place to serve as a basis for its development and implementation. This basis comprises hygiene programs, also called prerequisites, which, as their name suggests, must be implemented before HACCP. Therefore, two parts of the HACCP system can be distinguished:

- Prerequisite programs (PRPs)
- HACCP-based procedures

In order to carry out a HACCP study, it is first necessary to develop the PRPs. Once the PRPs have been developed, the HACCP team can start the HACCP study. When these two components of the HACCP system are added with quality management elements, such as those defined in ISO 9000, the three parts make up what is known as a food safety management system (FSMS).

For a food operator with the PRPs implemented, HACCP provides a higher level of control, like a plus consisting of analysis, vision, insight, the ability to anticipate problems, learn from mistakes, check results and updates, and continuous improvement. However, in some companies located at certain stages of the food chain or corresponding to specific sectors or types of establishments, applying flexibility criteria, it may be considered that the hygiene prerequisites are sufficient to control the identified hazards and that it would therefore not be necessary to continue with the development of a complete HACCP system. That means we can find food operators with only hygiene programs, without HACCP. However, we should never see the opposite: establishments that have developed HACCP without PRPs. To summarize it, as if it were a slogan, it can be said that:

“"There can be prerequisites without HACCP, but not HACCP without prerequisites.”"

Document CXC 1-1969: General Principles of Food Hygiene, adopted by the Codex Alimentarius Commission and currently in its fourth revision published in 2022, in addition to containing, as we have seen, the description of the HACCP system (Figure 4.1), is also the fundamental international reference for the implementation of PRPs by food companies. The following table shows the development of the different aspects related to food hygiene, which correspond to the eight sections that this document uses to describe each of the prerequisites (Figure 4.2).

HACCP System Auditing for Food Safety: Principles and Techniques, First Edition. Luis Couto Lorenzo.

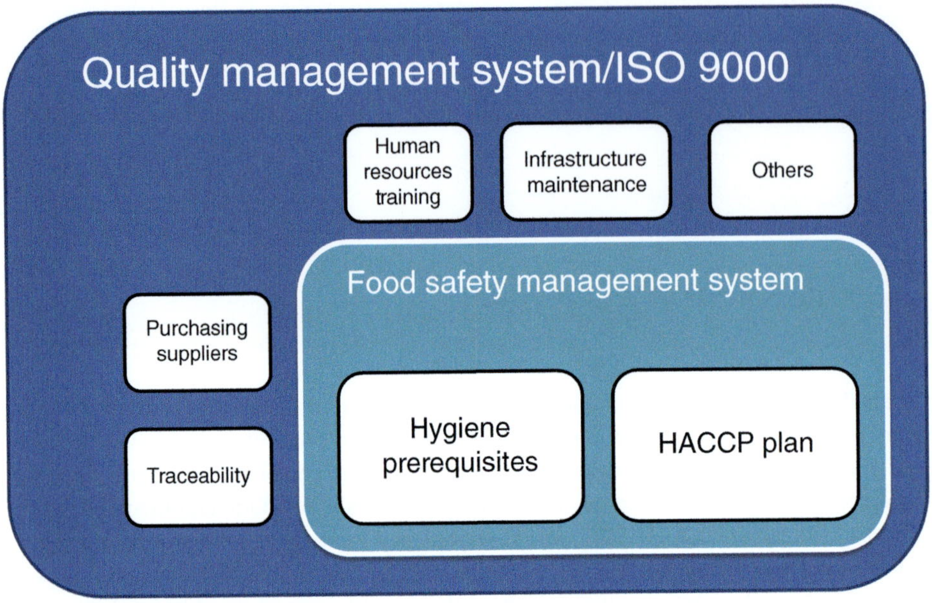

Figure 4.1 HACCP and other elements.

1. Primary production Environmental control Hygienic production Handling, storage, and transport Cleaning, maintenance, and personnel hygiene	**5. Establishment: personal hygiene** Health status Illness and injuries Personal cleanliness Personal behavior Visitors and other persons from outside the establishment
2.Establishment: design of facilities and equipment Location and structure Facilities Equipment	**6. Control of operation** Description of product and processes Key aspects of GHPs Water Documentation and records Recall procedures – removal from the market of unsafe food
3. Training and competence Awareness and responsibilities Training programs Instruction and supervision Refresher training	**7. Product information and consumer awareness** Lot identification and traceability Product information Product labeling Consumer education
4. Establishment maintenance, cleaning and disinfection, and pest control Maintenance and cleaning Pest control systems Waste management	**8. Transportation** General Requirements Use and maintenance

Figure 4.2 General principles of hygiene.

From a legal point of view, at the EU level, the most relevant provisions on prerequisites are set out in:

- The general hygiene requirements in Annex I to Regulation 852/2004 that apply to primary production and related operations.
- The general hygiene requirements in Annex II to Regulation 852/2004 that apply to the stages of the food chain after primary production.
- The specific hygiene requirements for food of animal origin in Annex III to Regulation 853/2004, some of which are related to primary production, e.g. in the eggs, raw milk, live bivalve mollusks, and fishery products sectors.

In the European Union, Regulation 852/2004 includes the following general hygiene requirements:

- Provisions concerning the layout, design, construction, location, and size of food premises
- Cleaning and disinfection
- Pest control
- Maintenance of facilities and equipment
- Raw materials (selection of suppliers)
- Food waste
- Water supply
- Personal hygiene
- Equipment requirement
- Wrapping and packaging requirements
- Heat treatment
- Transport
- Training
- Redistribution of food
- Food safety culture

For food companies under the jurisdiction of the Food and Drug Administration (FDA), the primary reference for developing hygiene requirements and good manufacturing practices (GMPs) is the Code of Federal Regulations (CFR), Title 21, Part 110, which contains the following subparts:

- General Provisions,
  - Definitions
  - Current good manufacturing practice
  - Personnel
  - Exclusions
- Buildings and Facilities,
  - Plant and grounds
  - Sanitary operations
  - Sanitary facilities and controls
- Equipment,
  - Equipment and utensils
- Production and Process,
  - Processes and controls
  - Warehousing and distribution
- Defect Action Level,
  - Natural or unavoidable defects in food for human use that present no health hazard

As we can see, depending on the source we consult, we can find some differences in terms of the number or denomination of the hygiene requirements; thus, if we review the National Advisory Committee on Microbiological Criteria for Foods (NACMCF) HACCP Guide, we can see a list composed of eleven prerequisites, compared to the eight sections of the Codex document, or the five of the CFR. In any case, these differences between the different organisms or geographical areas are essentially formal, and what matters is that the set of prerequisites includes all the aspects necessary to provide the environment and operating conditions for hygienic food production. This set of prerequisites represents the pillars of what has become famous as the "House of Food Safety." At the top is the HACCP plan, and at the base the most essential condition of all: the commitment of the company's managers to promote the implementation and maintenance of the system (Figure 4.3).

Depending on the characteristics of each food sector or the type of establishment in question, there will be situations in which some hygiene requirements do not apply. In order to decide when a specific requirement is necessary or appropriate in a particular case, a risk assessment should be carried out, preferably within the framework of the HACCP approach (FAO and WHO 2003).

The role of the prerequisites within HACCP is to control certain types of general hazards associated with the work environment, which can occur in many parts of the establishment, reducing the workload of the HACCP plan, which can thus focus its resources on controlling the most specific, highest-risk hazards related to a particular phase of the production process. That makes it possible to significantly reduce the critical control points (CCPs) to a more realistic and easily controllable number, thus improving the efficiency and operability of the entire system (Figure 4.4).

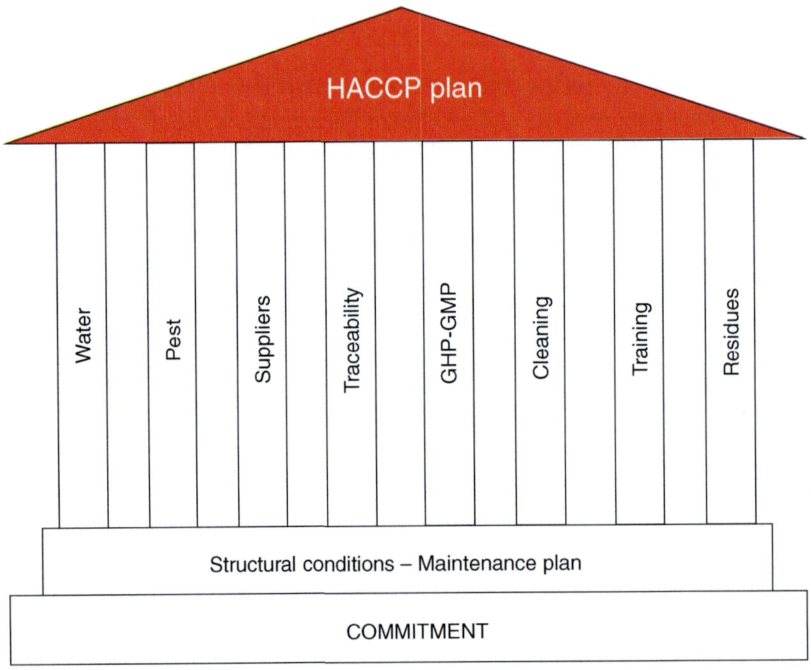

Figure 4.3 The House of Food Safety.

Hazards controlled by prerequisites	Hazards controlled by HACCP plan
General	Specific
Before hazard analysis	After hazard analysis
At many points or places	At one point
Working environment	Process
Normally related to contamination prevention or growth	Usually related phases in which hazards are eliminated
Associated with low risk levels	Associated with high risk levels
Non-priority for validation	Priority for validation

Figure 4.4 Specific and general hazards.

4.1 Type of Prerequisites

Depending on the scope or objectives covered by each of the PRPs, two broad categories of prerequisites can be distinguished:

- Prerequisites related to infrastructure, maintenance, and work environment.
- Prerequisites related to the product or production process.

The second group corresponds to the one that ISO 22000:2018 classifies as operational prerequisites, which we refer to when dealing with principles 1 and 2 of HACCP, and which it considers as a group of control measures with a more significant effect to avoid contamination or the proliferation of hazards in the product or the production process. However, determining the impact of a control measure on a hazard can only be done effectively after the hazard analysis or after conducting a risk assessment in the framework of HACCP.

Some examples help to understand better what we want to express. Thus, if we think about a cleaning and disinfection plan, it is easy to understand that cleaning the walls or floor of a finished product packaging room does not have the same effect on hazards as CIP (clean in place) cleaning of the pipes of pasteurized milk packaging equipment, where insufficient rinsing will lead to the presence of detergent or disinfectant residues in the final product. To give another example, the design of the type and frequency of cleaning is crucial to avoid cross-contamination in a production line where several products containing different allergens are produced. On the other hand, if we refer to a maintenance plan, it is also plain to see the difference in the effect on food safety between the monitoring of the maintenance of the lights in a container and packaging warehouse and that produced by a failure due to poor maintenance of the elements of an autoclave, which can cause an unsafe or nonconforming product, for not having received sufficient heat treatment.

Depending on the company's characteristics, size, and position in the food chain, each food operator must determine the control measures that it considers most effective to ensure the hygiene of its food products. Although in a HACCP system, we can distinguish its elements separately: PRP and HACCP plan, in practice, these elements are intertwined, and the food business operator

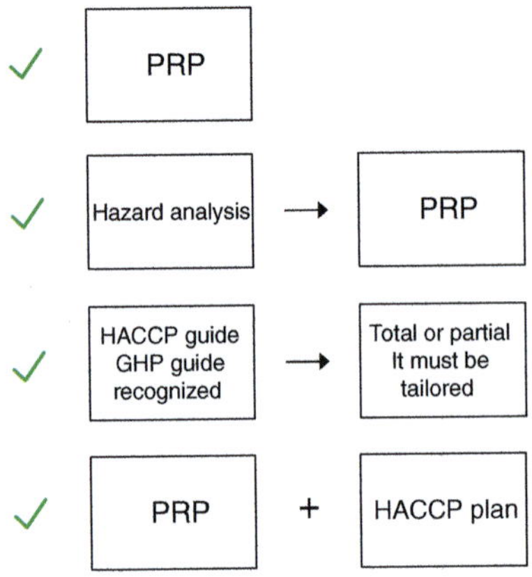

Figure 4.5 Flexibility options.

(FBO) must achieve hazard control through a combination of control measures that can be PRP, operational PRP (OPRP), and HACCP plan.

Applying different control measures according to the different levels of risk associated with hazards is the basis for flexibility. Flexibility is an inherent element of HACCP, so if we want it to be really helpful, it needs to be tailored to the characteristics of each food operator; one size does not fit all.

European regulations include several provisions on flexibility, which recognize that the food operator has several options to comply with the legal requirement to implement a self-control system; these options are as follows:

- Implement only PRPs when they can control all identified hazards. This option, in turn, presents two possibilities:
  - In some establishments, PRPs are assumed to be sufficient due to their characteristics, and no hazard analysis is required.
  - Facilities in which the hazard analysis concludes that the PRPs are sufficient for the control of identified hazards.
- Use a generic HACCP guide or a recognized guide on GHPs, which identifies hazards and control measures to be implemented in a particular food sector.
- Development of a complete HACCP system, applying all HACCP principles (Figure 4.5).

4.2 Flexibility in HACCP

Flexibility in HACCP is one of the issues that demand the most attention, both from food operators and from competent authorities and official control agents, which must evaluate the application of the flexibility criteria, also taking into account the purpose of enabling companies to continue their activity without having to deal with unnecessary hindrances.

The concept of flexibility within HACCP denotes the ability of food companies to adapt to different situations or needs without having to follow fixed rules and the faculty they have to reach an objective, but choosing between several ways, i.e. starting from the premise that the important thing is the goal rather than the way to achieve it.

4.2.1 Background

Again, at the EU level, the very article 5.1 of Regulation 852/2004, which requires FBOs to "put in place, implement and maintain a permanent procedure or procedures based on the HACCP principles" contains the basis for accepting a certain degree of flexibility for the application of HACCP by FBOs.

Recital 15 of this regulation already states that HACCP requirements "should provide sufficient flexibility to be applicable in all situations, including in small businesses. In particular, it is necessary to recognize that, in certain food businesses, it is not possible to identify critical control points and that, in some cases, good hygienic practices can replace the monitoring of critical control points."

Next, Article 5.2 of Regulation 852/2004 sets out the two main criteria that allow a food operator to adopt the flexibility of HACCP-based procedures: their nature and size.

What does the regulation mean when it mentions a procedure based on HACCP? The European Commission has clarified this concept, explaining that although HACCP is a valid model for the control of hazards in food, when this objective can be achieved by equivalent means, which replace in a simplified but effective manner some of those seven principles, the requirement set out in Article 5.1 of the Regulation 852/2004 must be considered to be fulfilled.

Therefore, the main idea supporting flexibility is that food operators can choose the way that best suits their characteristics to control the hygiene of their production, provided that they meet the established objective and that this does not entail an unacceptable increase in risk for consumers.

Since the debate on flexibility in applying HACCP arose, there has been no particular objection to recognizing the room for maneuver that companies need to comply with the legal requirement of self-control. However, it is necessary to address when and how to apply flexibility, in what situations, to whom, or to determine the extent to which it can be admitted.

During all these years, experience in implementing the HACCP system has shown that flexibility continues to be one of the concepts least understood by food operators and with most difficulties for its application, as well as for its interpretation and evaluation by official control agents. This lack of clarity demands competent authorities and other stakeholders for more practical guidance on what types of food operators can apply flexibility, under what conditions, and the extent to which it can be considered acceptable.

In order to meet this demand, in 2022, the European Commission published the Commission Notice on the implementation of FSMSs covering GHPs and procedures based on the HACCP principles, including the facilitation/flexibility of the implementation in certain food businesses (2022/C 355/01), which replaces the previous one from 2016. The purpose of this document is to provide practical guidance on:

- the relevant legislation, the link between GHP, PRP, OPRP, and HACCP-based procedures within a FSMS, the relationship with international standards and training, and the use of guides to GHPs;
- the implementation of GHP, including the flexibility provided for certain food establishments by EU legislation related to their implementation;
- the implementation of HACCP-based procedures, including the flexibility provided for certain food establishments by EU legislation related to their implementation; and
- the auditing of FSMS.

As a document limited to the European Union, readers from other continents may well skip the rest of this section; however, many of the guidelines provided on flexibility in HACCP have universal application and can be of great help to people who have to implement, evaluate, or audit the HACCP system.

The 2022/C 355/01 Commission Notice states that there are differences in risk depending on the nature of the activity, which should be considered when considering flexibility in applying GHP and HACCP. In order to ensure proportionality in administrative burdens, several requirements, such as documentation and recordkeeping, can be simplified in small businesses compared with larger ones carrying out the same activity. HACCP-based procedures should provide sufficient flexibility to be applicable in all circumstances. In summary, this document describes three ways to apply flexibility in self-control systems:

1. Flexibility in the implementation of GHP.
2. Flexibility in the implementation of HACCP-based procedures.
3. Flexibility in auditing of GHP and HACCP-based procedures.

4.2.2 Flexibility in the Implementation of GHP

Since all food production sectors must apply them, the requirements in Annexes I and II of Regulation 852/2004 have to be generally described. Therefore, the generic nature of these requirements implies a high degree of flexibility so that they can be complied with by very different food sectors.

We must bear in mind that not all GHPs apply to all establishments. Hence, a case-by-case assessment is necessary to identify the relevant GHPs for each establishment, depending on their nature and size.

In order to facilitate the implementation of GHPs in small FBOs, Regulations 852/2004 and 853/2004 contain several flexibility provisions, such as the following:

- The general nature of the provisions for primary production and related operations in Annex I to 852.
- The simplified general and specific requirements listed in Annex I to Regulation 852/2004 for premises and rooms intended for movable and/or temporary premises, premises used mainly as a private dwelling house but where foods are regularly prepared for marketing, and vending machines.
- Exclusions from the scope of Regulation 852/2004, for example, the direct supply by the producer of small quantities of primary products to the final consumer or to a local retail establishment directly supplying the final consumer.
- Exclusions from the scope of Regulation 853/2004, for example, the direct supply by the producer of small quantities of meat from poultry and lagomorphs slaughtered on the farm to the final consumer or to local retail establishment directly supplying such meat to the final consumer.
- Exclusion of most retailers from Regulation 853/2004.
- The possibility to adapt GHP under the legislation of the EU member states to enable the continued use of traditional methods or to accommodate the needs of FBOs situated in regions subject to special geographical constraints.

4.2.3 Flexibility in the Implementation of HACCP-based Procedures

According to the 2022/C 355/01 Commission Notice, flexibility in HACCP-based procedures is acknowledged in Article 5 of Regulation 852/2004, which requires the implementation of HACCP for food operators, as we can see in the following points of that article:

- The clear statement in Article 5(1) that the procedure must be **based on** the HACCP principles.
- The statement in Article 5(2)(g) that the need for establishing documentation and records must be **commensurate to the nature and the size of the food business.**
- Article 5(5) of the Regulation, which allows the adoption of arrangements to facilitate the implementation of the HACCP requirement by certain FBOs. These include **the use of guides for the application of HACCP-based procedures**.

The generic guidelines could suggest hazards and control measures common to certain food businesses and facilitate developing HACCP control procedures based on a generic hazard analysis and proper recordkeeping. When generic HACCP guides are used, the FBO should ensure that all the activities of the FBO are covered; otherwise, the FBO should develop its own procedures based on HACCP principles for the additional activities.

Concerning the 12-step approach of the HACCP system, the 2022/C 355/01 Commission Notice includes several examples of flexibility for the preliminary steps and for applying HACCP principles.

Preliminary Steps

- **Assembling of a multidisciplinary HACCP team:** In small businesses, HACCP activities might be carried out by one person who is (temporarily or regularly) assisted by external expertise, but it is essential that the FBO takes ownership of its FSMS. When the HACCP system has been done with the help of an expert, FBOs must ensure that their staff is familiar with it and knows how to implement it. Being so basic and common sense, FBO staff's unfamiliarity with HACCP is a frequent issue in implementing HACCP.
- **Description of the product**: Where there is no processing or other manufacture, the product description may be limited to the information available on the label (prepackaged foods) or other information about the food from reliable sources.
- **Flow diagram**: In certain establishments such as retailers, flow diagrams can be very simple and can be based on models such as those included in the 2020/C 199/01 Commission Notice providing guidance on FSMSs for food retail activities, including food donations.

HACCP Principles

- **Hazard analysis (Principle 1)**: In small FBOs, it might be sufficient that the hazard analysis describes in a practical and simple way the methods to control hazards without entering into detail on the nature of the hazards. Nevertheless, such analysis should cover all significant hazards in a business and should clearly define procedures to control these hazards and the corrective action to be taken in case of deviations.
- **Identification of CCPs (Principle 2)**: Following the Codex guidelines, flexibility in this regard is based on the possibility of using different methods to determine CCPs, including different decision trees, risk matrices, or assessment through the use of other criteria, as long as a logical and structured approach is applied. On the other hand, it should also be noted that in some cases, due to the nature of the operations, it is unnecessary to identify CCPs and that the PRPs are sufficient to control the hazards. However, it must be stressed that flexibility in identifying a hazard as significant or not should not be directly linked to the size of the company but to the risk involved in certain foods and/or processes, which are also present in small companies.
- **Critical limits at CCP (Principle 3)**: Critical limits can be determined based on:
  - Experience (best practices).
  - International documentation, such as Codex standards setting critical limits for several operations, e.g. food canning, pasteurization of liquids.
  - Advice on specific processes or operations in the good practice guides.
  - Scientific publications.
  - EU legislation and European Food Safety Authority (EFSA) opinions.

  It is not always possible, nor necessary, to set a numerical value for critical limits, particularly when monitoring procedures are based on visual observation, for example:

- The fecal contamination of carcasses after dressing in a slaughterhouse.
- The boiling temperature of liquid food.
- The change of physical properties of food during processing (e.g. the thorough cooking of food).

- **Monitoring procedures (Principle 4)**: In line with what has been said for critical limits, monitoring is not always fulfilled by measuring; sometimes, it may be a simple procedure such as the observation of processes or changes in products.

The type of CCP determines the frequency of monitoring. A long series of favorable results makes it possible to reduce the frequency of monitoring.

- **Corrective action (Principle 5):** When a critical limit is exceeded, it is not always possible to identify the root cause of the incident; in such cases, generic corrective actions can be implemented to cover several suspected causes, at least until more information is obtained if the incident occurs again.
- **Validation and verification procedures (Principle 6):** In some cases, the verification may be limited to verifying that the monitoring is being carried out in the manner set out in the procedures. Simple verification procedures may include a physical audit or check on the monitoring and physical audit or review of monitoring records, including checking corrective actions whenever a noncompliance has been recorded. In very small FBOs where very few people are involved in implementing the FSMS, the owner or manager can conduct visual inspections to verify that the system is working as intended. Another option for verifying HACCP compliance in small businesses is outsourcing it to an external company.

Another example of flexibility in Principle 6 is using examples of verification procedures included in the generic HACCP guides. However, when the control measures described in those guides are modified, the food operator should validate its control measures for the identified hazards.

- **Documentation and recordkeeping (Principle 7)**: This point is where food operators demand more flexibility to reduce the administrative burden involved in documenting and managing the system's records, which small companies may consider excessive or unnecessary in some cases.

The fundamental concept is that records play a helpful role in HACCP. However, every food operator should assess the advantages and disadvantages and restrict themselves to only those crucial for ensuring food safety. Records are indispensable but not an end in themselves. The general guidelines given to make documentation and recordkeeping more flexible are as follows:

- A simple documentation and recording system can utilize existing models and administrative procedures implemented by the company's employees. For instance, delivery invoices are a useful means to record the product's conformity upon receipt and its temperature.
- Where generic HACCP guidelines are available, the hazard analysis documentation, determination of CCPs, determination of critical limits, validation activities, and possible modifications of the FSMS can replace the individual documentation on HACCP-based procedures. These guidelines should also specify when records are necessary and how long they must be kept.
- Monitoring is more important than recording; it is easier to accept flexibility in records than in monitoring activity, for example, by changing its frequency. Monitoring procedures may stipulate

that only exceptions or noncompliant outcomes are recorded, following the principle of recording by exception.

- A diary or checklist may be a simple and efficient method for recordkeeping. FBOs can tick boxes to indicate their actions or provide more detailed information by writing in text boxes to demonstrate compliance with control points. Daily recordkeeping entails confirming opening and closing checks with a tick and signature to confirm the employment of safe methods. An example of the use of diaries is the one proposed in the English guide "Safer Food, Better Business," recognized as an example of good practice in self-control. This diary model is simple and easily accepted by workers to record some routine tasks. The FBO should define a list of controls to be executed at the start and end of each activity. The operator will conduct the listed checks, record them in the daily register upon commencing and finishing work, and then sign off. Deviations or changes must be noted and described in detail, along with the corrective action taken for each case. Throughout the day, the FBO will be able to identify recurrent problems. Recording this information will facilitate the identification of solutions to these incidents through a more detailed analysis of the root causes (Figure 4.6).

Finally, one question to consider is: What is the limit of flexibility? It is common sense to locate this limit at the point wherein the safety of food products can no longer be guaranteed or where there is an unacceptable increase in risk to the consumer. Therefore, the solution to this question must emerge from assessing the risks and validating control measures, following a logical and scientifically based methodology that establishes acceptable criteria based on the available information in each case. When an FBO wants to use control measures that are not the usual recognized methods, it must validate those control measures. Essentially, the level of flexibility is determined by validation. If the measures can be validated, then they are acceptable. However, if they cannot be validated, they are unacceptable.

4.3 Development and Implementation of Hygiene Prerequisites

The PRPs, which form the basis of the HACCP system, attempt to cover all aspects that are necessary to ensure a hygienic working environment and suitable manufacturing conditions.

As we saw above, the fundamental reference for the development of the prerequisites is the Codex Alimentarius General Principles of Hygiene. At the EU level, the legal reference for the PRPs is in the annexes to Regulations 852/2004 and 853/2004 of the Hygiene Package. Furthermore, there are other rules governing various aspects of the PRPs that the auditor must also consider when evaluating the content of these programs, such as the regulations on drinking water, by-products, traceability, use of biocides, and labeling, among others.

As a starting point, every PRP must attain a clear objective connected to controlling hazards or diminishing risks in a specific aspect of food production. In order to achieve this objective, each of the hygiene prerequisites must contain minimum elements that are considered operationally necessary for the plan to be effective.

Some of the elements included in the PRPs are specific and characteristic of each of them, depending on the nature of the aspect of production they cover; thus, a pest control plan must describe the preventive measures and, where appropriate, the treatments required to prevent or control the presence of pests in an establishment; meanwhile, a training plan identifies the

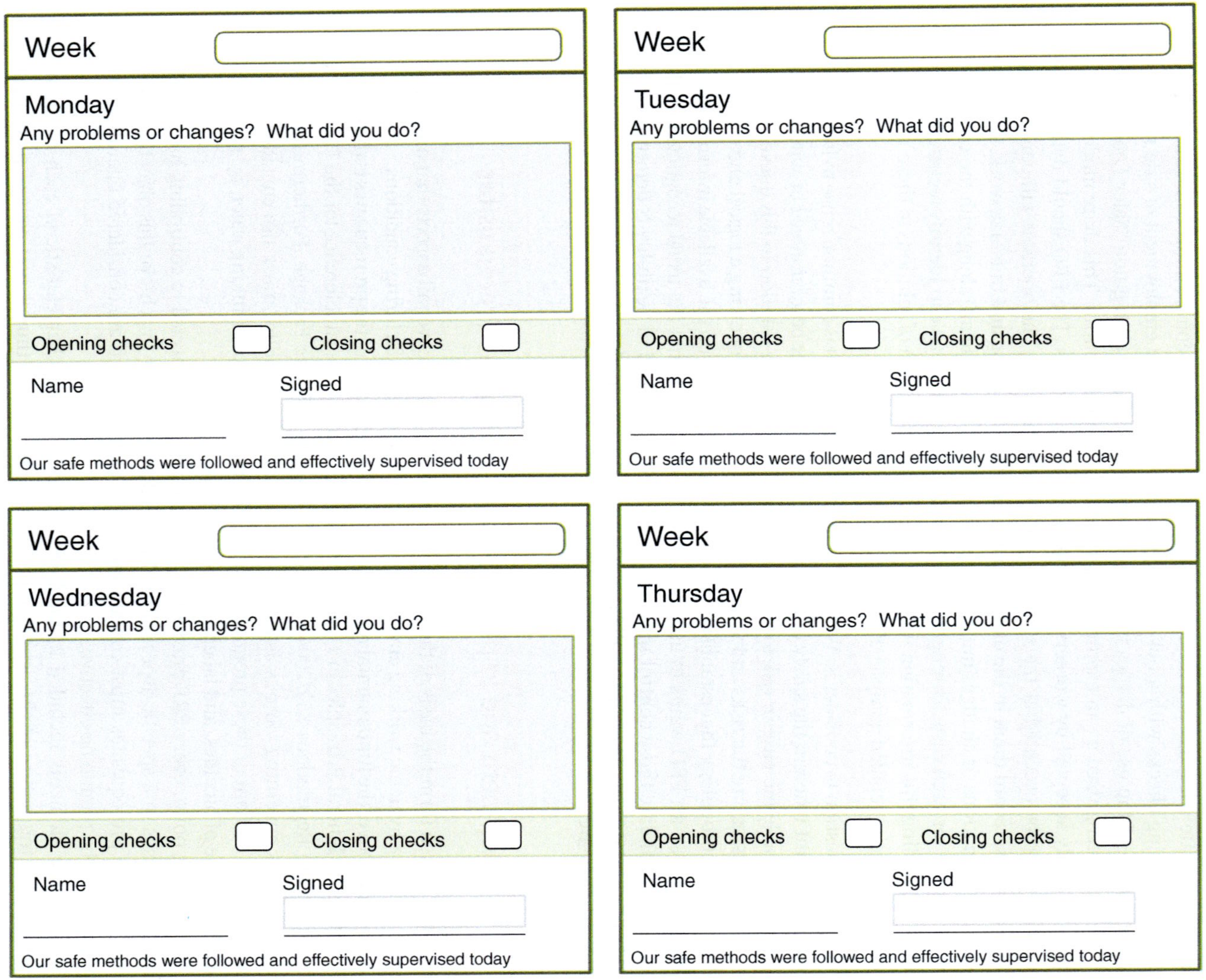

Figure 4.6 Diary. Source: Adapted from Food Standards Agency (2020).

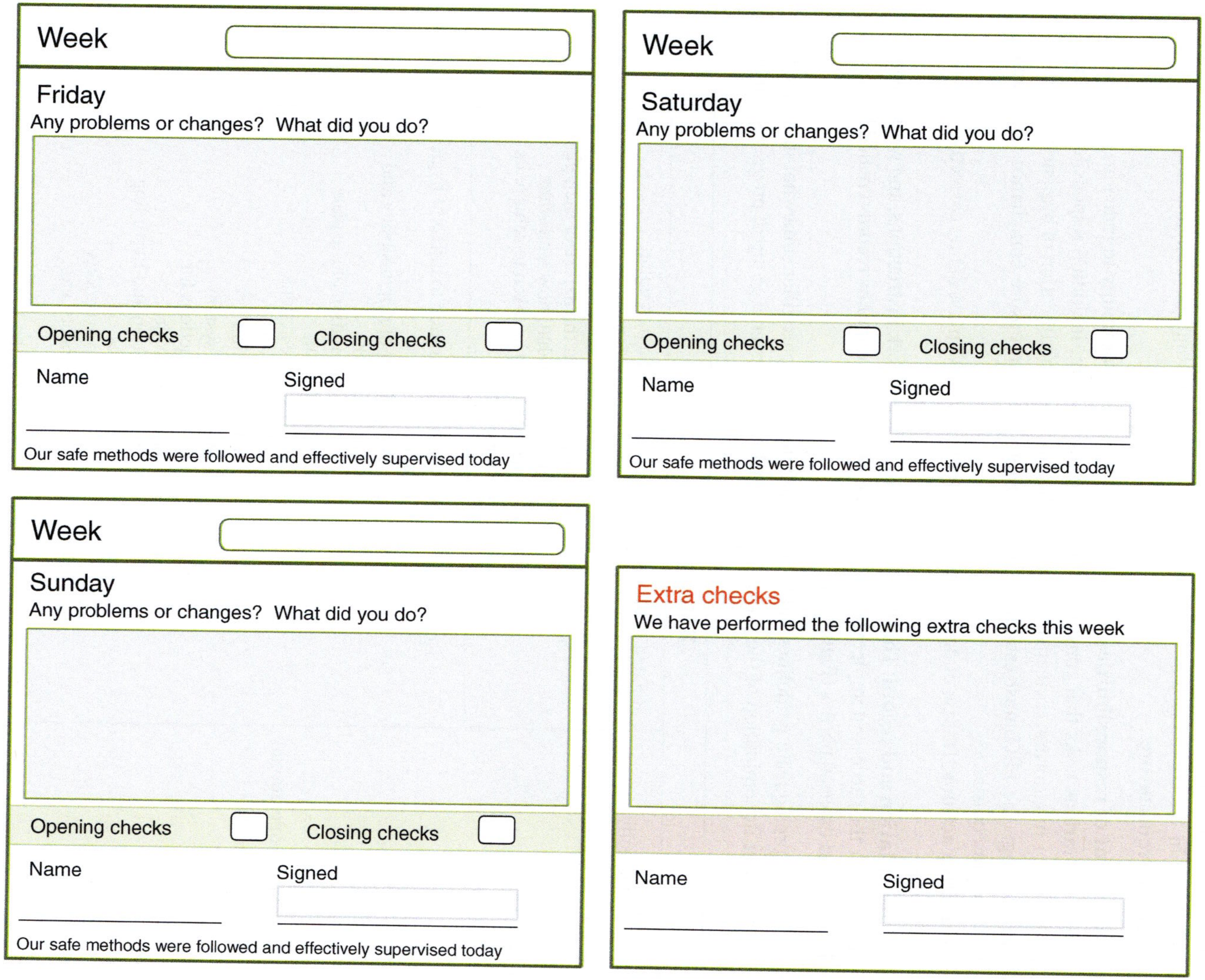

Week

Friday
Any problems or changes? What did you do?

Opening checks ☐ Closing checks ☐

Name ____ Signed ____

Our safe methods were followed and effectively supervised today

Week

Saturday
Any problems or changes? What did you do?

Opening checks ☐ Closing checks ☐

Name ____ Signed ____

Our safe methods were followed and effectively supervised today

Week

Sunday
Any problems or changes? What did you do?

Opening checks ☐ Closing checks ☐

Name ____ Signed ____

Our safe methods were followed and effectively supervised today

Extra checks
We have performed the following extra checks this week

Name ____ Signed ____

Figure 4.6 *(Continued)*

training requirements for employees in various positions within an organization, ensuring adequate training is scheduled to meet those needs. Therefore, within the contents of each hygiene prerequisite, we can identify the specific components that determine how to achieve each plan's particular objective, along with other aspects related to the implementation and management of the plan within the company that are common to all PRPs. Among the elements shared across different prerequisites are:

- **Assignment of responsibilities**: It is crucial to allocate responsibility to individuals for executing the various tasks that constitute the self-control system, including responsibilities and authority for monitoring, corrective actions, verification, and modification of hygiene plans.
- **Monitoring**: The FBO must establish a monitoring system to ensure compliance with each hygiene prerequisite.
- **Corrective actions**: In case of deviations or incidents, the FBO must define corrective actions to be taken.
- **Documentation and record**: This includes the procedures, work instructions, and worksheets required for the successful implementation of each PRP. It also contains records created by both internal and external parties (Figure 4.7).

The hygiene prerequisite guidelines published by different agencies differ somewhat in terms of the number or denomination of the programs they include; however, a group of programs can be

HACCP prerequisite		
Objective:	Related to the control of the hazards of a particular aspect of food production	
Prerequisite operational elements	Specific	Depending on the nature of each plan, they define what, how, and when actions are to be taken to achieve its objective

	Common	Elements common to different hygiene plans
		• Assignment of responsibility and authority for
		  • Implementation of the plan
		  • Monitoring
		  • Corrective actions
		  • Verification
		  • Modification
		• Monitoring system
		• Corrective actions for nonconformities
		• Documentation and Recording
		  • Procedures
		    • Work instructions
		    • Worksheets, etc.
		  • Records
		    • Internal
		    • External

Figure 4.7 Prerequisite structure.

considered typical and present in all of them. Each of these hygiene plans contains the elements that are considered to be the minimum necessary to be effective, but this does not preclude the existence of other requirements; for example, in companies where a standard or certification scheme has been implemented, the requirements contained in each of these standards must also be considered.

4.4 Water Control

This PRP aims to ensure that water used in food businesses is not a source of contamination, either directly or indirectly, to the food being processed. This is an aspect regulated in different countries; therefore, some requirements for water control plans are prescribed by law. Although drinking water must legally be provided, recycled and non-potable water can be used for specific purposes if distributed through separate pipes. Moreover, specific companies like ships and mollusk purification centers are allowed to utilize clean seawater for various activities.

The water control plan must comprise the basic elements necessary for ensuring the safety of water used in the food industry. These include water supply, distribution, maintenance, monitoring, and, where required, treatment. Adherence to these elements will guarantee the water used in food processing is safe and does not contaminate the final product. In addition, there are other legal requirements set out in the drinking water regulations which the auditor will also take into account when assessing the compliance of this control plan.

The first point to consider in assessing the content of this plan is the origin of the water supply:

- **Obtaining from own water sources**: The plan must describe the treatment and its component elements. In this case, an administrative authorization or concession is required from the corresponding body.
- **Supply by external manager**: There must be a supply contract with the manager.
- **Supply by cisterns or mobile tanks**: A favorable report from the health administration is required.

The plan should cover the following, depending on the type of supply and the characteristics of each establishment:

- The point of entry or catchment of the water.
- The treatment(s) carried out, indicating the nature of each (mechanical, physical, or chemical), the composing phases, equipment used, and the nature of the substances involved.
- **Piping**: Typically, a diagram or plan illustrating the distribution of potable water is necessary, and when applicable, a map of pipes for non-potable water, along with information on the pipe material.
- **Tanks**: Number, location, size, and usage.

Another aspect to consider within the plan is the cleaning and maintaining the facilities and equipment. Maintenance activities include pumps, dosing equipment, filters, tanks, and ozone or ultraviolet treatment equipment, among others.

Cleaning tasks should be scheduled for all parts of the distribution system and water tanks, specifying the method and frequency required to prevent them from becoming a source of water pollution.

The monitoring system of the plan must describe the controlled elements, method, frequency, and responsible person for each monitoring task. Among the monitoring activities are:

- Organoleptic examination.

- Residual disinfectant detection (measuring free and combined chlorine levels at designated frequencies and sampling points).
- pH determination.
- Turbidity measurement.
- Checking the correct functioning of the treatment equipment.
- Verification of compliance with cleaning and maintenance tasks.

The corrective actions to be taken in the event of incidents or noncompliance with any of the elements of the water control plan must be defined in advance, as well as the allocation of responsibility for carrying them out.

Verification of this plan's effectiveness is typically attained through water analysis. The frequency and type of analysis depend on the characteristics of the supply system and the volume of water used in the establishment, in accordance with the requirements laid down in the drinking water regulations of various countries.

Depending on the number of parameters analyzed, regular or thorough control analyses can be conducted. Thus, a routine control analysis typically includes assessments of odor, taste, color, turbidity, conductivity, pH, ammonium, coliform bacteria, *Escherichia coli*, and disinfection byproducts. If there is suspicion that the installation pipes contain any of these materials, additional testing for copper, chromium, nickel, iron, or lead should be conducted. If an intermediate deposit is found, then assessments for aerobic bacteria count at 22 °C, *Clostridium perfringens*, and aluminum or iron when used as flocculants should be conducted (Figure 4.8).

Parameter	Parametric value
Odor	3–25 °C Dilution rate
Taste	3–25 °C Dilution rate
Turbidity	1 NTU: tank outlet
Color	15 mg/l Pt/Co
Electrical conductivity	2500 µS/cm −20 °C
pH	>6.5–9.5
Ammonium	0.50 mg/l
Escherichia coli	0 cfu/100 ml
Coliforms	0 cfu/100 ml
Free chlorine	≤1 ppm
Colony count 22 °C (tankoutlet)	<100 cfu/1 ml
Clostridium perfringens (tank outlet)	0 cfu/100 ml
Aluminum (when used as a flocculant)	200 µg/l
Iron (when used as a flocculant)	200 µg/l

Figure 4.8 Water parameters.

As a guideline, the required analyses can range from a more basic control analysis at the beginning of the activity for establishments with an external manager and no intermediate tank, to a full analysis at the beginning of the activity and then periodically, depending on consumption, for establishments with their own supply.

Among the documentation and records associated with the water control plan, the most common are:

- Plan or diagram of the distribution system and water treatment.
- Water supply contract with the external manager.
- Water consumption bills.
- Technical data sheet of the disinfectant used.
- Record of chlorine determinations and organoleptic analyses.
- Report of the analysis of the own water supply.
- Water analysis reports provided by the external manager.
- Records of the cleaning of pipes and tanks.
- Records of maintenance operations.
- Records of incidents and corrective actions.

4.5 Cleaning and Disinfection Plan

The objective of the cleaning and disinfection plan is to remove dirt and reduce microbial contamination in the premises, facilities, equipment, and tools to levels that ensure an appropriate hygienic environment for conducting activities.

The plan can control three main types of hazards:

- **Microbiological**: The plan aims to prevent food contamination by contact with dirty or undisinfected equipment, tools, or surfaces.
- **Chemicals**: The plan must take into account the possibility of contaminating food with residues from cleaning and disinfection products and establish preventive and surveillance measures to prevent this from happening.
- **Allergens**: If a production line manufactures multiple foods with varying allergen content, it may be required to conduct intermediate cleaning between the different preparations, depending on the sequence of production, in order to avoid cross-contamination by allergens between a product and the one being processed below.

The design of the operational elements of the cleaning and disinfection plan should be based on a risk assessment, which must determine the most appropriate cleaning methods and frequency for each of the areas, units, or equipment involved in the process, depending on the microbiological, chemical, or allergenic hazards identified. For example, in establishments where the risk of contamination by *Listeria monocytogenes* may be significant because of the environmental conditions or the nature of the products, a specific cleaning and disinfection plan must be developed to reduce the presence of this germ to acceptable levels, taking into account the areas where it is most likely to be found and its ability to form biofilms, which requires a specific method of removal.

The cleaning and disinfection plan should contain the following elements:

- **What**: it is necessary to include the different dependencies, facilities, equipment, and utensils that need to be cleaned and/or disinfected.

- **How**: The description of the cleaning methods should specify:
  - Phases that make up the cleaning operations.
  - Parts or elements of equipment to be disassembled.
  - Cleaning supplies.
  - Products used.
  - Concentration.
  - Temperature of use.
  - Times for each task, or, for example, the contact time of the disinfectants.
- **When**: the frequency and, if applicable, the time at which each area, equipment, or process tool should be cleaned or disinfected.
- **Who**: Determine who is responsible for carrying out the tasks included in the cleaning and disinfection plan.

The plan should describe the monitoring system and who is responsible for checking that cleaning and disinfection methods are being applied in the prescribed manner and with the prescribed frequency. In the event of incidents or noncompliance with the cleaning and disinfection plan, it is necessary to have defined the corrective actions to be taken.

Sometimes, the food operator must validate the effectiveness of the cleaning and disinfection methods they intend to use in their establishment. In such cases, the validation documentation should contain the data related to the cleaning and disinfection methods chosen, the chemicals used, the pathogens that may cause problems, and the tests carried out to demonstrate the effectiveness of the cleaning and disinfection plan.

The cleaning and disinfection plan should also specify the frequency and the person responsible for the verification activities. Ways to achieve this include:

- Visual inspection.
- Microbiological analysis of surfaces using swabs, sponges, or contact plates.
- Verification tests for chemical residues in the rinse water by measuring the pH.
- Rapid tests: using bioluminescence to detect the presence of ATP in organic debris.

In addition, at the EU level, Regulation 2073/2005 establishes that food operators who produce ready-to-eat foods that pose a risk of *L. monocytogenes* must include in their sampling plan the analysis of surfaces of equipment and production areas to detect the presence of this bacterium.

As for the documents and records associated with the cleaning and disinfection plan, the most common are:

- Delivery notes for cleaning and disinfection products.
- Technical data sheets for cleaning and disinfection products; under current regulations, some disinfectants other than bleach must be authorized and registered or labeled as suitable for use in the food industry. For some products, it is also necessary to take into account the safety measures indicated by the manufacturer of the cleaning product, the need for protective equipment, and whether training is required for its use.
- Records of the implementation of the cleaning and disinfection plan.
- Checklists for monitoring the plan.
- Records of surface analysis.
- Documentation of validation of cleaning and disinfection methods, if applicable.

4.6 Pest Control Plan

The pest control plan aims to prevent the contamination and spoilage of the food produced by insects, rodents, and other types of undesirable animals that can act as vectors of human diseases (typhoid, salmonellosis, diarrhea, etc.).

The hazards associated with the implementation of this plan are:

- **Biological and microbiological hazards**: they can lead to cross-contamination of food with different microorganisms and/or become vectors for certain diseases.
- **Physical hazards**: caused by unwanted animal remains such as insects, rodents.
- **Chemical hazards**: resulting from the inappropriate use of pesticides and biocides in the prevention and treatment of pests.

The significance of this PRP within the food industry varies considerably, depending on the establishment's location, infrastructure design and condition, type of food, and processes. Thus, in some types of companies, pest control poses hardly any problems, while in others, it requires much more effort and attention to keep pests at a level that does not pose unacceptable risks to food.

On the other hand, it should be noted that the effectiveness of the pest control plan is closely linked to the implementation and proper management of other prerequisites of the HACCP system, such as cleaning and disinfection, maintenance of equipment and facilities, waste management, and good handling practices.

A pest control plan is much more than just a periodic review of the baits placed at various points in the establishment. A comprehensive approach is necessary to achieve pest control effectiveness, wherein preventive and monitoring measures take priority while limiting pesticide use to necessary scenarios.

The following elements should be taken into account when developing a pest control plan:

- Diagnosis of the situation.
- Preventive measures:
  - Physical barriers.
  - Active methods.
  - Hygienic measures.
- Monitoring system.
- Corrective actions:
  - Pest treatment.
  - Other actions.

4.6.1 Diagnosis of the Situation

Before developing the pest control plan, the first thing to do is to carry out a diagnosis of the situation in order to analyze the risks of access and settlement of the different types of pests in the facility. Depending on the location and environment in which the premises are situated, it is necessary to study what type of animals can cause problems, the possible entry routes, the places in the plant where they can settle and proliferate, and the practices that can favor them.

4.6.2 Preventive Measures

As mentioned above, the most effective measure to control pests is to prevent them from entering the premises and, if they do enter, to ensure that they do not find favorable conditions for their settlement or proliferation. Preventive measures include a range of actions that must be taken to address the factors that may influence the presence of pests in a food business. These measures include:

- **Physical barriers**: this is a passive measure, but its importance is crucial, and it can be said that its condition is what makes the difference between pest control from one factory to another. The aim is to seal the factory to the entry of pests through its different routes. To do this, it is necessary to check the condition of all installations: walls, floors, ceilings, openings to the outside, pipes, etc., and cover or protect all entry routes, for example, by:
  - Mosquito nets on windows.
  - Curtains on doors and passageways.
  - Weatherstripping on doors.
  - Siphons in the drains.
  - Sealing of pipes and ducts.
- **Active methods**: they aid in eliminating animals in the factory and prevent their proliferation, particularly when they are not at high levels; additionally, these methods help detect and monitor the presence of pests in the establishment. Active methods used include:
  - Traps with pheromones, glues, or baits.
  - Insect repellents on doors and windows.
  - Insect lamps.
  - Air curtains on doors to prevent flies and mosquitoes from entering.

Bait traps should be correctly placed and maintained to prevent product contamination. The usage of toxic baits should be avoided whenever feasible in areas where food is not protected during production or storage. Insect lamps should be positioned to prevent insect or fragment contamination of food and should not be placed in areas where they attract insects. They should also be positioned in areas where the lamp's attractive effect is not compromised by overly bright surroundings. These are key details for the auditor to observe during their tour of the plant.

- **Hygiene measures**: these include several types of measures of a more general nature that also have an impact on pest control, such as:
  - Appropriate storage and disposal of waste to prevent contamination and pest settlement.
  - Cleaning and disinfecting all areas of the establishment, paying particular attention to places where conditions may be favorable for undesirable animals and damp places or hotter areas due to the presence of engines, which can cause some pests to nest.
  - Prevent food from being accessible to pests, in open containers or in direct contact with the ground.
  - On receipt of goods, inspect raw materials that may contain some pests, such as sacks of vegetable products or other types of packaging, to prevent them from entering the premises.
  - Observe good handling practices, such as not eating in work areas, storing food in lockers, wearing appropriate clothing.
- **Monitoring system**: The food operator must establish a monitoring system and appoint a responsible individual or team to verify the correct implementation of the pest control plan. This includes reviewing the condition of physical barriers, assessing the effectiveness of active methods and compliance with hygienic measures to prevent pests, and searching for any signs

of unwanted animals or early infestations in order to be able to take corrective actions as soon as possible, preventing animal population levels from increasing excessively, and then becoming more difficult to eradicate or requiring more drastic actions.

The monitoring of the pest control plan can be carried out by the food company's own staff or by professionals from a pesticide service company. However, when hiring a pest control service, a frequent mistake is to delegate all the management of the plan to it, neglecting the tasks of monitoring and the application of the rest of the prevention measures. The reason is that the contracted company is only present in the facilities sporadically, so many times it will not be able to detect the first signs of the presence of pests in time, and also because the pest control plan requires, as we saw, the application of other hygienic measures and prerequisites that are the food operator's responsibility and cannot be delegated to external companies.

- **Corrective actions**: The pest control plan requires corrective actions which can be broadly categorized into two types of causes:
  (1) Noncompliance with the preventive measures of the control plan such as physical barriers, active methods, or hygienic measures.
  (2) Elimination treatment for the presence of pests at a level that requires it.

  If pesticide treatment is necessary, it is crucial to comply with the relevant legal requirements and adhere to safety measures regarding permitted use, handling, and safety deadlines. If the treatment is carried out by a pest control company, then:
  - The company must be authorized and registered in the register required by the regulations on biocidal establishments and services in various countries.
  - Staff must provide appropriate training for the type of product used.
  - The products used must be authorized and registered in the corresponding register of biocides, in compliance with current legal regulations.

  The food company's staff can perform the treatment if they have the necessary training for handling those products.

  If the complete pest control plan is successful and there is no need for elimination measures, then the food operator may solely manage the pest control plan without hiring a pest control company or utilize their services only when eradication treatment is necessary.

  Documentation and records: The auditor usually asks for certain documents and records to check the pest control plan. These include:
  - Pest control plan prepared by the pesticide services company.
  - Professional cards or training certificates for pesticide applicators.
  - Technical data sheets or safety data sheets for the biocides used.
  - Plan of the premises showing where the baits and other means of action against pests are located.
  - Monitoring records or pest inspection sheets of baits and the condition of the installations.
  - Treatment certificates stating the type of pest treated, the product used, and the safety measures adopted, among others.
  - Records of incidents and corrective actions.

4.7 Training Plan

The training plan is the most essential prerequisite for successful HACCP implementation. In a system developed and operated by people on a day-to-day basis, training forms the foundation of the entire system, affecting the performance of the rest of the prerequisites and elements of the self-control system. However, it is a plan that is more challenging to evaluate, as it is related to intangible aspects of human nature, such as the attitude, commitment, or motivation of the workers.

The aim of a training plan is to ensure that all employees involved in food hygiene and the implementation of the FSMS have the necessary competence to carry out their tasks correctly.

The training of company personnel follows a distinct approach from education in other sectors of society. In the workplace, training serves the purpose of equipping employees with the necessary knowledge, skills, and abilities to effectively carry out their designated tasks. This is contingent on their job requirements, disregarding any other factors related to one's comprehensive personal development, which falls under the purview of the education system.

The training of employees in the company should be aimed at improving the performance of the employees. If the performance of the workers in the company is adequate and it is observed that they carry out the assigned tasks correctly, then it can be concluded that the training plan is effective and, therefore, there is no need to plan any training activities.

Performance in the workplace is influenced by many factors, of which knowledge is just one, and not always the most crucial. It is frequent to encounter individuals who perform their tasks proficiently despite a lack of specific knowledge or understanding of why and how they are doing it. Sometimes, workers understand what to do, but factors like insufficient time, unfavorable working conditions, work pressure, lack of motivation, or scarce resources impede them from executing their tasks correctly.

For Frank Yiannas, a widely recognized figure in food safety culture, the equation is simple: food safety equals behavior. Thus, if we want to enhance food safety, we must work to improve behavior. According to Yiannas, better integration between behavioral and food sciences is necessary to create a FSMS firmly rooted in behavior (Frank Yiannas 2009). Training will often be the most important tool to promote a sound food safety culture (Figure 4.9).

Training for food handlers is mandatory in many countries. At the EU level, food operators are required to create a training plan for their staff, as per Regulation 852/2004. Despite the significant importance of training in the food industry, there is a lack of guidance on how to develop a training plan and the necessary conditions that should be met in this area, except for specific food sector's legislation.

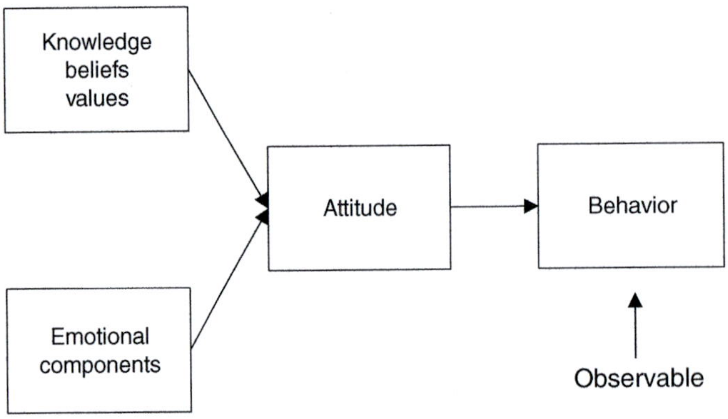

Figure 4.9 Beliefs, attitude, and behavior.

In this way, the training of workers is the full responsibility of the business operator, who, in accordance with Regulation 852/2004, must ensure:

1. that food handlers are supervised and instructed and/or trained in food hygiene matters commensurate with their work activity;
2. that those responsible for the development and maintenance of the procedure referred to in Article 5(1) of this regulation or for the operation of relevant guides have received adequate training in the application of the HACCP principles; and
3. compliance with any requirements of national law concerning training programs for persons working in certain food sectors.

When a training plan is implemented, it is very easy to obtain training records showing that the company's personnel have received or have followed courses, sometimes not in person, on food hygiene in general or other specific aspects. However, this is not the end of a training plan; it is only one means, and not the only one, of achieving the real objective of the plan: that employees acquire the necessary skills to perform correctly the tasks assigned to them that affect food safety. This point requires proper analysis and planning.

The development of a training plan differs considerably depending on the size and characteristics of the organization; in medium-sized to large companies, there is likely to be a human resources manager or a mid-level position responsible for planning the training of employees over a given period; in small or family-run companies, the training plan may be much less formal and based on the perceived needs of the manager or the employees themselves to attend specific courses or on their own interest in acquiring knowledge on aspects related to their work that can have a positive impact on the company's performance or development prospects. In both cases, it is possible to identify the advantages and disadvantages associated with each of these approaches.

Ideally, for a training program to be successful, a methodical and organized approach should be adopted, comprising four stages (Figure 4.10):

1. Identification of the training needs.
2. Design and development of the training.
3. Providing the training.
4. Evaluating the training outcomes.

The first step in developing a training plan is to identify training needs. In order to identify the training needs of employees, the person responsible for the plan needs to know the characteristics

Figure 4.10 Training cycle.

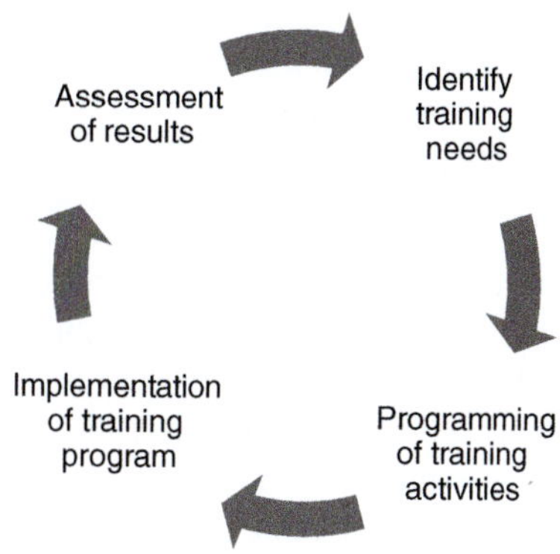

of each job, the list of tasks it contains, and the knowledge, skills and abilities required to perform them correctly. The next question that the person responsible for drawing up the training plan must ask is: Does the employee perform the assigned tasks in the manner indicated? If the answer is "yes," then no training is required. If the answer is "no," then there may be a need for training.

A performance gap is the difference between what a worker has to do to achieve the required level of performance and what he actually does. However, this performance deficiency may be due to several causes: it may be due to lack of knowledge, and in that case, it can be solved by training; it may be due to lack of practice, and it can also be solved by hands-on training. Nevertheless, it may also arise from other reasons, such as inadequate motivation, attitude, work environment, excess workload, and other related issues. In these cases, training alone may not address the problem.

Once the training needs have been identified, the next step is to plan the most appropriate training actions to address the identified competence or performance gaps. Among the aspects to be specified in the training program are the following:

- Specific objectives.
- Target group.
- Training actions:
  - Number.
  - Duration.
- Timetable.
- Training provider:
  - External.
  - Internal.
- Method:
  - Face-to-face.
  - Distance learning.
  - Group workshop.
  - Self-training: delivery of guides or manuals.
  - On-the-job training.
- Resources.
- Follow-up or evaluation.

After implementing the training plan, it is necessary to evaluate the results of the training. Evaluation is another essential element of the training plan, used to check whether the objectives set have been achieved. The evaluation of training can focus on two aspects:

- Checking that the activities that make up the training plan have been carried out according to the timetable set.
- Checking whether performance deficiencies have been eliminated or whether the level of competence of the employees has been improved.

In order to test this second aspect, it is often necessary to allow a period of time for staff to be observed in the workplace to be able to assess the extent to which the knowledge and skills acquired lead to a change in the behavior of the workers and an improvement in the tasks they perform.

On the other hand, it is necessary to identify the person(s) responsible for managing the different stages of the training plan. The food safety auditor may be interested to know what mechanisms or criteria are used by the person in charge of the plan to identify training needs or how to check whether the training activities had the expected effect.

Documentation and records related to the training plan include the following:

- Scheduling or timetable of training activities.
- Training certificates issued by an external entity.
- Records of internal training activities.
- Records of delivery or receipt of guides or codes of good practice for new operators.
- Checklists.
- Training results evaluation reports.
- Incidents and corrective actions.

In addition to reviewing the specific elements that make up the training plan, such as compliance with the scheduled training actions or the training records of the workers, the food safety auditor is in an optimal position to assess the outcome of the training on the food operator. As it is a prerequisite that affects the performance of the operators in a transversal way, by observing the handling or manufacturing practices, or by interviewing the personnel, during their stay in the plant, the auditor will be able to get a fairly well-founded idea of the overall competence of the workers, their degree of commitment, or the level of maturity of the management system and thus deduce the real effectiveness of the training plan.

4.8 Supplier Control

The objective of the supplier control plan is to prevent the introduction of hazards originating upstream in the food chain into the food produced by the operator. It is essential to bear in mind that hazards present in raw materials or ingredients that are not being controlled in the previous steps will have to be eliminated during the food operator's production process or, ultimately, by the action of the final consumer prior to consumption. In many food safety incidents, the source of the problem has been found to be in the raw materials or ingredients coming from a supplier.

For each raw material or each supplier, the FBO must analyze the risk of microbiological, chemical, and physical hazards and the contamination or presence of undeclared allergens that must be controlled in its process.

Typically, the control of risks associated with suppliers and raw materials is done through the approval and monitoring of approved suppliers. Types of suppliers include the following:

- Suppliers of raw materials:
  - Ingredients.
  - Additives.
  - Adjuvants.
  - Spices.
- Suppliers of machinery and equipment in contact with foodstuffs.
- Suppliers of containers and packaging.
- Suppliers of cleaning and disinfecting products.
- Service suppliers:
  - Pesticide service companies.
  - Maintenance services.
  - Laboratory.

- Water supply.
- Transport.
- Waste manager.

The supplier approval system is based on the fulfillment of requirements set by the food operator, and these requirements can be of various kinds:

- **Legal requirements**: in some countries, various authorizations or registrations are required, so satisfying these requirements is the first thing a supplier must comply with:
  - Sanitary Authorization to operate: an administrative requirement for FBOs to be able to operate.
  - Registration or, where applicable, accreditation of food testing laboratories.
  - Registration of pesticide service companies in the corresponding register.
  - Authorization of the waste manager.
- **Supplier evaluation**: in the initial evaluation, based on the supplier's knowledge, the FBO must assess the supplier's capacity to supply the requested product or service with the agreed characteristics and conditions.
- **Product specifications**: significant for the effectiveness of the supplier control program, the FBO must define all the factors that may affect the quality or safety of the product supplied.

Having an authorization or registration is a legal requirement that must be fulfilled, but it is often not very indicative of the supplier's real guarantees to supply products with the required conditions to ensure the safety of foodstuffs. For this reason, the food operator must develop a system for assessing suppliers, which may be based on a combination of the following methods:

- **Historical**: when the FBO has been working with a supplier for many years, without any incidents, and the product supplied is well known.
- Certifications in food safety standards such as Brand Reputation Compliance Global Standards (BRCGS), International Featured Standards (IFS), ISO 22000.
- Questionnaires.
- Product analysis.
- Certificates of analysis.
- Visit to the supplier.
- Audits of the supplier's FSMS.

In addition to evaluating the supplier's characteristics and conditions to receive products with necessary guarantees, the FBO can agree with its supplier to comply with the purchasing specifications that it considers necessary to ensure the safety and quality of its production process. Purchasing specifications cover aspects such as:

- Composition of raw materials and ingredients.
- Microbiological specifications.
- Labeling requirements.
- Type of packaging and packing.
- Storage and distribution conditions.
- Specifications concerning traceability and documentation accompanying incoming consignments.

The FBO must have an up-to-date list of approved suppliers. This list must include identification details such as company name, address, email, etc., the legally required authorization or registration, and the list of products or services supplied. The operator must designate a person responsible for approving suppliers and, where necessary, removing them from the list for any reason, such as noncompliance with agreed conditions or other commercial reasons.

Once suppliers are approved, there should be a monitoring system in place to check whether the conditions established for the supply of the products or services contracted are being met over time. Part, but not all, of these monitoring activities are carried out at the time of receipt of the goods. For this reason, the monitoring system, the frequency, the person responsible, and the corrective actions are often described at the reception stage of raw materials and ingredients or in the HACCP control chart.

Documents and records related to the implementation of the supplier control plan include the following:

- Updated supplier list.
- Initial assessment form.
- Questionnaires.
- Copy of the health license or registration in the relevant registers.
- Copy of certificates of food safety standards such as BRCGS, IFS, ISO 22000, etc.
- Supplier audit reports.
- Raw material specification sheet.
- Product data sheet.
- Declarations of conformity for materials in contact with food.
- Records of monitoring on receipt of goods.
- Records of incidents and corrective actions.

4.9 Traceability Plan

Traceability is an essential tool for the management of food safety by the food operator and the control of public health risks through all stages of the food chain. Traceability alone does nothing to control hazards in food, but it is a prerequisite for effectively controlling food safety problems for the public.

At the EU level, traceability is a legal requirement for FBOs, established by Regulation 178/2002, which defines it in Article 3(15) as "the ability to trace a food, feed, food-producing animal or substance intended or expected to be incorporated into food or feed through all stages of production, processing and distribution."

Regulation 178/2002 requires that the FBO must be able to identify who has supplied it with a product and to whom it has been supplied, i.e. it is based on the "one step up" and "one step down" approach in the food chain. This obligation means that FBOs must put in place systems and procedures to identify the suppliers and the direct recipients of their products, other than the final consumer, and that this information must be made available to the competent authorities on request (Figure 4.11).

Depending on the nature of the business activity and the direction of traceability, three types of traceability can be recognized:

- **Upstream, also known as supplier, traceability**: the capability to locate the source and traits of a product at any stage of the food chain; in other words, it involves identifying which products have been supplied by which suppliers.
- **Internal traceability**: it is crucial for food operators to connect input products such as raw materials, ingredients, and packaging to the output of final goods and the production

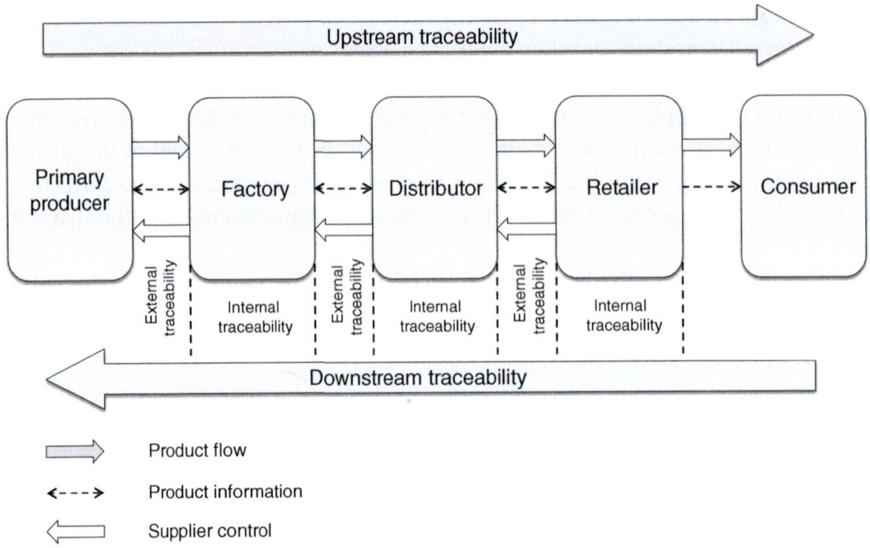

Figure 4.11 Supply chain traceability scheme.

process's features and parameters. Although not required by law at the EU level, it is highly recommended. This practice enables the more accurate identification of products to be withdrawn from the market in case of production failures.

- **Downstream or customer traceability**: it refers to the capability of identifying the precise customer who received the products and promptly locating the items that have been distributed within the market according to specific criteria. In simpler terms, it involves specifying the products supplied to each individual customer.

Another classification for traceability distinguishes between external traceability – the relationship established between operators in the food supply chain, which serves to trace a product's history and location through all or part of the chain; and internal traceability – the system that enables food operators to gather all the information pertaining to the ingredients and processes of the batches manufactured on their premises. Internal traceability specifies traceability within a single stage of the food chain.

The development of a traceability system can vary considerably depending on the food sector, the type of product, the characteristics of the process, customer requirements, and the technology used to collect, manage, and transmit the data. In the beef sector, for example, it is perfectly possible to establish traceability throughout the food chain so that the farm of origin of a piece of meat served in a restaurant can be easily identified, as required by EU legislation. On the other hand, in the case of products that are distributed in bulk, such as milk, cereals, or oil, which are collected in silos or storage tanks, sometimes huge ones, traceability is lost, and it is no longer possible to identify the origin with the same precision.

To be effective, a traceability system must meet three fundamental requirements (FSA 2020):

- Identification of units of all ingredients and products.
- Information on when and where they have been moved or processed.
- The method by which all this information can be linked together.

The unambiguous identification of the different product groups is one of the essential elements on which the traceability system is based. There are three types of unit or group of products that can be traced:

- Batch.
- Consignment.
- Logistical unit.

The definition and identification of the batch is a necessary element in traceability procedures and is the type of unit most commonly used by food operators. A batch is defined as a set of sales units of a food product produced, manufactured, or packaged under virtually identical circumstances.

Consignment refers to the commercial grouping of units shipped by one operator to another at one time.

A logistic unit is any combination of trade units or other objects put together in a case, on a pallet or truck, established for the purpose of transport or warehousing.

The type of grouping, the criteria, and the code used to identify the grouped product units must be clearly defined in the traceability procedure. The food safety auditor will certainly check that the batches or groupings used by the operator for the different products placed on the market match the coding system described in the procedure.

Another decision to be made by the FSMS responsible is the size of the batch or grouping to operate in the food chain, taking into account the type of product, the characteristics of the production process, and the total volume of production. In general, the smaller the batch, the lower the cost of the amount of product to be segregated or withdrawn from the market and the lower the risk to the operator of customer complaints or damage to the company's image. However, the smaller the batches, the more the data to manage. In addition, the type of product or the nature of the process, e.g. bulk products, limits the ability to reduce the number of units or the amount of product per batch, or using a small batch size would imply a reduction in process efficiency, leading to shorter production cycles, with more time lost in downtime or for equipment start-up. In the end, therefore, the food operator must find a trade-off between the increased management work and costs associated with smaller batches and the risk it assumes for the recall of larger batches.

A traceability system should be able to achieve the following objectives:

- Determine the origin and history of a product.
- Identify the location of a product on the market.
- Facilitate the withdrawal or recall of a product.
- Provide appropriate information to stakeholders, organizations, and consumers.
- Help the FBO to control its processes and identify the causes of failures.
- Facilitate compliance with food safety and quality objectives (Figure 4.12).

To meet these objectives, the elements that make up a traceability system must allow the collection and maintenance of the data necessary to answer the following questions:

- **Who**: the operators involved throughout the food supply chain.
- **What**: the products and the information transmitted.
- **Where**: location of the products.
- **When**: to determine when a food product was produced and its subsequent movements.
- **Why**: to find out the causes of failures or nonconforming products.

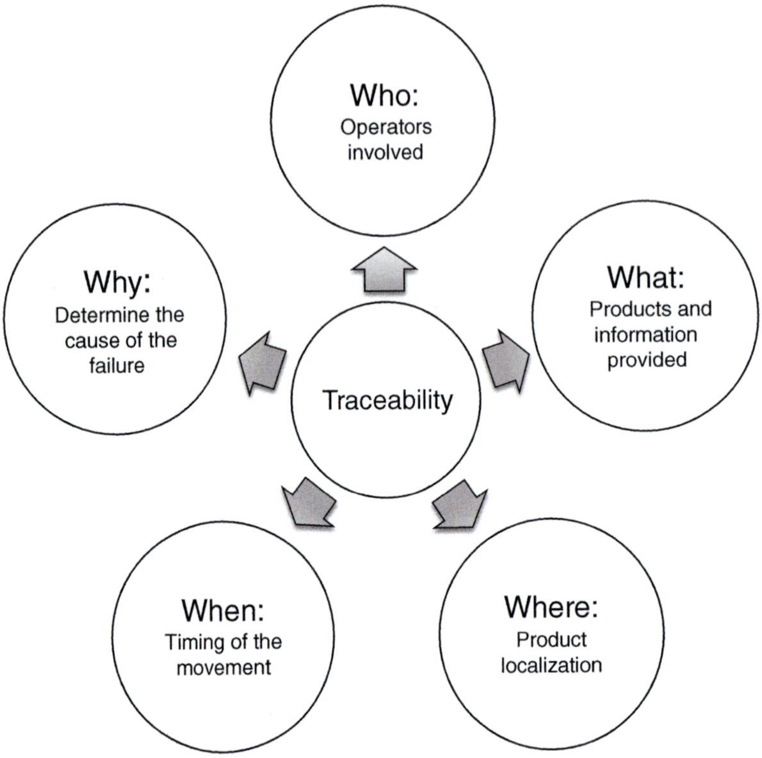

Figure 4.12 Traceability objectives.

Food operators are free to choose the tools or methods that best suit their circumstances and needs; however, to be effective, the traceability program must be:

- **practical**: easy to implement;
- results oriented;
- verifiable;
- compliant with general and sector-specific legislation; and
- be affordable for the operator.

There are a variety of methods and technologies available to FBOs to manage their traceability system, ranging from simple paper or computerized records such as a spreadsheet, to more advanced technologies capable of automatically recording, storing, and sharing information on food products throughout the food chain.

For small businesses where the amount of data to be managed is very limited, a system based on manual paper records or the use of a spreadsheet may be sufficient to meet the needs and objectives of the traceability plan.

However, in larger companies where the amount of data is much larger, or simply because they want to reduce the burden of collecting and managing traceability data and increase its accuracy and security, there is a growing number of devices such as smartphones, tablets, sensors, radio frequency identification (RFID) tags, vehicles, cameras, drones, and machines that allow the use of increasingly advanced identification and monitoring technologies (IMT). The emergence of the Internet of Things (IoT) and rapid advances in information and communication technologies will drive a transformation in food traceability systems that we are likely to see in the coming years.

The goal of the IoT is to connect diverse and heterogeneous smart devices without human intervention, from data collection, sensing, and identification to data analysis, processing, and storage for better decision-making. An IoT-based food traceability system goes through a series of steps in which different technologies and platforms coexist. These technologies can be classified into three main categories: IMT, communication technologies (CT), and data management technologies (DMT), which include the use of 5G as a new CT and blockchain as a new peer-to-peer technology (DMT) (Mehannaoui et al. 2023).

IMT Include

- **Barcodes**: It is widely used by the food industry, easy to use, and cost-effective. Barcodes are machine-readable optical labels used to record product information, ideal for inventory management and cash registers. Barcodes can be one-dimensional or two-dimensional, such as Quick Response (QR). QR can store more information than one-dimensional barcodes and can be read by a barcode reader, smartphone, and webcam.
- **RFID**: RFID is a form of contactless (wireless) data communication technology in which users can send radio frequency signals to the tag via a reader and receive information from the tag in response. It is now a widely applied technology for food identification and traceability. RFID is also employed for monitoring with a wireless sensor network (WSN) and locating food products with GPS. RFID consists of three components: a tag, a reader, and a data management system.
  - **Tag**: it is a chip containing the information with an antenna to communicate with the reader. Tags can be passive, active (equipped with a battery to transmit a signal), or semi-passive.
  - **Reader**: it is an active device that emits radio frequencies to activate the tags by passing in front of them or by supplying them with the necessary energy at a short distance.
  - **Data management system**: it allows the data transmitted by the reader to be recorded for sorting history and data management.
- **DNA barcoding**: It is widely adopted in various food sectors to authenticate raw materials and track them through the supply chain as a way of ensuring food traceability and minimizing food piracy.
- **WSN**: A WSN is a network consisting of a large number of small sensor nodes, either stationary or mobile, with wireless communication between these nodes. These sensor nodes are distributed in the environment to be monitored (place or object), far away from the user. The collected data (temperature, humidity, light, pressure, etc.) are routed via multi-hop routing to a collection point called a receiver node (base station or sink) and from there to the user via the Internet. It is an ideal system for real-time monitoring of critical safety and quality parameters and for improving food traceability, making it very useful in the HACCP system.

Communication technologies: The majority of IoT applications rely on wireless technologies because of the need for accessibility and mobility; the most common wireless communication technologies used are Proximity (NFC, RFID), WPAN (Bluetooth, Zigbee, Z-wave, and 6LoW-PAN), WLAN (Wi-Fi, Wireless HART), WMAN (WI-mAX, NWave), and WWAN (5G, NB-IoT, WRANE, etc.).

Data management technologies cover data analysis, integration, storage, sharing, and security. Some of the most reliable and promising new technologies that can be used in IoT-based traceability systems are Cloud Computing, Big Data, and Blockchain.

The food safety auditor will find a variety of traceability systems in food businesses, from the simplest to the most complex, but whatever the traceability tool used by the food operator, the important thing is to check that it works. And in this sense, the traceability program is easy to evaluate, it is enough to conduct one or more traceability exercises, by selecting some of the operator's batches, to see if the plan is valid and correctly implemented. Many food safety standards consider a reasonable time for the food operator to obtain the required information on the batches or articles concerned to be around four hours.

Taking into account the objectives to be achieved by a traceability program, the operator must be able to collect and manage the following data:

- On backward traceability:
  - Who the supplier is.
  - What was received.
  - When.
  - Quantity received.
  - Batch received.
- In internal traceability:
  - What products it makes.
  - Which ingredients are used.
  - How it mixes or combines them.
  - How it processes them.
  - When.
- In forward traceability:
  - Who they supply the product to.
  - What product he delivers.
  - When.
  - Quantity delivered.
  - Batch or consignment delivered.

At the EU level, Regulation 178/2002 requires that if a FBO considers, or has reason to believe, that a product it has placed on the market is not in compliance with food safety requirements, it must immediately withdraw the product from the market and inform the competent authorities and other businesses with which it has traded the product. This implies that there should be a procedure in the traceability program or elsewhere in the FBO's management system that defines the responsibilities for the effective withdrawal or recall of products from the market and the communication channels for informing the competent authorities and the operators concerned.

As with the other prerequisites we have seen, the FBO should specify what the monitoring system is and who is responsible for checking that the traceability plan is implemented correctly. Monitoring activities should ensure that products are identified as specified and that records containing traceability information are completed with the required data.

To verify the traceability plan, the operator may, inter alia, carry out traceability exercises or simulations at specified intervals to verify that the traceability of the processed products can be established from raw materials, ingredients, auxiliary products, and packaging materials to the final product, including the corresponding quantities or mass balance of the process.

4.9.1 Documentation and Recordkeeping

In addition to the records specific to the traceability plan, such as records of incoming raw materials and outgoing final products, there are other records related to the food operator's activities that contain data on the identification and quantities of raw materials used and the subsequent manufacturing processes that can also be taken into account to establish the traceability of food products. Records that can be used to establish traceability in the food business include:

- Invoices or delivery notes from suppliers of raw materials, ingredients, and supplies.
- Invoices or delivery notes for outgoing or dispatched products.
- Reception register.
- Production order.
- Product file.
- Production report.
- Production sheet.
- Stock records.
- Verification records.
- Traceability tests or simulations.
- Incidents and corrective actions.

4.10 Maintenance Plan

The purpose of the maintenance plan is to ensure that the facilities are maintained in a satisfactory state of hygiene and that the equipment is working properly so as not to compromise the safety of the foodstuffs processed in the establishment.

Implementation of the facilities and equipment maintenance plan also affects the effectiveness of other HACCP prerequisites such as the pest control plan, cleaning and disinfection, maintenance of the cold chain, GMPs, and the food manufacturing process itself.

The hazards that can be controlled by the maintenance plan are:

- Microbiological hazards:
  - **Contamination**: Poorly maintained equipment can make cleaning more difficult, encourage pests that can contaminate equipment and food, and lead to an unhygienic working environment, for example with increased humidity or dust.
  - **Growth**: Poorly functioning refrigeration equipment will result in temperatures not being maintained at levels required to prevent microbial growth.
  - **Survival**: Malfunction of certain equipment, e.g. heat treatment equipment such as heat exchangers or autoclaves, or failure to calibrate equipment measuring process parameters, may result in food not receiving the intended treatment to eliminate target microorganisms.
- **Physical hazards**: Deterioration of equipment can lead to contamination by foreign bodies from walls or ceilings in poor condition, chipped or broken glass, rust particles, or poorly maintained equipment that can lose parts such as screws, nuts, sawdust, blades. Conversely, the failure of a metal detector due to lack of maintenance prevents the removal from the production line of food that has been contaminated with metallic objects during the process.
- **Chemical hazards**: For example, contamination by non-food-grade lubricants or disinfectant residues due to failure of the CIP system, etc.
- **Allergenic hazards**: From food-grade lubricants that are suitable for use with food, but whose allergenic status is not known.

The maintenance plan includes different tasks depending on the infrastructure of the establishment, the nature of the processes, and the equipment used by the FBO. The importance of these tasks within the self-control system varies considerably according to the element to be maintained, but can be critical in the case of equipment where a malfunction would result in a nonconforming or unsafe product for human consumption. To give an example, it is easy to see that the maintenance of the walls or floor of a finished product warehouse cannot have the same impact on food safety as the maintenance of certain elements that directly affect the operation of an autoclave in the canning industry, such as the steam purge in a steam autoclave or the fan in a steam–air autoclave. In the case of a steam–air autoclave, the correct operation of the fan to maintain a steam–air mixture throughout the autoclave is critical to ensure a homogeneous distribution of treatment temperatures within the autoclave, thus avoiding products being treated below the specified thermal lethality time F0. This is not always detectable by the autoclave temperature probe (Figure 4.13).

In the maintenance plan, two types of activities can be distinguished according to the cause of their occurrence and the time at which they occur:

- Corrective maintenance
- Preventive maintenance

Corrective maintenance: this is carried out after a breakdown, to repair a malfunction or to overhaul equipment. It is therefore reactive and unplanned maintenance.

Preventive maintenance: this is the set of activities carried out on a scheduled basis on the various pieces of equipment in order to ensure their optimal operation and prevent breakdowns due to poor maintenance. It is planned and proactive maintenance.

Both types of maintenance are complementary and are necessary for the operator to achieve the objectives set for this HACCP prerequisite. However, it is the preventive maintenance plan that, by its very nature, is most closely related to the HACCP system's ability to anticipate problems, not just react after they have occurred.

On the other hand, the way of evaluating the two types of actions is also different; in the case of corrective maintenance, it can be said that its noncompliance is obvious; therefore, the mere observation or inspection of the equipment can be sufficient for the auditor to get an idea of the state of maintenance of the equipment during his stay in the plant, by locating the defects or breakdowns that remain uncorrected. If, on the other hand, we refer to the preventive maintenance plan, it is

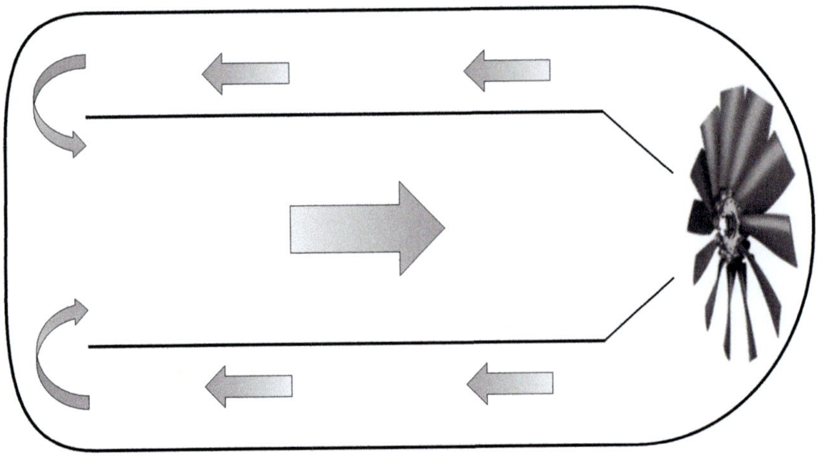

Figure 4.13 Steam–air sterilizer. Source: Adapted from Alserco, Galicia.

necessary first to review the schedule of maintenance tasks for the various pieces of equipment and then to check that they are being carried out according to schedule. In other words, you have to look for evidence in a different way.

The maintenance plan is a common cause of noncompliance or nonconformity in both inspections and audits. Optimal maintenance of plant and equipment requires constant attention and resources, and the operator does not always have sufficient time or financial resources to complete corrective maintenance tasks as quickly as would be desirable. For this reason, when assessing compliance with the corrective maintenance plan, it is necessary not only to note each deficiency but also to take into account its number and its impact on the working environment or production conditions and at the same time to check whether the operator has recorded it in his incident log and whether it is scheduled to be repaired within a foreseeable timeframe.

The preventive maintenance plan includes the planned maintenance activities, for the preparation of which it is necessary to carry out a prior study of the installations and all the equipment in the plant, to determine which are the ones that require preventive maintenance actions, and finally, to define what these actions consist of. The main elements of the plan are therefore as follows:

- List of equipment.
- Maintenance operations for each piece of equipment or element.
- Frequency.
- Person responsible.

Depending on the size and characteristics of each food business, the number of items of equipment requiring preventive maintenance may vary considerably. In some businesses, this list can be considerable and it is necessary for the FBO to prioritize preventive maintenance activities for equipment that is directly involved in the production process and for equipment where a failure could pose a higher risk to food safety.

There is a clear trend toward increasing automation in the food industry; processes that were previously carried out manually are more often performed by machines or robots; this reduces the need for manpower in certain processes, but brings with it new maintenance requirements and the need for suitably trained personnel to carry out the maintenance tasks on these machines or to outsource their maintenance.

In addition to the list of equipment, the preventive maintenance workload is also determined by the frequency of the activities to be carried out on each piece of equipment, so the program must be realistic and its planned activities determined by criteria such as:

- History of breakdowns or incidents
- Manufacturer's maintenance instructions
- Risk of failure
- Condition of the equipment.

One way to develop a risk-based preventive maintenance plan is by using failure mode and effects analysis (FMEA). FMEA is a tool specifically designed to systematically analyze the failure modes that can occur in equipment and the consequences that can result from those failures. In FMEA, the level of criticality or risk is calculated using matrices similar to those used in HACCP; combining probability of failure × severity of effect, to which the ease of detection of the failure is added.

Once the level of risk has been defined, priorities can be set and resources focused on those actions that address the most critical aspects of the equipment and have the greatest impact on the safety of the processed food.

Often, companies include the calibration and/or verification program of measuring equipment in the maintenance plan. A calibration and/or verification program should at least include:

- List of equipment to be calibrated or verified, identified by a code.
- Frequency of calibration or verification.
- Responsibility.
- Recording system.

Calibration consists of the comparison of the values obtained by a measuring instrument with the measurement corresponding to a reference standard and shall be carried out by external bodies issuing a calibration certificate in accordance with ISO/IEC 17025:2017.

Maintenance plan MP CS-08 Rev.1

Maintenance schedule log

Equipment: Freezer nº _____
Model: ___________________

Item for inspection and/or maintenance	Schedule date/Frequency	Performed date____	Responsible
Rubber doors			
Evaporator and compressor pressure cleaning			
Compressor inspection/ replacement			
Refrigerant pressure/flushing and draining			
High- and low-pressure check			
Oil level check/addition			
Fan inspection			
Thermometer review/ calibration			
Drain and evaporator check			

Incidents/breakdowns

Corrective actions/repairs

Figure 4.14 Maintenance schedule log.

Verification consists of comparing the values of calibrated equipment and the measuring equipment to be checked and may be carried out internally by the FBO. The result of the verification may be satisfactory, or it may require adjustments to the measuring equipment or repairs, or the equipment may have to be discarded. In any case, the company should determine the tolerance associated with its measuring equipment (Figure 4.14).

The monitoring tasks of the maintenance plan include:

- Periodic review of the condition of facilities and equipment to identify areas, rooms, or equipment in need of refurbishment or repair.
- Checking that corrective maintenance tasks are carried out within the timescales specified in the corrective action plans.
- Checking that preventive maintenance tasks are carried out in accordance with the established maintenance program.
- Checking that the frequency of the calibration and/or verification plan for measuring equipment is adhered to and that appropriate action is taken on the basis of the results obtained (Figure 4.15).

4.10.1 Documentation and Recordkeeping

Some of the records that the auditor may examine to verify the implementation of the maintenance plan for plant and equipment include

- Invoices or delivery notes from companies providing the external maintenance service.
- Invoices or delivery notes for supplies and spare parts.
- Calibration certificates for measuring instruments.
- Internal verification records for measuring equipment.
- Work reports.
- Records of incidents and corrective actions.

4.11 Good handling Practices

The objective of good handling practice is to prevent people who have direct or indirect contact with food from becoming a source of contamination for the food.

GHPs are a set of preventive measures or hygiene practices of a general and basic nature, but they are no less important; on the contrary, they are a fundamental pillar of the FSMS.

The hazards that can be controlled through the application of GHPs are all those in which the operator is involved, either as a direct source of contamination, as a potential carrier of germs, or through cross-contamination due to inappropriate behavior; these include microbiological hazards, physical hazards, through the use of personal items, and allergenic hazards, by moving from areas with a particular presence of allergens to others or by bringing food to work.

In the literature and in some GHP guides, GHPs are sometimes described together with GMPs, which is a perfectly valid approach; however, it may be equally useful to distinguish between non-specific hygiene practices common to different types of establishments, which are more related to the basic hygienic behavior of the operators, and specific operations linked to the different stages of the production process, which deal with the "know-how" of the operators.

PREVENTIVE MAINTENANCE PLAN: COLD STORE ENGINE ROOM (EXTRACT)

Equipment	Operation	Duration (h)	Frequency	Worker	Total hours	Responsible	Piece or material	Tools or equipment
Compressor	Oil filter replacement	3	Every 4 months	2	6 h every 4 months	Maintenance technician/assistant	Compressor oil filter	Open-end wrench/adjustable wrench
	Oil change	4	Every 4 months	2	8 h every four months	Maintenance technician/assistant	XXX ISO 68 Oil	Vacuum pump/open-end wrench/ adjustable wrench/
	General review	–	Every 25000 working hours	–	–	External company	Wearing parts found during review	–
	Rectify alignment	2	Annual	1	2 hours every year	Maintenance technician	–	Plate micrometer
Condenser	Cleaning of water pipes	3	Every 6 months	1	3 hours every 6 months	Maintenance technician	–	Caustic soda/anti-algae
Oil cooler	Cleaning of coolers	5	Every 3 months	2	10 hours every 3 months	Maintenance technician/assistant	Degreasing agent	Hydrojet/fixed spanner/ wire brush
Cooling tower	Coil inspection and cleaning	1	Monthly	2	2 hours every month	Maintenance technician/assistant	Degreasing agent	Hydrojet/screwdriver/pliers
	Suction filter maintenance	1	Monthly	2	2 hours every month	Maintenance technician/assistant	Water	Water source/spatula
	Cleaning the internal filling	4	Monthly	2	8 hours every month	Maintenance technician/assistant	Water	Hydrojet
Oil pump	Oil filter replacement	3	Every 3 months	2	6 hours every 3 months	Maintenance technician/assistant	Oil filter	Open-end wrench/ adjustable wrench

Figure 4.15 Maintenance plan of cold store room. Source: Adapted from Yessica M. Aguanes.

Since their description by the Codex, the content of GHPs has usually been divided into different aspects related to:

- **Health status**: This includes provisions related to persons who are known or suspected to be ill, or who may be carriers of pathogens, and who should not be allowed to work in areas where they may contaminate food.
- **Illness and injury**: If a food handler shows signs of the following illnesses or injuries, he should be examined to assess the possibility of temporary exclusion from food handling:
  - Jaundice.
  - Diarrhea.
  - Vomiting.
  - Fever.
  - Sore throat with fever.
  - Visibly infected skin lesions (boils, cuts, etc.).
  - Discharges from ears, eyes, or nose.
- **Personal hygiene**: This includes provisions for personal hygiene, the use of protective clothing for exclusive use (headgear, footwear, mask, instructions for use of gloves, etc.), the need to cover wounds and cuts with appropriate waterproof bandages, and instructions on how and when to wash hands.
- **Personal behavior**: This refers to actions that must be avoided by food handlers, such as smoking, spitting, chewing or eating, and sneezing or coughing on unprotected food. Personal items such as jewelry, watches, pins, or other items such as false nails/eyelashes should not be worn or brought into food handling areas if they pose a threat to the safety and suitability of the food.
- **Visitors to food businesses**: Visitors, including maintenance staff, particularly to food production, processing or handling areas, should be instructed and supervised as appropriate, wear protective clothing, and comply with other personal hygiene requirements for staff. Visitors should be briefed on the hygiene policy of the establishment prior to visits and encouraged to report any illness/injury that may raise cross-contamination issues.

As with other HACCP prerequisites, the GHP should specify the monitoring system and the person responsible for checking that handlers are following good handling practices. Although the size of the company must be taken into account, the persons designated to carry out the monitoring should be able to act without pressure and independently of the activities being monitored.

A responsible person with sufficient authority should also be appointed so that, in the event of noncompliance with GHPs, he can take appropriate corrective action to correct the poor practice and prevent its recurrence. As we have seen, training may be one of these actions, but not the only one, depending on the situation and work environment of each company.

4.11.1 Documentation and Recordkeeping

In addition to observing workers' handling practices, the auditor may ask for records related to the monitoring of GHPs, such as the following:

- GHP monitoring lists.
- Records of incidents and corrective actions.

4.12 Good Manufacturing Practices

The aim of GMPs is to determine how manufacturing activities should be carried out in order to reduce or eliminate hazards identified at different stages of the manufacturing process.

Unlike GHPs, GMPs do not refer to general aspects of food handling, but to the specific control measures to be applied in each of the operations that make up the manufacturing process of a food product. GMPs are related to the "know-how" of food handlers, not their basic hygienic behavior, and therefore, their content is quite different from that of GHPs.

Due to their specific nature, GMPs cannot be defined in advance; to develop them, it is first necessary to know the production process of each foodstuff, to draw up its flow diagram, to carry out the hazard analysis and, once these steps have been taken, to define the most appropriate control measures for the hazards identified at each stage. In this sense, GMPs can be equated to the role of operational prerequisites (OPRP) within the management system.

In any case, the FBO can choose the way in which the GMPs are developed and so they can sometimes be documented as a GMP program or as technical or work instructions within their management system.

GMPs should be described in sufficient detail so that operators know exactly how to carry out an operation so that the process is correct. By way of example, the type of information that may be described in GMPs includes aspects such as:

- Specific ways of carrying out an operation.
- Use of tools and equipment for each operation.
- Times for each stage.
- Working temperatures.
- Appropriate values of process control parameters.
- Instructions for handling equipment.
- Correct dosing of ingredients, additives, auxiliary raw materials, etc.

Depending on the type of foodstuff, the processes to which it is subjected, the degree of mechanization, etc., the content of the GMPs may vary considerably from one company to another, which means that the auditor's professional judgment becomes even more important in the absence of reference elements in order to assess whether the GMPs are sufficiently developed to achieve the objectives of the self-control system. To assess the content of the GMPs, the auditor may consider the following criteria:

- They are described for all stages or operations in the production process, or at least for those where significant hazards have been identified following a hazard analysis.
- The level of detail is sufficient to define the correct way of manufacturing.
- They are effective for the control of the hazards identified.
- There is evidence of validation by the food operator.
- Operators are aware of them and apply them as specified.

As with the other prerequisites, the FBO must define the monitoring system and the person responsible for verifying that the GMPs are correctly applied.

In the event of noncompliance with GMP or when the parameters of the manufacturing process are not as specified in the GMP, corrective action must be established. Products that have been manufactured under conditions where GMPs or OPRPs, when considered as such, have not been met should be assessed for the level of risk posed by that product and, if deemed necessary, managed as if they were nonconforming products (ISO 2018). The FBO must then decide whether such products can be released for consumption, whether they require reprocessing or further treatment, or whether they should be destroyed or disposed of as waste if the risk is unacceptable.

4.12.1 Documentation and Recordkeeping

As it depends on the characteristics of each production process, the records related to manufacturing may vary considerably from one company to another, but in any case, the records that the auditor may request to verify the application of GMP include:

- GMP checklists or GMP control lists.
- Production sheets.
- Records of process values.
- Incident and corrective action records.

4.13 Waste Control Plan

The purpose of this plan is to prevent waste generated in a food business from directly or indirectly contaminating food, contaminating the production environment or favoring the presence of pests on the premises.

The hazards that can be prevented by the waste control plan are mainly:

- **Biological hazards**: due to the presence of organic matter and waste that may favor the presence and growth of microorganisms and cross-contamination of food and equipment.
- **Physical hazards**: remnants of containers, packaging (staples, seals, adhesives, plastics, etc.), and other waste from the production process.

The FBO must identify the different types of waste generated in its business and establish guidelines for their disposal outside the areas where raw materials, ingredients, or processed foods are handled and stored. For some types of waste, it is necessary to take into account the requirements of specific legislation, such as the management of animal by-products or frying oils, which in both cases must be disposed of by authorized waste management companies.

In general, the waste covered by this plan can be classified as follows:

- **Municipal solid waste**: those that can be collected by the municipal cleaning services, such as packaging remnants, glass.
- **Animal by-products**: the most important from a health point of view, they must be managed in accordance with the health regulations in force in the different countries or regions on animal by-products not intended for human consumption.
- **Chemical waste**: disposal of chemical waste and their packaging to an authorized collector.
- **Other waste**: depending on the food sector, waste can be very diverse, to give a few examples: frying oils in the catering sector, stems and diatomaceous earth in wine production, vegetable residues and waste in the vegetable processing industry.

The waste control plan should describe aspects such as:

- Waste disposal routes.
- Identification of areas or containers for waste collection.
- Identification and, where required by law, sorting of waste.
- Use of appropriate containers: easy to clean and disinfect, with lids, etc.

- Frequency of disposal.
- Final destination or delivery to the authorized waste manager.
- Responsibility for disposal and collection.

At the EU level, Regulations 1069/2009 and 142/2011 require FBOs to classify animal by-products into three categories, which must be stored and disposed of separately:

- **Category 1**: specified risk materials (SRMs), by-products from animals treated illegally, by-products containing residues of certain environmental contaminants in excess of permitted levels, catering waste from international transport, and mixtures of category 1 material with other categories, among others.
- **Category 2**: By-products containing residues of authorized substances in excess of permitted levels, products of animal origin unfit for human consumption due to the presence of foreign bodies, mixtures of categories 2 and 3, by-products other than categories 1 and 3, etc.
- **Category 3**: carcasses and parts of slaughtered animals fit for human consumption but not intended for human consumption for commercial reasons, by-products derived from products intended for human consumption, aquatic animals and parts thereof showing no signs of disease, aquatic animal by-products from establishments preparing products for human consumption, mollusk shells and eggs and egg by-products, provided that they come from animals showing no signs of disease, and catering waste, except in the case of international transport.

The FBO must identify the by-products by means of legends or color codes and keep them separate until they are handed over to the approved by-products manager.

The monitoring system and the person responsible for checking that the handling, storage, and disposal of waste are carried out in accordance with the guidelines set out in the waste control plan must be established.

4.13.1 Documentation and Recordkeeping

Records generated by the implementation of this plan may include:

- Delivery notes for hazardous waste.
- Commercial documents resulting from the delivery of animal by-products.
- Records of monitoring the implementation of the plan.
- Records of incidents and corrective measures.

References

Europe, European Parliament and Council (2002). Regulation (EC) No 178/2002 of the European Parliament and of the Council of 28 January 2002 laying down the general principles and requirements of food law, establishing the European Food Safety Authority and laying down procedures in matters of food safety. *Official Journal of the European Union* 1–24.

Europe, European Parliament and Council (2004a). Regulation (EC) No 852/2004 of the European Parliament and of the Council of 29 April 2004 on the hygiene of foodstuffs. *Official Journal of the European Union* 139 (1): 1–54.

Europe, European Parliament and Council (2004b). Regulation (EC) No 853/2004 of the European Parliament and of the Council of 29 April laying down specific hygiene rules for food of animal origin. *Official Journal of the European Union* 55–205.

Europe, European Parliament and Council (2009). Regulation (EC) No 1069/2009 of the European Parliament and of the Council of 21 October 2009 laying down health rules as regards animal by-products and derived products not intended for human consumption and repealing Regulation (EC) No 1774/2002 (Animal by-products Regulation). *Official Journal of the European Union* 1–33.

European Commission. *Commission Regulation (EU) No 142/2011 of 25 February 2011 implementing Regulation (EC) No 1069/2009 of the European Parliament and of the Council laying down health rules as regards animal by-products and derived products not intended for human consumption and implementing Council Directive 97/78/EC as regards certain samples and items exempt from veterinary checks at the border under that Directive.* 2011.

European Commission (2022). Commission Notice on the implementation of food safety management systems covering Good Hygiene Practices and procedures based on the HACCP principles, including the facilitation/flexibility of the implementation in certain food businesses (2022/C 355/01). *Official Journal of the European Union* 1–58.

FAO and WHO (2003). *General Principles of Food Hygiene.* Codex Alimentarius Code of Practice, No. CAC 1-1969. Codex Alimentarius Commission. Rome.

FAO and WHO (2022). *General Principles of Food Hygiene.* Codex Alimentarius Code of Practice, No. CXC 1-1969. Codex Alimentarius Commission. Rome.

Food Standards Agency (2020). *Safer food, better business for caterers.* www.food.gov.uk.

International Standard for Organization (2017). *ISO/IEC 17025:2017: General Requirements for the Competence of Testing and Calibration Laboratories.*

International Standard for Organization (2018). *ISO 22000:2018: Food Safety Management Systems – Requirements for any Organization in the Food Chain.*

National Advisory Committee on Microbiological Criteria for Food (NACMCF) (1997). *Hazards analysis and critical control point principles and application guidelines.* https://www.fda.gov/food/hazard-analysis-critical-control-point-haccp/haccp-principles-application-guidelines/ (accessed 25 November).

Mehannaoui, R., Mouss, K.N., and Aksa, K. (2023). *IoT-based food traceability system: architecture, technologies, applications, and future trends. Food Control* 145: 109409.

Yiannas, F. (2009). *Food Safety Culture. Creating a Behavior – Based Food Safety Management System.* New York: Springer.

5

HACCP as a Management System

5.1 From Words to Work

Once the hygiene requirements have been developed and the 12 steps that make up the hazard analysis and critical control point (HACCP) study have been completed, we have a document: the HACCP plan. But does that mean we are done? The answer is no. On the contrary, the most important task now begins: implementation. This means that the people in the organization have to put the rules and procedures of the system into practice in the day-to-day running of the business. To use an analogy, when the team completes the HACCP study, it is as if we have the plans of what our "food safety house" will look like, but the building has yet to be constructed.

In the overall process of implementing the HACCP system to which we referred earlier, we are at a point where the first two phases have been completed: preparing and carrying out the HACCP study, but the next two are still to be completed: implementing and maintaining the system. It is also important to remember that the HACCP plan is a living document that needs to be revised and updated from time to time. Sometimes, when the HACCP system is implemented, it is found that the procedures need to be adapted to the real situation of the production process, which may be different from what was expected when the study was carried out. On the other hand, one of the requirements of the system established by principle 6 is precisely to modify the elements of the system whenever there is a change in the processes, products, or organization, in order to keep it up to date and ensure that it faithfully reflects the reality of the company.

The Codex Alimentarius Guidelines for the implementation of HACCP contain the necessary information for any food operator to carry out the HACCP study in a satisfactory way. However, they provide little guidance on the next steps to be taken to implement and maintain HACCP in practice.

Years ago, when HACCP began to be used in the food industry, it soon became apparent that the food operator had difficulties in putting into practice what was on paper in his HACCP manual, and this often led to poor implementation and results that were not as expected. In analyzing the causes of this failure in implementation, which is quite common in food businesses, professionals and experts in the field have come to the conclusion that other elements are needed that are not included in the HACCP principles and guidelines but which are really necessary for the implementation and management of the HACCP system to be put into practice in an effective way. So, what are the elements we are talking about? What is required to implement HACCP?

HACCP System Auditing for Food Safety: Principles and Techniques, First Edition. Luis Couto Lorenzo.

5.2 What is Required to Implement HACCP?

To answer this question in a logical order, it is not difficult to see that the first thing that is needed is what we have already mentioned above: the commitment of the organization to implement the system. It is therefore important to remember that without a real commitment from the organization, which must start at the highest level of the company, the implementation of HACCP will be very difficult. If the commitment of the organization is real, it will be easy to understand that the next step is to provide adequate human resources, in terms of both the number of staff and their training and competence to carry out the tasks assigned to them, without overlooking the issues related to the creation of a food safety culture, where leadership, communication, motivation, and involvement of the employees play a fundamental role. HACCP is first developed by a team of people and then the rest of the organization has to apply it, so human resources management is fundamental to the success of the HACCP system.

In addition to human resources, the organization must have an adequate infrastructure, which includes the necessary facilities, equipment, and services to perform the processes correctly.

Before proceeding with the rest of this section, it is useful to recall the definitions of "process" and "procedure" from ISO 9000: 2015 Quality management systems – Fundamentals and Vocabulary, in order to better understand how they are integrated into HACCP:

Process: set of interrelated or interacting activities that use inputs to deliver an intended result.
Note: The result of a process may be called an output, a product, or a process, depending on the context in which it is used.

Procedure: specified way to carry out an activity or a process.
Note: Procedures may or may not be documented.

The HACCP team should identify all control and monitoring activities resulting from the completion of the HACCP study. These activities should be defined by processes, which describe the sequence of operations necessary to achieve an intended result. In order for these processes to be controllable and reproducible, they must always be carried out in a specified manner, which is achieved by developing procedures that specify what, how, who, when, and where a particular task or activity is to be carried out. In many cases, it is also desirable for the procedures to be documented to facilitate their consistent execution by all operators. This is the essence of a management system (Figure 5.1).

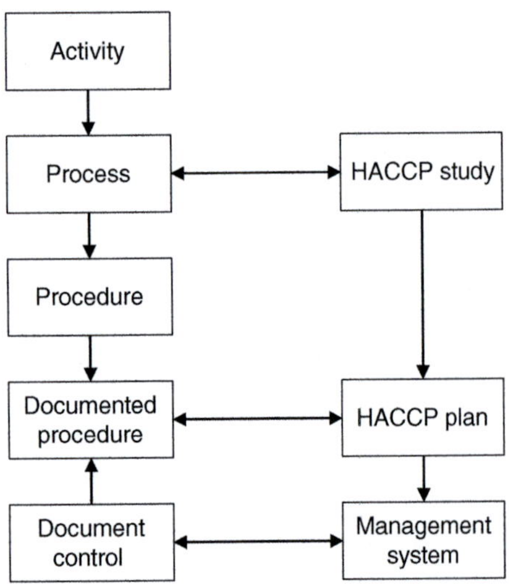

Figure 5.1 Activity, process, procedure, documented procedure of a HACCP study, HACCP plan and HACCP system.

5.3 Building a Food Safety Culture

In 2015, the conviction of the Global Food Safety Initiative (GFSI) that food safety must go beyond formal regulations and live within a company's culture led it to launch a technical working group with the idea of providing food companies with guidance on how to promote a food safety culture within their organizations. According to the GFSI, "Unlike the rule of law, culture draws its power from the unspoken and the intuitive, from simple observation and from beliefs as basic as 'this is the right thing to do' and 'we would never do this'. Rules state facts; culture lives through human experience" (GFSI 2018).

Later, in 2020, the revision of the General Principles of Food Hygiene (CXC 1-1969) introduced the implementation and maintenance of a food safety culture as a fundamental element of a well-functioning food safety management system (FSMS). Since then, several food safety standards and certification schemes have included food safety culture as a requirement for food business operators (FBOs). Likewise, in 2021, European Regulation 2021/382 introduced a specific chapter on food safety culture, identifying the components of food safety culture. Therefore, at the EU level, FBOs carrying out activities other than primary production and related operations must comply with this requirement.

Food safety culture can be defined as the shared values, beliefs, and norms that influence food safety attitudes and behaviors within, across, and throughout an organization. With this concept in mind, we need to identify the elements that support the food safety culture in a food company:

- **commitment** to safe food production and distribution by management and all employees: employee commitment is the perception of the level of engagement and involvement in food safety of all employees in the FBO;
- **leadership** to set the right direction and involvement of all personnel in food safety practices: leadership can be defined as the perception of the extent to which the FBO's leader(s) are able to motivate staff to achieve food safety performance and compliance to meet food safety requirements and to ensure appropriate responses to risks, deviations and changing risks, and deviations and changing circumstances;
- **awareness** of the importance of food hygiene by all food business personnel: awareness is the perception of the extent to which all staff in an FBO are aware of and control the food safety risks relevant to their tasks and have these under control;
- open and clear **communication** between all food business personnel, within an activity and between successive activities, within a production site or between different sites of an FBO, including communication of deviations and expectations; and
- the availability of **sufficient resources** to ensure the effective operation of the food hygiene system: sufficient resources is defined as the perception of the extent to which the material and nonmaterial means necessary to operate in a food-safe manner are available in the FBO (e.g. time, personnel, infrastructure, education/training, and procedures).

Although most of the components of food safety culture are subjective as they are based on the perception of the above elements, some tools have been developed for the objective measurement of the food safety culture of an FBO. An example of the tools available to assess and analyze food safety culture are surveys in which employees are asked to express their agreement or disagreement with certain indicators or statements on a given scale, e.g. 1–5, for each of the culture components. This tool can also be used as a basis for verifying the food safety culture of an FBO during an audit (European Commission 2022).

5.4 Management System Concept

We come to the concept of a management system from all the elements we have mentioned. A management system is defined as a set of interrelated or interacting elements of an organization to establish policies and objectives, and processes to achieve those objectives. These elements in turn determine:

- The structure of the organization.
- Roles and responsibilities.
- Planning.
- The documentation of the system: procedures, work instructions, plans, organization charts, specifications, diagrams, records, etc.
- Human and material resources.
- Products and/or services and their characteristics.
- Customers and their requirements.
- Operations.
- Rules of the organization
- Practices to achieve established objectives.

A management system may deal with a single discipline, such as an FSMS, or it may cover several disciplines, such as a quality and environmental management system.

The elements that make up a management system can be identified in all organizations. Even if they have not been previously defined, or if there is little awareness of their existence, in any organization, however small, most of these elements could be recognized or established to guide the operation of the organization toward the achievement of specific objectives (Figure 5.2).

The main reference for these management elements, which are considered necessary for the implementation and maintenance of HACCP, are the standards of the ISO 9000 series. Since the original formulation of HACCP by the Codex, it has been recognized not only that the two systems are compatible but also that HACCP works best as part of a structured management system and integrated into the overall management activities of the organization.

When the elements of ISO 9000 are incorporated into HACCP management, the system can benefit in areas such as

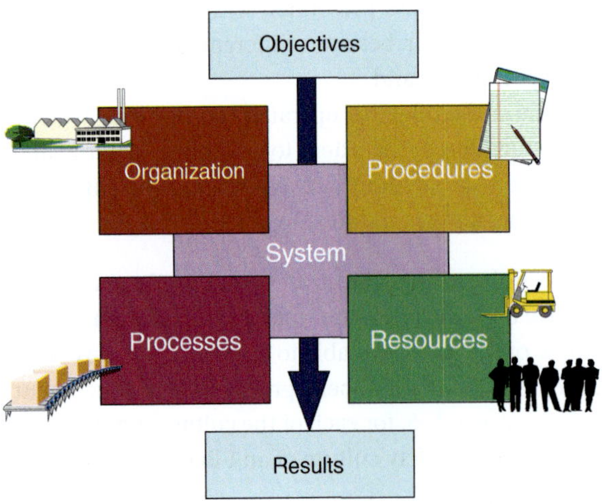

Figure 5.2 Management system elements.

- Use of calibrated equipment.
- Management of nonconforming products.
- Control of documentation.
- Control of records.
- Roles, responsibilities, and authority.
- Staff competence and awareness.
- Communication internally, with customers and other interested parties.
- Monitoring, measuring, analyzing, and evaluating.
- Internal audit.
- Management review.
- Continuous improvement.

5.5 The Concept of a Food Safety Management System

An FSMS can be defined as a set of elements that interact to manage food safety in a food business. The generally recognized key elements that need to be combined to ensure food safety are as follows:

- HACCP plan.
- Prerequisite programs (PRPs).
- Management elements of the system.

Some other components of FSMSs can be found in the literature and in the publications of various reference bodies, such as interactive communication, traceability, food recovery system, and compliance with legal requirements. In any case, the three main parts mentioned above are present in all food safety standards, and what varies is the relative weight of each of them in the overall system (Figure 5.3).

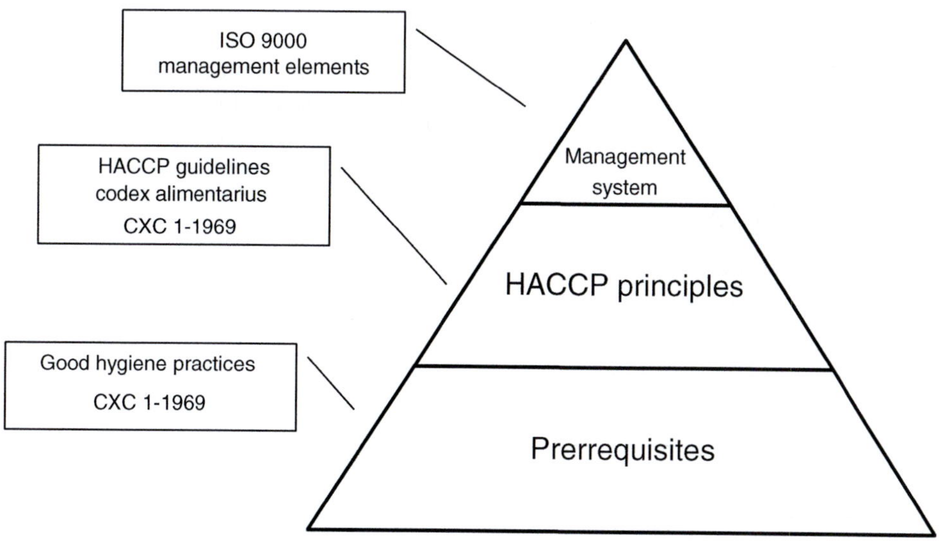

Figure 5.3 Components of a food safety management system.

5.6 System Documentation

An essential element of management systems is the documentation that supports the whole system. Each organization shall determine what documentation is required to ensure the proper development and implementation of the system. The resulting documents include:

- Statement of policy and objectives.
- Manual.
- Procedures.
- Work instructions.
- Records.
- Forms.
- Drawings.
- Specifications.
- Organization charts.
- Process maps, process flowcharts, and/or process descriptions.
- Documents containing internal communications.
- Production schedules.
- Quality plans.
- Strategic plans.

5.6.1 Document Hierarchy

The documentation system is usually organized in a pyramid structure, with the highest level and most general documents at the top of the pyramid and more specific and detailed documents defining specific aspects of the management system as one moves down the pyramid. Each organization develops its own documentation system to suit its own characteristics and needs, but the hierarchy of documentation can be seen in all of them (Figure 5.4).

5.6.2 Document Control

Document control is a characteristic requirement of all management systems. In the case of HACCP, its application overlaps with principle 7: establishing a system of documentation and

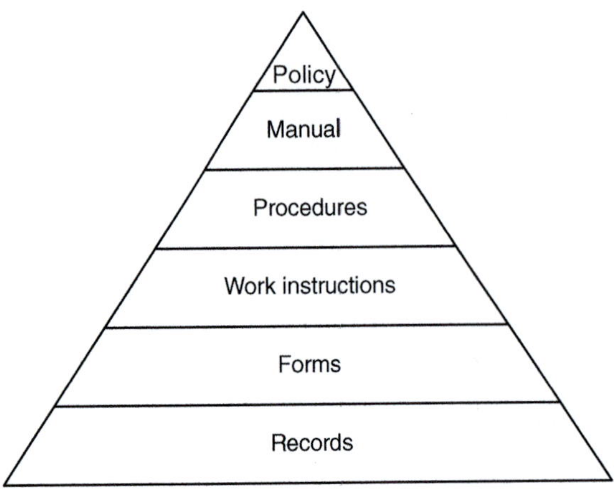

Figure 5.4 Document hierarchy of a management system.

Company logo	**Corrective and preventive action management procedure**	**P-CA-08** Page 1 of 8 **Edition 2 May 2023**
Table of content	1 Purpose	
	2 Scope	
	3 References	
	4 Definitions	
	4.1 Corrective action	
	4.2 Preventive action	
	5 Requirements	
	5.1 Responsibility	
	6 Process	
	6.1 Review of nonconformances	
	6.2 Determine causes	
	6.3 Evaluate need for action	
	6.4 Implement action	
	6.5 Effectiveness verification	
	6.6 Closure	
	7 Annexes	
	8 Records	
Revision history		
Last revision	Modification	
Edition 2	Point 7: Annexes Point 8: Records	
Reviewed by Quality manager Signed Date	Approved by the Manager Director Signed Date	

Figure 5.5 Example of corrective and preventive action management procedure.

recordkeeping. The synergy between ISO and HACCP is evident at this point; document control serves to improve the effectiveness of the documented system by making it more robust and reliable.

Document control consists of a set of provisions that require the operator to:

- Approve documents before they are distributed to the organization's personnel.
- Review, update, and reapprove documents as necessary.
- Ensure that only the latest and most up-to-date versions of documents are used.
- Ensure that changes to documents are identified and that the current status is known.
- Ensure that relevant versions are available where they are used.
- Prevent the unintended use of out-of-date documents.
- Identify documents of external origin.
- Ensure that documents remain legible and easily identifiable (Figure 5.5).

5.6.3 Control of Records

This is another typical management system requirement. Records are considered a special type of document, and the operator shall ensure that they remain legible, easily identifiable, and retrievable. The necessary controls shall be defined for their identification, storage, protection, retrieval, retention period, and disposal.

5.7 Food Safety Standards

Shortly after the HACCP system became widely accepted and mandatory in some countries, the first food safety standards began to emerge. Probably motivated by the need for independent assessment of self-control systems and the possibility of obtaining certification of the system, similar to quality management systems based on ISO 9000, which at that time were in full expansion, the first initiatives to develop food safety standards or certification schemes were taken in some countries. For example, Safe Quality Food (SQF) was introduced in Australia in 1994, the first version of the Dutch standard "Requirements for a HACCP-based Food Safety System" appeared in 1996, and the Danish standard 3027 followed a little later in 1997.

We will look briefly at the characteristics of some of the most commonly used standards in the following sections.

5.7.1 Safe Quality Food (SQF)

SQF was originally developed by the Western Australian Department of Agriculture in 1994 in response to demand from farmers and small food producers for a quality assurance system that would enable them to meet food safety and quality requirements for their food. In 2003, SQF was purchased by the Food Marketing Institute (FMI) in Washington and is now administered by the FMI–The Food Industry Association.

SQF is a food safety standard recognized by the Global Food Safety Initiative (GFSI). The GFSI has approved SQF as one of the benchmarked schemes that meet its requirements for FSMSs. SQF provides food safety and quality codes for all aspects of the food supply chain, from primary production to retail and food service.

SQF's Food Safety Programme is a food safety and quality program designed for companies wishing to meet the GFSI requirements of their retailers and buyers. The SQF Food Safety Programme comprises 13 codes designed to meet industry, customer, and regulatory requirements from all sectors of the food supply chain. These codes provide sites with a HACCP-based approach to ensuring their products meet food safety requirements.

The Food Safety Code is currently in its ninth edition and has gained support over time, and today, companies certified in one of its various schemes and codes can be found in more than 40 countries.

5.7.2 Dutch HACCP Code

The Dutch HACCP (Requirements for a HACCP-based Food Safety System) was one of the first internationally recognized food safety certification schemes. Since 2004, it has been managed by the Foundation for Safety, Quality and Sustainability (SCV) and has come to an end on January 1, 2021.

The Dutch HACCP Code covered all operators in the food chain, but focused primarily on the stages of preparation, processing, manufacturing, packaging, storage, distribution, and sale of food, excluding suppliers of services to the food industry, such as packaging materials, equipment, and cleaning services. This standard combined the HACCP principles of the Codex Alimentarius with the elements of ISO 9000 management systems and required the existence of hygiene programs in accordance with the Codex General Principles of Food Hygiene to be in place and operational before the implementation of the management system.

5.7.3 The Danish Standard DS 3027

The first version of this standard was published in 1997: DS 3027:1997 "Food safety according to HACCP – Requirements for Food Processors and their Subcontractors." Like the Dutch standard, it integrated the principles of HACCP into a quality management system. DS 3027 was organized as a set of requirements that supplemented the principles of HACCP with the elements of a management system in order to obtain certification of the system. The first version followed the chapter structure of ISO 9000:1994, which was in force when the Danish standard was developed.

In 2002, DS 3027 was revised and replaced by DS 3027:2002, which in turn was replaced by DS/EN ISO 22000:2005, the current version of which is dated 2018. The extinct Dutch HACCP Code and the withdrawn DS 3027 are considered to be the precursors of ISO 22000:2005.

5.7.4 Brand Reputation Compliance Global Standards (BRCGS)

Brand Reputation Compliance Global Standards (BRCGS) is one of the world's most widely used standards, used in more than 130 countries. BRCGS started in 1996 when the British Retail Consortium (BRC), in response to industry needs, developed the BRC Food Technical Standard, first published in 1998.

The Global Standard for Food Safety was used to assess manufacturers of retailers' own brand food products. And it was developed to help retailers and brand owners produce food safely and consistently, which would help with a "due diligence" defense in the event that they were the subject of a prosecution by the enforcement authorities. Now in its ninth edition, BRCGS was the first standard to be benchmarked by the GFSI and to introduce food safety culture requirements.

The success and widespread use of the Global Standard for Food Safety led BRCGS to publish other global standards, such as the first edition of the Packaging Standard in 2002, followed by the Consumer Products Standard in 2003, the Storage and Distribution Standard in 2006, the Agents and Brokers Standard in 2014, the Retail Standard in 2016, and the Ethical Trade and Responsible Sourcing Standard in 2019.

The requirements of the Global Standard for Food Safety are structured in the following way:

- Senior management commitment
- The food safety plan – HACCP
- Food safety and quality management system.
- Site standards.
- Product control.
- Process control.
- Personnel.
- Production risk zones – high risk, high care, and ambient high care.
- Requirement for traded products.

The audit protocol included in this standard provides three audit options:

- Announced audit (with mandatory unannounced audit every three years).
- Blended announced audit – two-part unannounced audit.
- Unannounced audit.

In this way, food manufacturers and suppliers can achieve a higher audit grade in an unannounced BRCGS food safety audit. An independent, unannounced audit gives customers added confidence in a company's ability to consistently meet quality standards.

5.7.5 International Featured Standards (IFS)

The first version of the IFS Food Standard, a standard for quality and food safety audits of food products, was drafted in 2002 and published in 2003 by the Hauptverband des Deutschen Einzelhandels (HDE) with the aim of having a common auditing standard for food suppliers. In 2003, the French Fédération des Enterprises du Commerce et de la Distribution (FCD) joined the project, and together they published the next edition of the standard. In the following years, the Italian distribution associations Associazione Nazionale delle Cooperative di Consumatori (ANCC), Associazione Nazionale Cooperative fra Dettaglianti (ANCD), and Federdistribuzione also joined the IFS. The IFS standard is currently at version 8 as of April 2023.

The IFS is a HACCP-based food safety and quality management protocol that has been developed for producers of all types of food products. The IFS standard has been specifically designed to meet the needs of distributors and retailers, focusing on the requirements for suppliers of private label products to ensure food safety at all stages of the supply chain under the responsibility of the distributor. Like the BRCGS, the IFS does not apply to primary producers, but does apply when the product is handled or packaged.

The content of the IFS Food Standard is structured as follows:

Part 1 – IFS Food Certification Protocol
Part 2 – IFS Food Audit Checklist (list of requirements for the IFS Food Audit)
Part 3 – Requirements for accreditation bodies, certification bodies, and auditors
Part 4 – Reporting, IFS software, and IFS database

The IFS Food Audit requirements specified in Part 2 are laid down as follows:

1. Governance and commitment.
2. Food safety and quality management system.

3. Resource management.
4. Operational processes.
5. Measurements, analyses, and improvements

An IFS certification is a product and process certification for which the auditor questions the audited companies on the audit checklist specified in part 2 to determine the level of compliance of processes and products.

Unlike the BRCGS, the IFS protocol uses a scoring system to assess compliance with the requirements of the system. There are six possible points and the option of not applying. Points are assigned for each requirement according to the following table:

Result	Explanation	Points
A	Full compliance	30 points
B (deviation)	Almost full compliance	15 points
C (deviation)	Part of the requirement is not implemented	5 points
D (deviation)	The requirement is not implemented	−20 points
Major (nonconformity)	A Major nonconformity can be issued to any regular requirement (which is not defined as a KO requirement) Reasons for Major rating are: • There is a substantial failure to meet the requirements of the standard, which includes but is not limited to food safety and/or the legal requirements of the production and/or destination countries • A process is out of control which might have an impact on food safety	Major nonconformity will subtract 15% of the possible total amount; the certificate cannot be issued
KO requirement scored with a D (nonconformity)	The requirement is not implemented	KO nonconformity will subtract 50% of the possible total amount; the certificate cannot be issued.
N/A Not applicable	The requirement is not applicable. N/A can apply to any requirement, except for most of KO requirements. The auditor shall provide an explanation in the report.	Not included in the calculation of the total score

The IFS Food Standard also includes options for announced and unannounced audits, as decided by the certification body, but ensuring that at least once every third IFS Food Audit is performed unannounced.

In addition to the IFS Food standard, the IFS has developed other standards related to food, products, and services, such as:

- IFS Logistics.
- IFS Broker.
- IFS Wholesale/Cash and Carry.

- IFS HPC (household and personal care products).
- IFS PACsecure (packaging).

The IFS is also a GFSI-recognized standard and one of the most popular certification standards, particularly in Europe, with a total of 47% of food accreditations being the IFS. Meanwhile, the United Kingdom and the Netherlands are most certified in the BRCGS, which represents 39% of the total and is the main leading trade association for UK retailers.

5.7.6 ISO 22000

ISO 22000 Food safety management systems – Requirements for any organization in the food chain was published at the end of 2005 with the aim of establishing the requirements that a FSMS must meet in order for the organization to ensure that the food produced is safe and to demonstrate to third parties its ability to control food safety hazards.

When ISO 22000 was published, there were already a large number of standards in the agri-food sector, such as EurepGAP, BRC, IFS, Dutch HACCP Code, SQF, each of which applies to different areas or sectors of the food chain. This means that it can be too complicated and time-consuming for any FBO to comply with the different requirements of each of these standards and to undergo as many audits as the standards of each customer's choice. In this context, ISO 22000 was created with the intention of becoming a universal standard, applicable to any organization, regardless of its size, valid for all sectors, and due to its vertical nature, extendable to all links in the food chain.

ISO 22000 now comprises a family of standards consisting of the following standards:

- -ISO 22000: 2018 Food Safety Management Systems – Requirements for any organization in the food chain.
- -ISO TS 22002-1: 2009 Prerequisite Programmes on food safety – Part 1: Food manufacturing.
- -ISO TS 22002-2: 2013 Prerequisite Programmes on food safety – Part 2: Catering.
- -ISO TS 22002-3: 2011 Prerequisite Programmes on food safety – Part 3: Farming.
- -ISO TS 22002-4: 2013 Prerequisite Programmes on food safety – Part 4: Food packaging manufacturing.
- ISO TS 22002-5: 2019 Prerequisite Programmes on food safety – Part 5: Transport and storage.
- -ISO TS 22002-6: 2016 Prerequisite Programmes on food safety – Part 6: Feed and animal production.
- ISO 22003-1: 2022 Food Safety – Part 1: Requirements for bodies providing audit and certification of food safety management systems.
- -ISO 22003-2: 2022 Food Safety – Part 2: Requirements for bodies providing evaluation and certification of products, processes, and services, including an audit of the food safety system.
- ISO 22005: 2007 Traceability in the Food and Feed Chain – General principles and basic requirements for system design and implementation.

ISO 22000:2018 uses the Annex SL format, which is a high-level structure that is common to all of the ISO management standards and is therefore divided into 10 parts as follows:

1. Scope
2. Normative references
3. Terms and definitions
4. Context of the organization
5. Leadership
6. Planning

7. Support
8. Operation
9. Performance evaluation
10. Improvement.

As indicated in the introduction to this standard, the elements that ISO 22000 considers key to the implementation of an FSMS are:

- Interactive communication.
- System management.
- -PRPs.
- HACCP principles

ISO 22000 is far from being the universal standard that was envisaged when it was created. However, due to its structure and comprehensive description of management system requirements, it can be considered as the main reference model of an FSMS for all actors involved in food safety: operators, consultants, official inspectors, auditors, etc.

5.7.6.1 FSSC 22000

Food Safety System Certification (FSSC) 22000 is a comprehensive, GFSI-recognized certification scheme for FSMSs, based on ISO 22000 and technical specifications for sector-specific PRPs, with additional requirements to ensure consistency and integrity, governance, and agility in addressing emerging food safety issues.

FSSC is currently in its sixth edition, published in April 2023, and consists of three components:

- ISO 22000.
- PRP specifications (ISO/TS standards and BSI PAS).
- Additional requirements.

The scheme consists of five parts, two appendices, and five annexes, the five parts being the following:

Part 1 Overview
Part 2 Requirements for organizations to be audited
Part 3 Requirements for the certification process
Part 4 Requirements for the certification bodies
Part 5 Requirements for the accreditation bodies

There are currently more than 33500 FSSC-22000-certified food companies worldwide (Figure 5.6).

5.7.7 Global Food Safety Initiative

The GFSI was founded in 2000 by the heads and representatives of the world's leading food companies at the Consumer Goods Forum, promoted by the Comité International d'Enterprises à Succursales (CIES), with the aim of finding solutions to common food safety problems, avoiding duplication of audits and associated costs – due to the existence of numerous certification standards – and improving trust in relations between operators in the food chain.

Under the motto "Once certified, recognized everywhere," the main objective of the GFSI is to avoid recognized audits by establishing common procedures and criteria for accreditation and certification bodies, so that when a company is certified to a GFSI-recognized standard, it is accepted

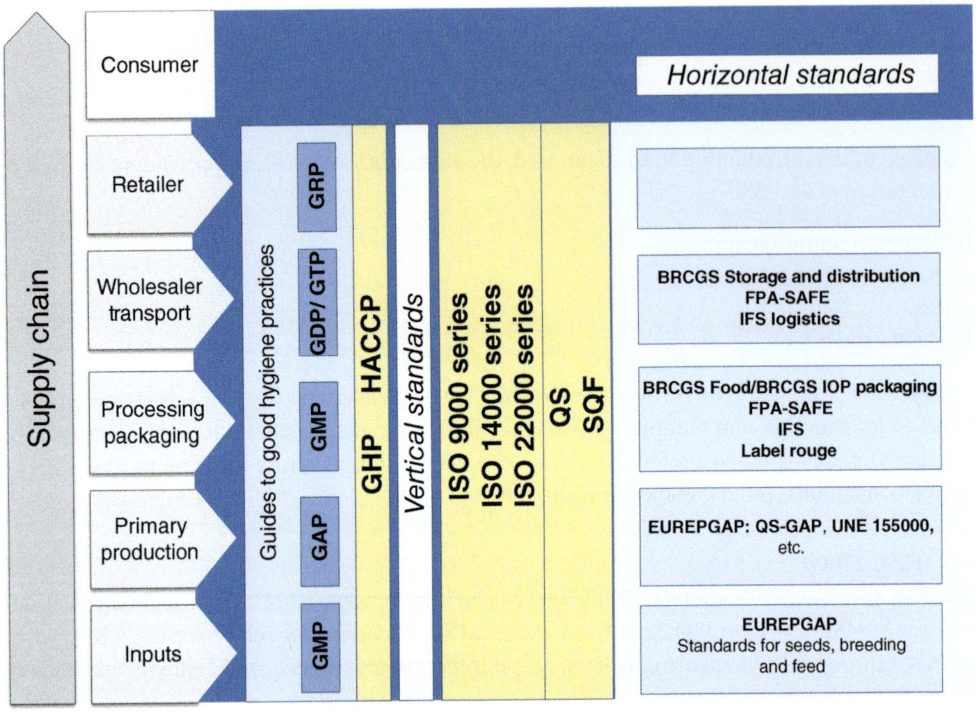

Figure 5.6 Horizontal and vertical standards in the food supply chain. Abbreviations: GAP: good agricultural practices; GMP: good manufacturing practices; GDP: good distribution practices; GTP: good transportation practices; GRP: good retail practices; FPA-SAFE: Food Products Association Safe Program; SQF: Safe Quality Food; QS-GAP: Qualität und Sicherheit GmbH – good agricultural practices. Source: Adapted from Will and Guenther (2007).

by all members of the GFSI community. This avoids suppliers facing multiple audits with different certification criteria for different customers.

In 2001, the GFSI published its first draft of the Benchmarking Requirements, formerly known as the GFSI Guidance Document. This model is not a food safety standard, and companies cannot be audited or certified against it, but is intended to provide guidance, set out how the assessment process should be conducted, and specify the requirements to be met by owners of food safety certification schemes wishing to be recognized by the GFSI. The certification schemes currently recognized by the GFSI are:

- BRCGS
- CanadaGAP
- Freshcare
- FSSC 22000
- GRMA: Global Retailer and Manufacturer Alliance
- GRMS: Global Red Meat Standard
- Global Seafood Alliance
- GlobalG.A.P.
- IFS: International Featured Standards
- JFSM: Japan Food Safety Management Association
- ASIAGAP: Japan GAP Foundation

- PrimusGFS Standard
- SQF.

It is estimated that there may be around 150.000 GFSI-recognized food safety standard certificates around the world, illustrating the extent of this initiative.

References

European Commission (2021). *Commission Regulation (EU) 2021/382 of 3 March 2021 amending the Annexes to Regulation (EC) No 852/2004 of the European Parliament and of the Council on the hygiene of foodstuffs as regards food allergen management, redistribution of food and food safety culture.*

European Commission (2022). *Commission Notice on the implementation of food safety management systems covering Good Hygiene Practices and procedures based on the HACCP principles, including the facilitation/flexibility of the implementation in certain food businesses* (2022/C 355/01).

International Featured Standards (2023). *IFS food version 8*. Berlin.

International Organization for Standardization (ISO) (2015a). *ISO 9000:2015: Quality management systems – fundamentals and vocabulary.*

International Organization for Standardization (ISO) (2018). *ISO 22000:2018: Food safety management systems – requirements for any organization in the food chain.*

Will, M & Guenther, D (2007). Food quality and safety standards as required by EU law and the private industry. 2. Communities.

6

Audit: Concept, Types, and Personnel

6.1 Audit Concept

In colloquial language, thinking of the average citizen, the term "audit" is usually associated with the review of companies' financial statements and, in general, with an activity related to the financial world. However, since the last decades of the last century, audits were extended to other aspects of organizations than accountants, such as the effectiveness and efficiency of processes; thus, from 1970, they began to generalize audits of quality management systems and, somewhat later, environmental management systems, and already at the end of the 1990s, its usefulness in the field of food safety has started to be recognized.

Commonly used dictionaries of different languages define audit as "official examination" or "careful examination"; besides, we can consider it as a synonym for "supervise" or "assess" or even link it to the term "inspection." Etymologically, it comes from the Latin word *audire*: to hear, and thus "auditor" would be one who listens, something that seems to be related to the existence in the antiquity of a hearing procedure during the review of the accounts.

An auditor's job is to collect data and analyze information obtained to check whether the system meets requirements or even for another agreed-upon purpose.

Auditing is a systematic process that consists of several activities carried out in a particular order; first, the planning or preparation; then the on-site audit, after which auditors issue a report with the results; and finally, the audited organization responds with an action plan that is subject to follow-up and closure of the corrective actions (Russell 2013).

In addition, auditing is considered a profession, so auditors must abide by a code of ethics and behave professionally in the exercise of their activity.

An audit is a versatile tool, adaptable to different areas and purposes; it can be applied to assess conformity with the requirements established for products, processes, or systems, whether quality management or food safety, among others; so, what varies in each case are the subject of the audit and the requirements against which it is audited, but the system used is very similar and follows the same principles.

For people who work with management systems and for professionals involved in the field of audit, it is necessary to be familiar with some of the definitions of audit established in the quality standards or other applicable regulations, such as the case of Regulation (EC) 2017/625 on official controls of European Parliament and of the Council:

Audit: a systematic and independent examination to determine whether activities and the related results of such activities comply with planned arrangements and whether these arrangements are applied effectively and are suitable to achieve the objectives.

HACCP System Auditing for Food Safety: Principles and Techniques, First Edition. Luis Couto Lorenzo.

This definition is the most direct reference for the agents involved in the control feed and food officer, food industry operators, and other stakeholders such as consultancies, laboratories. It comes from the one in ISO 8402:1994: "Quality management and assurance, vocabulary"; its advantage over others that we can find in the literature is that it allows us to know intuitively what the audit consists of and its purposes.

We can find a somewhat different definition included in the most recent editions of ISO 9000:2015, Quality Management Systems – Fundamentals and Vocabulary, and ISO 19011:2011 Guidelines for the audit of management systems. This more recent definition of audit, contained in these two standards, is not as intuitive as the one we have just seen because it uses two terms: "objective evidence" and "audit criteria", the definitions of which must be known in advance in order to fully understand its meaning, as we will see later.

In any case, based on this definition of Regulation 2017/625, we can explain some of the elements that are essential to describe audits:

- **Systematic**: systematic examination means that the auditors must conduct the process in a structured, planned way, following a scheme or order of activities according to a series of phases.
- **Independent**: another keyword that refers to an established audit principle: auditors' independence from the audited organization and its activities, which is one of the requisites we must meet as much as possible, avoiding or limiting the effect of various factors that can undermine it.
- Check if the audited organization meets the planned arrangements; verify whether the organization applies what is established in the documented system. The objective is to check how the system operates in practice and whether it complies with the requirements and arrangements of its documented system.
- Determine if they are adequate to achieve the objectives: it is essential to highlight this aspect within the definition because it reminds us that management systems aim to achieve specific objectives and results. So audits, if helpful, must determine whether the documented system complies with the requirements and whether the system works effectively.

In addition to the aspects explicitly mentioned in the above definition, we can add two others:

- **Documented**: in the course of the audit, different types of documents such as audit plan, checklist, audit report, and plan of corrective actions, among others, serve as a means of communication and a record of the entire process. In this way, the information about what happened during the audit can be retrieved and the trail of its different phases can be followed to monitor or inform other persons or stakeholders who did not intervene directly in the audit.
- **By sampling**: due to time constraints or because it is not strictly necessary, auditors do not review all the elements or items of the system but select certain parts or items. The extent and manner of the sampling depend on what the auditor deems sufficient in each case to make a judgment or an objective assessment of the audited element.

This distinctive aspect of audits gives rise to some degree of uncertainty that is recognized and accepted as usual. Indeed, auditors can mention it during the opening meeting or include it in the final report stating that the judgments made relate exclusively to the audited elements and that we should not extend them to items or parts we did not audit.

Anyone intending to work in the field of audits cannot ignore the following definition, included in both ISO 9000: Fundamentals and Vocabulary and ISO 19011:2018 and that contains a more technical description than the previous one because it includes two terms: objective evidence and audit criteria, defined below, which are also essential to understand the function of the auditor.

Audit: systematic, independent, and documented process for obtaining objective evidence and evaluating it objectively to determine the extent to which the audit criteria are fulfilled.

Audit criteria: set of requirements used as a reference against which objective evidence is compared.

Objective evidence: data supporting the existence or verity of something.

6.2 Why Audit?

Once we know the concept of auditing and some of its distinctive aspects, we can ask ourselves: Why should we audit? What for? What are the benefits expected from audits? As we will see below, several reasons justify conducting audits, which may have more or less importance depending on the person or party promoting the audit in each case.

- The first reason may be intrinsic or inherent in management; once the system has been developed and implemented, it is necessary to check how it works in practice, at different levels of the organization, and among the people involved in its implementation. During the initial phases of implementing a system, it is usual to monitor compliance with the planned arrangements, to check if the personnel involved know and apply procedures as expected, or to detect problems or failures that need to be corrected. Once this initial phase has finished, if the degree of implementation is satisfactory, the frequency of these verifications or audits can be decreased until it becomes a routine activity.
- Check if the food operator has adequately developed, documented, and implemented the elements of the system.
- Check as time goes by if deviations, failures, or problems start to happen due, among other reasons, to the fact that there may appear a trend to relax their compliance among workers and those responsible for enforcing the management system. There is a saying in the world of work that "employees only do the tasks monitored by superiors." Without assessing how true it is, the reality is that it is a factor to consider in monitoring the management system. Conducting regular audits, besides correcting deviations at the same time they are detected and then avoiding that they go further, is supposed to be a way of visibly showing management support and commitment and has a positive effect of psychological reinforcement, in the sense that the staff involved in implementing the system can see how important is their role and that this is the right way to achieve the expected outcomes. If an operator notices that no one monitors or follows up on how he applies the procedures or work instructions that correspond to him, the message implicit that one receives is that, in the end, his work should not be so necessary for the system's functioning or the smooth running of the business, as instructed.
- To update the system, it is usual that management systems, once designed, change when implemented due to new activities, renewal of equipment and machinery, changes in products, new data or information, or simply the experience gained, which justifies the need to modify some element. The audit, in this sense, is an excellent tool for detecting the existence of these changes against the documented plan and thus carrying out the necessary updates so that the system does not end up becoming something obsolete, which does not faithfully reflect the actual activity of the organization, which ends up for influencing its effectiveness.
- Principle 6 of hazard analysis and critical control point (HACCP) requires the existence of procedures for the verification of control measures and the system as a whole; therefore, the mere application of the HACCP principles of the Codex already shows that there must be a verification of the system, and auditing is one of the recommended ways to carry it out (Figure 6.1).
- A fundamental principle of quality intrinsically linked to the tasks of verification or audit is continuous improvement, also known as the Deming cycle or the plan–do–check–act cycle (PDCA).

This cycle consists of four phases:

- **Plan**: Establish the action plan, procedures, and resources according to the objectives of the system.
- **Do**: Carry out the planned arrangements.
- **Verify**: Track processes, products, and results obtained to check if they correspond to the objectives and planned activities.
- **Act**: Based on the information obtained in the verification, establish the actions needed to improve performance.

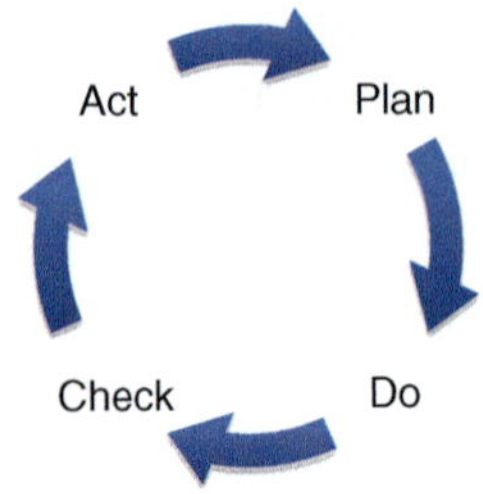

Figure 6.1 PDCA cycle.

Conducting audits corresponds to the third step of this cycle.

For the wheel to turn and move toward continuous improvement, it is necessary to carry out the verification; therefore, this means that audits are a management tool fundamental to effective implementation and for the continuous improvement of performance in organizations.

- For the approval and control of suppliers, together with other measures, supplier auditing allows evaluation of the supplier's capacity for the supply of products (raw materials, ingredients, packaging, etc.) according to the agreed features or specifications. Once the supplier is approved, periodic audits verify that they maintain the supply-stated conditions.
- For certification or registration purposes, the company requests an independent organization to conduct an audit to verify that the operator's management system complies with the requirements set out in a standard or certification scheme and thus obtain the certificate, registration, authorization, or corresponding license.

6.3 Persons Involved in an Audit

An audit involves the client, the auditor, and the auditee; the role that each of them plays will determine the development and results of the audit (Figure 6.2).

In addition to this central core, which cannot be missing in any case, there may be another vital part, not so visible, but which has a central function in the planning and management of audits: the direction of the audit program, that is, the organization that coordinates and directs the auditors and on which they depend.

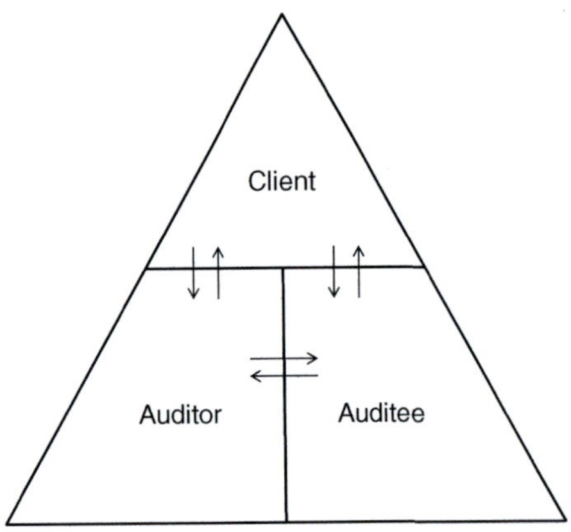

Figure 6.2 The triangle of audit parties.

These parts can be defined as follows:

- **Audit client**: organization or person requesting the audit.
- **Auditor**: person who conducts an audit.
- **Auditee**: organization as a whole or parts thereof being audited.
- **Audit program management**: Organization or part of an organization responsible for planning, directing, and conducting a series of audits. Among its functions is the selection and coordination of the auditors who report to it.

During the audit process, other persons with different roles may be present, such as those listed below:

- **Auditor in training**: auditors present at the audit to cover the period of practice required to complete their training.
- **Technical expert**: person who provides specific knowledge or expertise to the audit team but does not act as an auditor.
- **Guide or companion**: person designated by the auditee who attends and accompanies the audit team.
- **Observer**: individual who accompanies the audit team but does not audit and whose presence was previously agreed upon by the main parties.
- **Consultant**: person acting on behalf of the auditee and likely to have participated in developing and implementing the organization's management system.

6.3.1 Client

- Chronologically, the client is the first person or party involved in the audit; he is who decides to start the whole process for any of the reasons previously mentioned, to carry out an audit either in their or another organization.

From that moment, the client has to select the audit organization or the auditors responsible for conducting an audit, or a series of them, according to the audit program.

The client is the person who will determine the purpose, scope, and objectives of the audit and who will pay for all expenses incurred. The client is not usually present during the audit process but could be at the final meeting if desired. It is the one who will receive the audit's final report – among others – and decide the actions to carry out after the audit, either the follow-up of the implementation of the corrective actions or another.

6.3.2 Audit Team

The audit team consists of one or more auditors; when there is more than one person, one must exercise the lead auditor role, assuming management and communication functions above the rest of the team members. All team members must act in a coordinated manner, all rowing in the same direction and with one voice, without any divergence of views or opinions, particularly in the presence of the auditee; this would significantly undermine the credibility and professionalism of their work.

6.3.2.1 Lead Auditor

The lead auditor role includes the following functions:

- Establish initial contact with the auditee and convene the audit.
- Responsible for communication tasks with the client, the auditee, and the director of the audit program.

- May intervene in the selection of auditors for a specific audit.
- Plan audit activities according to the scope and approach requested.
- Lead the opening and closing meetings of the audit.
- Assign tasks or responsibilities to the rest of the team members and coordinate their work.
- Audit the part that corresponds to them.
- Resolve any conflicts of interest or other situations that may occur during the audit process and which may affect the achievement of objectives of the audit.
- Prepare the audit report and take care of its distribution to the client and other recipients.
- Maintain confidentiality about the data and information obtained during the audit.

6.3.2.2 Auditor

The functions of an auditor are as follows:

- Prepare the audit.
- Follow the audit plan according to the established scope and approach.
- Audit the part that corresponds to them, according to the task distribution among the auditors.
- Attend the opening and closing meetings.
- Cooperate and loyally support the lead auditor during the audit, especially in difficult times, for example, when giving bad news to the auditee is necessary.
- Prepare audit notes and findings so the lead auditor can prepare the report with reliable data.
- Maintain the confidentiality of the information and data accessed in the audit.

After attending the opening meeting and likely auditing some parts together, the most common for auditors is to act separately and thus make more efficient use of the team resources. For this reason and better coordination among auditors, effective communication mechanisms should be established between auditors and team meetings planned to review the progress of the process and, thus, decide changes or adjustments in the audit plan that may be necessary to achieve the objectives. If audits last several days, these meetings can occur at the end of the day.

6.3.2.3 Trainee Auditors

As part of the audit team, there may also be auditors-in-training; a fully established principle recognized by ISO 19011:2018 is that achieving the competencies and skills to be an auditor is not possible only with theoretical training, but requires passing through a phase of previous experience, quantified in a certain number of days or audits as a trainee auditor.

A question arises: To what extent can people without the necessary qualifications or experience participate in the process? ISO 19011:2018 does not specify the tasks of the trainee auditor within the team, nor does it deny that they can participate in the process. Therefore, the lead auditor, responsible for coordinating the team, must decide the type of tasks that can be performed by the auditors-in-training, always remaining in the background and preventing them from negatively influencing the evidence collection, sometimes caused by how they ask questions or relate to auditees.

6.3.2.4 Technical Expert

ISO 19011:2018 defines a technical expert as someone who provides specific knowledge or expertise to the audit team. In addition, the two notes annexed to the definition add that specific knowledge or expertise relates to the organization, the activity, process, product, service, discipline to be audited, or language or culture and that a technical expert to the audit team does not act as an auditor.

Therefore, the need to incorporate an expert in the audit team lies in the complexity or difficulty that we can find in some sectors or companies, by the type of processes, by the technology used, or by operations or areas that may require specialized technical knowledge.

The technical expert's job is to advise or assist the audit team to the extent required. The audit team must know how to limit the intervention of the technical expert to his role and prevent him from acquiring too much prominence or interfering in the process in an unwanted way.

Suppose the chosen expert needs to gain knowledge or experience about the methodology and audit principles. In that case, it must be explained to him his role of reporting objectively on the aspects consulted and the duty to maintain confidentiality about the areas that are being subject to review.

6.3.3 Guide

The guide is the person designated by the auditee to accompany and assist the audit team. The need for a guide can be discussed and confirmed during the opening meeting.

The functions of the guide are:

- Facilitate access to the facilities and the different departments of the auditee.
- Help the auditor locate and identify the people selected for an interview during the audit.
- Provide documentation of the self-control system and help locate records requested by the audit team.
- Where necessary, provide the auditors with the personal protective equipment necessary to access certain rooms and monitor compliance with safety rules in force.
- Provide information and mediate in any situation that may affect the audit process.

Another essential role of the guide is to witness the facts, circumstances, or findings that occur during the development of the audit. Their presence will facilitate the knowledge and acceptance by the auditee of specific facts that could be considered a finding or a nonconformity.

Often, the companion is a person who usually has responsibility in the management system, typically in a middle or intermediate position within the company's organization chart, such as a technical manager, a quality manager, or a manager with surveillance or supervision functions.

The guide should not influence the process of the audit, for example, by setting the pace of the audit route, interfering with the selection of the sample, or because of their position in the company, affecting the conducting of the interviews; therefore, the audit team must be attentive to these situations and try to minimize their effect as much as possible.

There may be situations in which the guide is only sometimes accompanying the audit team throughout the audit or others when there are no companions designated for each of the auditors acting separately after the division of work between them. In these cases, if an auditor considers it necessary for a guide or a representative of the auditee to observe or verify a particular situation or circumstance that is relevant to the audit objectives, what they must do is request their presence in that place so that someone from the auditee can witness that evidence.

6.3.4 Observer

There may be persons who, individually or on behalf of an entity, are interested in being present during the audit process. These persons are observers who accompany the audit team but do not audit, nor should they interfere in the audit process.

Internal audit	External audit	
	Supplier audit	Third party audit
Sometimes called first party audit	Sometimes called second party audit	For legal, regulatory and similar purposes
The system audited is owned or managed by the audit client	Client-supplier relationship	Performed by an independent entity in which the audit client has no ownership or management responsibility or client-supplier relationship.

Figure 6.3 Types of audits depending on the parties involved.

Observers may be appointed by the audit client, the auditee, any regulatory authority, or others interested in the process. The parties involved in the audit can agree on the observer's presence at the initial contact with the auditee and confirm it before the audit's announcement.

At the opening meeting, all participants are introduced, including the observers, and are reminded of their duty to keep the confidentiality of the information obtained during the process and to follow the safety rules and the need for personal protective equipment in specific areas.

6.3.5 Consultant

Another member who may be present during the audit is the consultant appointed by the auditee and acting on his or her behalf.

In small and medium-sized companies, it is common to ask for the help of an external advisor or consultant to develop, implement, and maintain the HACCP system.

In this sense, their presence is fully justified since it facilitates the flow of information and the development of the audit; in addition, witnessing the facts provides them with better knowledge of the audit findings so that consultants can prepare the corrective action plan more efficiently (Figure 6.3).

6.4 Types of Audit

We can classify audits in several ways depending on the criteria applied. The most commonly used criterion is based on the relationship between the parties involved in the audit.

In this way, audits are classified into two broad categories:

- Internal audits.
- External audits.

What defines an internal audit is that the client and the auditee are the same party. It is the same organization that owns or is responsible for the food safety management system that decides to audit its own system for one of the reasons we discussed above. In external audits, the client is not the auditee. The audit client is not responsible nor the owner of the management system but is the one who decides to audit the system due to the existence of a customer–supplier relationship, for certification purposes by an independent body, or due to legal requirements such as in the case of official control.

This classification that we have just referred to is the one that appears in the introduction to ISO 19011:2018.

When comparing the classification of internal or external audits with the classification of first, second, or third parties, there can be some confusion due to the origin of the auditors. The point is that the fact that the auditors are from the audited organization itself or from the client, or that they are auditors hired from external entities does not affect these classifications.

Thus, an organization that performs an internal audit can choose to do so with auditors belonging to the own organization or, in case of not having sufficient and adequately trained human resources or because it cannot guarantee the necessary independence from the audited party, may decide to hire auditors from an external body. Moreover, in a second-party audit, from client to supplier, it is possible to do it with auditors who belong to the client's organization or to hire auditors from an independent body. In the latter case, three organizations are involved: the client, the auditee, and the auditing body. However, it is still a second-party audit because what is decisive is the client–supplier relationship and not the origin of the auditors.

According to this classification, the audits conducted by the official control agents to the food operators are external third-party audits done for regulatory reasons, in which there is no link with the system's owner, and neither does a customer–supplier relationship. In this case, the audit client is the health administration that, as part of its official control programs, has to plan audits of food companies to verify the functioning and efficiency of their self-control systems.

The auditors the administration has to carry out the audit program may be official control agents, i.e. public health inspectors who, after undergoing an appropriate period of training, must be sufficiently prepared to carry out this task effectively; or, in other cases, can be external auditors from an auditing company.

If, rather than using the relationship between the participants in the audit, we use the audit purpose as a criterion of classification, we can distinguish the following types:

- product audit
- process audit
- system audit.

6.4.1 Product Audit

The product audit aims to check whether a product or service follows the requirements, standards, or specifications set by the client.

The product audit is done once the production process has finished when it has already passed the final inspection and is ready to be delivered to the customer. Therefore, in addition to the intrinsic characteristics of the product, the audit can consider other aspects such as packaging, labeling, or documentation accompanying the product, among others.

In a service audit, the approach is different since there are no physical properties to examine, but there are other aspects to check in each case.

6.4.2 Process Audit

Process auditing aims to verify that a process is performed following the established procedures and work instructions or remains within the limits set.

This type of audit can be better understood if it is analyzed taking into account the quality management principle of the process approach, included in ISO 9001:2015, Quality Management Systems – Requirements, by which it is necessary to examine inputs, process, and outputs to verify whether the activities correspond to the provisions and if the expected results are obtained.

During the audit, the discovery of deviations, weaknesses, or a lower-than-expected performance level should serve to implement corrective actions or improvements in the process.

6.4.3 System Audit

The system audit verifies the functioning of a management system. A system is a set of things working together as parts of a mechanism or an interconnecting network, like a complex whole.

In the case of a management system, these elements are the organization, activity, processes, procedures, and human and technical resources that interact between them to achieve specific objectives. So, the operators, following established procedures and using the necessary equipment and machinery, carry out a process within the company's overall activity, which has specific objectives or purposes.

System audits are conducted on the whole set of elements that make up the system, and that is why they have a broader scope than the previous two types we have seen. Moreover, a process audit can be understood as a system audit limited to verifying a selected process; the audit techniques are the same, and what changes is the narrower scope of the process.

A system audit aims to verify that the system has been designed, implemented, and maintained under established requirements, such as those included in a quality standard or resulting from applying the HACCP system.

System audits are usually done against a reference standard. These standards include several requirements that indicate "what" must be done but not "how" it must be done. It is up to the system's owner to find the most effective way to fulfill the "what" according to the characteristics of each organization. This circumstance is even more evident in the HACCP system because its development results from applying principles rather than on the specific points covered in a standard.

The term "scope," to which we are referring, is defined in ISO 19011 as follows:

Audit scope: extent and boundaries of an audit.
Note: The audit scope generally includes a description of the physical locations, organizational units, activities, and processes, as well as the time period covered.

Depending on the scope agreed upon, we can establish two other types of audit: total or partial. There are cases in which, due to lack of resources or time or because the audit client wants to give priority to specific parts of the system, they prefer to do partial audits, focusing only on some elements (locations, services, or functions), versus the total audits in which they want to examine the entire system.

All these types of audit mentioned according to the different classification criteria are not mutually exclusive; in practice, we can find represented different classes in the same audit. Here are some examples:

- Second-party audit, supplier, focused on a product.
- Third-party, certification, and system audit.
- First-party, internal, and partial audit since it focuses only on specific services.

6.4.4 Research Audit

In some cases, when problems or repeated failures in production appear, the food operator may commission an investigative audit. This type of audit is a thorough examination of an activity or process to discover the root cause behind the occurrence of production failures and thus establish the relevant actions to prevent their recurrence.

For some authors (Scott Parsowith 1995), this type of audit would be the highest rank or hierarchy, depending on the scope of it, maybe because the limits or extent of the audit is not fixed in advance but can be extended depending on the data obtained during the investigation of the cause of the problem.

6.4.5 Follow-up Audit

Once the audit has finished and the report delivered to the auditee, if there is any noncompliance, the audited organization has to submit an action plan to remedy nonconformities.

The follow-up audit is an activity of the post-audit phase. Its purpose is to verify that the auditee has made the necessary corrections and corrective actions to resolve nonconformities and that these have been effective. The scope of the follow-up audits is usually limited to the nonconformities included in the previous audit report. Other times, depending on the resources or for reasons of the cost involved in the audits, the verification of the closure of nonconformities can be arranged to take place with the next inspection, in the case of official control, or with the next scheduled audit.

After seeing all these types of audits, we can better understand one of the most typical properties of the audit: its versatility. This ability to adapt to different users, systems, and objectives, depending on the interests and needs of each organization or each client, makes it an irreplaceable management tool, both for the maintenance and continuous improvement of management systems and, in general, to add value to the organization.

References

Europe, European Parliament (2017). Regulation (EU) 2017/625 of the European Parliament and of the Council of 15 March 2017 on official controls and other official activities performed to ensure the application of food and feed law, rules on animal health and welfare, plant health and plant protection products (Official Controls Regulation). *Official Journal of the European Union.*

International Organization for Standardization (ISO) (2015). *ISO 9000:2015 Quality Management Systems – Fundamentals and Vocabulary.*

International Organization for Standardization. *ISO 19011: 2018*(2018). *Guidelines for Auditing Management Systems.* Geneva: International Organization for Standardization.

Russell, J.P. (ed.) (2013). *The ASQ Auditing Handbook*, 4e. Milwaukee, WI: ASQ Quality Press.

Scott Parsowith, B. (1995). *Fundamentals of Quality Auditing.* ASQC/Quality Press.

7

The Food Safety Auditor

7.1 Competence of the Auditor

The ability of an audit, or audit program, to achieve planned objectives depends mainly on the auditors' competence. Without forgetting other people who are also involved in the planning and management of the program, it can be said, without exaggeration, that an audit is only as good as the auditors who carry it out.

Thus, an audit can be beneficial and productive for the parties involved, or it may be frustrating when performed by personnel without adequate training and experience for several reasons: by not achieving the expected objectives because they cannot detect system failures or situations that may pose a risk to the organization, reaching erroneous conclusions, or even creating discomfort among the company's staff due to poor human interaction by the audit team.

The ISO 19011:2011 introduced the concept of risk in the audit of management systems, which did not exist in the 2002 edition, and defines it as:

Risk: Effect of uncertainty on objectives

The approach taken by the standard refers to the risks of the audit process in achieving its objectives, such as the risk of the audit interfering with the activities and processes of the auditee. The people who manage the program and the auditors should be aware of the risks associated with audit activities, for example, lack of resources, inadequate selection of auditors, failure to respect the confidentiality of information, etc., to try to prevent these risks to the extent that is possible.

The competence to act as an auditor in food safety relies on having a series of knowledge and skills, among which are the following:

- Knowledge of the hazard analysis and critical control point (HACCP) system.
- Education in hygiene and food technology.
- Knowledge of food legislation.
- Knowledge of quality management systems.
- Education and training in auditing techniques.
- Previous work experience, preferably in the food sector.
- Development of personal skills and qualities to conduct the audit efficiently per audit principles (Figure 7.1).

More than knowledge, the food safety auditor has to master HACCP, manage its methodology with ease, and fully understand how to apply each HACCP principle so that the result is coherent and, this way, obtain a valid system for managing food safety in each company.

HACCP System Auditing for Food Safety: Principles and Techniques, First Edition. Luis Couto Lorenzo.

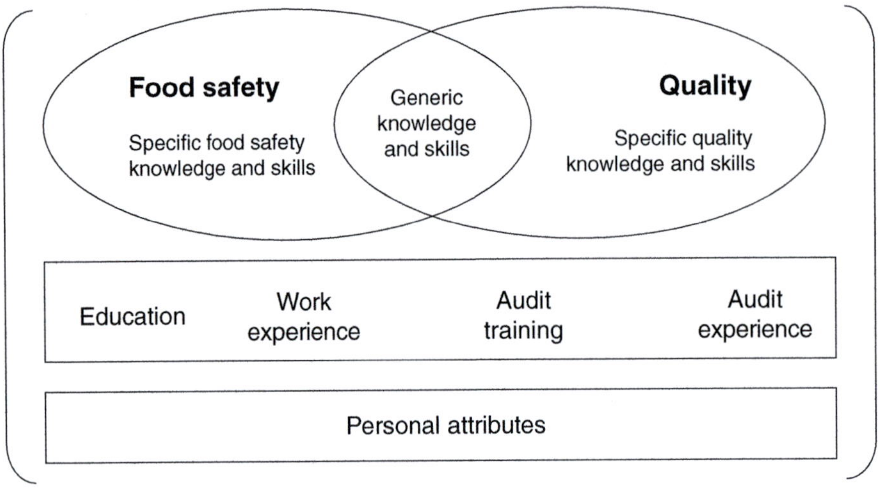

Figure 7.1 Elements of auditor competence.

The food operator is responsible for developing their HACCP system according to the nature and characteristics of their organization. Food companies are different in terms of activities, production processes, technology, position in the food chain, size, and organizational structure; therefore, it is logical that the self-control system has to adapt to these circumstances and thus adopt different formats. This variability requires the food safety auditor to know how to recognize the structure and application of HACCP principles regardless of each system's format and structure. HACCP system requirements define "what" must be fulfilled, but "how" is up to each food operator, who must choose the way that best suits their interests and needs.

In addition to the HACCP system, the food safety auditor must possess a solid background in hygiene and food technology. Disciplines such as bromatology, microbiology, food technology, toxicology, zoonoses are fundamental to competently evaluate a HACCP plan, which leads us to a professional profile of people with higher education, completed in several degrees related to health sciences or experimental and biological sciences. There are also other subjects, such as those we saw in the "HACCPverse" figure, which can be more or less familiar to the auditor, and that auditors may find helpful to evaluate management systems throughout their professional careers.

Likewise, the auditor must know the legislation that affects each food sector. Food legislation includes mandatory aspects for operators and refers to criteria, requirements, and hazards that the food operator must consider in the design of its food safety management system and for carrying out their activity. In the development stage of a HACCP system, legal standards also provide a frame of reference for operators to evaluate which hazards may be significant in their activity and to determine the acceptable level for controlling each identified hazard.

The fact that HACCP, in practice, is implemented as a management system requires the food safety auditor to be familiar with quality principles and management systems. ISO 19011:2018 describes the knowledge and skills in quality management that auditors must have with several examples, such as the following:

- Terminology related to quality, documentation, and audit processes.
- Management system approach.
- Leadership and management for the success of an organization.
- Participation of people, human factors, competence, training, and awareness.

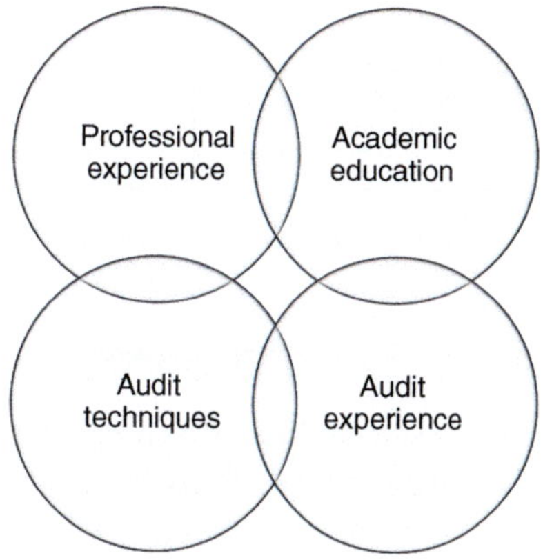

Figure 7.2 Both education and professional and audit experience constitute the auditor's competence.

- General business and management concepts, processes, and related terminology, including planning, budgeting, and people management.
- The needs and expectations of relevant stakeholders that have an impact on the management system.
- The type of organization, its governance, size, structure, functions, and relationships.
- The relationships and interactions between management systems processes.
- Understand the types of risks and opportunities associated with the audit and the principles of the risk-based approach to audit.

Audit implementation requires a certain degree of maturity in the people who carry it out. During the audit, auditors will have to face different situations, interact with people at various levels, make decisions, and many times, know how to act under pressure; for this reason, it is pretty convenient that the auditors have the experience and the emotional skills acquired by spending several years in the world of work. Preferably, that experience should be in positions with a technical or managerial function involving problem-solving, decision-making, and communication skills. When this work experience is in food companies, the result is even more valuable due to the increased understanding of the characteristics and requirements of the food sector (Figure 7.2).

7.2 Achieving the Auditor's Competence

Based on the candidate's academic education and previous work experience to be an auditor, the training process consists of an initial course in audit, followed by a traineeship period as an auditor in training.

The initial course in auditing techniques has a recommended duration of 40 hours, and its content includes aspects such as the concept and principles of auditing, methods of gathering information, use of working documents, documentation of findings, preparation of nonconformities, and drafting of reports. The practical part of this course usually includes role-playing games and simulation techniques of the real situations that occur in audits. The training period as a trainee auditor must be under the supervision of a lead auditor or one sufficiently experienced. This learning period should last long enough for the trainee auditor to acquire the minimum fluency required to act effectively in audits; as a guideline, participation in four complete audits or 20 days of audits is considered an adequate period.

This initial training process, consisting of the theoretical course plus the period as an auditor in practice, is normal among private certification bodies. In the case of official control, this process

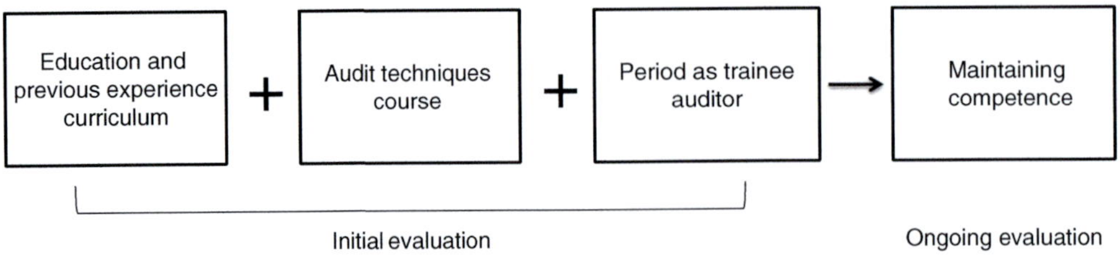

Figure 7.3 Process for gaining auditor competence.

is not so standardized, which may result in auditors not always having sufficient theoretical and practical preparation before starting to act in audits.

At the end of the training period, the auditor must be evaluated to verify that he meets the criteria established to participate in the audit programs. In the case of a lead auditor, the qualification process requires additional knowledge and skills, which allow him to lead the audit team and manage the entire audit process. The lead auditor must acquire this additional competency by working under the supervision of an experienced lead auditor (Figure 7.3).

There is a saying in the auditing world: auditing is learned by auditing. This means that after initial training, an auditor should continue acquiring knowledge and striving to maintain and improve their professional performance. Maintaining competence is achieved through regular participation in audits. In addition, more aspects can raise the need for updating knowledge, such as changes in laws and regulations, technology, or work processes.

Those responsible for the audit program should establish mechanisms to evaluate the performance of the auditors involved in the program's implementation, such as through review of reports, interviews, observation, or peer evaluation. Maintenance and continuous improvement of performance is a process that lasts throughout the professional life of the auditor.

7.3 Audit Principles

Planning and practical execution of audits depend on compliance with audit principles. These principles have been established for years in the general doctrine of audits and have been part of the content of ISO 19011 since its first versions.

These principles are considered as the prerequisites that must be taken into account both in the preparation of the audit program and in the process, in the methods used, and in the professional practice of the auditors during the conduct of the audits. The effectiveness of an audit program will depend to a large extent on the degree to which the different people involved adhere to compliance with these audit principles, which we will describe below.

The seven principles referred to in ISO 19011 are as follows:

- **Integrity**: auditors must conduct their duties honestly, in an impartial, even-handed manner, and maintain their professional judgment in the face of the pressures they may suffer.
- **Fair presentation**: based on the data and findings obtained, auditors must provide truthful and accurate information relevant to the objectives of the audit, complete and easy to understand for the parties involved, about the activities and results of the audit.
- **Due professional care**: Auditors must act with diligence and professional competence to achieve the objectives of the audit and respond to the trust placed in them. They must perform their work professionally during the audit to make reasoned judgments in all audit situations.
- **Confidentiality**: Auditors must maintain the confidentiality of the data and information they access during their duties. Audit information should not be used inappropriately for personal

gain by the auditor or the audit client or in a manner detrimental to the legitimate interests of the auditee. The audit team must protect the auditee's data and sensitive information.
- **Independence**: It is fundamental to the impartiality and objectivity of the audit conclusions. Auditors should not audit areas or activities for which they have direct responsibility. The audit body and auditors should be independent of the audited activity and free from bias and conflicts of interest.

Independence is not an optional merit or benefit in achieving audit objectives. The fact that it is part of the definition of audit is no accident; we must consider independence as a prerequisite, essential to the credibility and effectiveness of the entire process.

Nor should we understand independence as a binary condition, which is or is not, but as a continuum that can always be affected to some degree, between a minimum and a maximum. Independence does not come from heaven but must be actively protected and promoted by audit program managers at various levels, from the auditing body, which must give the auditors the appropriate authority through a clear and documented mandate, to the audit team itself, which must try to adopt an objective mental attitude, without bias and free of influences throughout the process.

- **Evidence-based approach**: Auditors must reach reliable and reproducible conclusions supported by objective and verifiable facts. The function of the auditor is not to provide ideas, opinions, or subjective assessments about the audited organization but to be able to make reasoned judgments based on the objective evidence obtained during the audit exercise.
  The reliability of the auditors' conclusions relies on the available information. Due to the lack of time or limited resources, searching for relevant information should be done systematically and through appropriate sampling to achieve the objectives. If the information provided is objective and sufficient, anyone should be able to accept the conclusions of the audit team without any problem.
- **Risk-based approach**: an audit approach that takes into account risks and opportunities. The risk-based approach should significantly influence the planning, conduct and reporting of audits to ensure that audit focus on matters that are important to the audit client.

7.4 Personal Attributes of Auditors

The competence to perform audits relies on developing specific attributes or personal qualities. As we have seen, being an auditor requires the participation of mature people who know how to act with professionalism, diligence, ethics, and equanimity, who are able to make decisions and respond appropriately to the different situations, not always comfortable, that may arise throughout the audit.

For the audit to progress appropriately, proactive management by the audit team is necessary. Auditors must ensure the process moves forward at the right pace based on available time. Team members must act kindly and diplomatically in their interrelationship with the auditees and with the firmness necessary to effectively direct audit activities to achieve the purposes established by the client.

In the section on personal behavior, ISO 19011 enumerates a list of the qualities that auditors must possess to act by the principles of auditing, which includes the following:

- ethical, i.e. fair, truthful, sincere, honest, and discreet;
- open-minded, i.e. willing to consider alternative ideas or points of view;
- diplomatic, i.e. tactful in dealing with people;
- observant, i.e. actively observing physical surroundings and activities;

- perceptive, i.e. aware of and able to understand situations;
- versatile, i.e. able to readily adapt to different situations;
- tenacious, i.e. persistent and focused on achieving objectives;
- decisive, i.e. able to reach timely conclusions based on analysis and logical reasoning;
- self-reliant, i.e. able to act and function independently while interacting effectively with others;
- acting with fortitude, i.e. able to act responsibly and ethically, even though these actions may not always be popular and may sometimes result in disagreement or confrontation;
- open to improvement, i.e. willing to learn from situations and striving for better audit results;
- culturally sensitive, i.e. observant and respectful of the auditee's culture;
- collaborative, i.e. effectively interacting with others, including members of the audit team and the auditee's staff.

When one reads or hears, for the first time, this enumeration of the personal qualities of the auditors, it is almost inevitable to react with astonishment and skepticism. Perhaps the reader, at this point, is smiling incredulously at such an accumulation of virtues required to be an auditor. However, we should not remove any items from this ISO 19011 list, since all of them have their importance and their practical application for the conduct of audits.

This question is easier to accept when we consider this enumeration not as the compendium of the attributes the auditors must possess but as the skills they should exercise throughout the audit and their professional practice, i.e. as a guide or a reference of the qualities that each auditor, to the extent of his possibilities, should try to apply to conduct the audit most effectively for the objectives of the audit and the interests of all parties involved in the process.

For auditors in training or those who are going to start doing audits, it is convenient that they know these personal qualities, think about their meaning, and then try to retain them as a helpful guide for their professional practice.

Later, when they are performing their work, there will surely be various situations that they have to face, and that is when they will best understand the usefulness of putting those skills into practice to conduct the audit most properly. Over time, it is easy for many of these qualities to be internalized and become part of the natural behavior, the unconscious competence of the auditor. This does not preclude the assertion that this professional competence is based much more on the experience acquired than on the alleged innate qualities of auditors.

If, from a general perspective, we understand the audit as a process, with some inputs, activities, and results, the essential purpose of audits is to provide helpful information for the management of organizations: either to assess whether system procedures are applied as intended, to verify that a supplier has sufficient capacity to deliver a product under specified requirements, or generically for informed decision-making for management or to direct organizational policy.

The degree to which the audit becomes worthwhile depends mainly on the auditor's ability to adopt the mindset and the most appropriate approach to reach reliable conclusions about audit objectives.

This state of mind to which we refer is so characteristic of this job; we can say that it is based, on the one hand, on being able to start from a position of independence for the audited activities, which allows the auditor to act without pressure and with the required freedom, and, on the other hand, on the need to provide information that is truthful, accurate, and objective about the operation or the actual state of the management system of an organization.

This mindset explains the commitment of auditors to look for evidence, to check things for themselves, and to investigate beyond the appearance and opinions or unverified ideas about audit matters. This approach aligns with the purpose of the audit; no one can replace the auditor in this task, and only from this attitude can audits achieve the expected utility.

From the point of view of the auditee, the announcement of an audit is usually received with a mixture of uneasiness and fear about the outcome and its consequences, perceiving the auditor's work often as something threatening, excessively focused on the negative and discovering failures. This reaction is easily predictable and has a logical explanation; we have a conflict of interest between the audit team and the auditee because they start from different positions and roles, and we do not all share the same ideas, beliefs, values, or criteria. Once the potential existence of this conflict of interest is recognized, the auditor's role is to manage it positively, avoiding confrontations and sterile discussions and trying to resolve it most favorably. In this sense, the role of the auditor should never become a personal confrontation to assert his criteria against the auditee, as if it were a boxing match, but, to follow the sports simile, it should be something much more like the chair umpire of a tennis match, with objectivity and distance; what he determines is whether the ball enters or falls out if it touches the net or in general if the rules of the game are met (Jon Basagoiti, personal communication) (Figure 7.4).

Figure 7.4 The auditor's proper role with the auditee. Source: andriyyavor/Adobe Stock.

A good auditor is one who, throughout the audit process, knows how to adopt an attentive attitude, investigates in depth, inquires, and tries to find out with professional care if things are done as planned. It can be challenging for the audited staff to understand this effort of the auditor to check things conscientiously. They may even think the auditor enjoys discovering flaws, but this is untrue. When things are done well, the auditor should have no qualms about recognizing it. However, when there is a flaw or a problem in the system, the auditor must report it. This role often involves giving bad news to the interlocutors, presenting them at the closing meeting, and including them in the final report, which is what we expect from a professional auditor and what gives meaning to the audit, not to hide or overlook the problems. To illustrate it better, we can make a simile with a person's attitude before a medical review. What do we prefer? They make us a superficial check, giving us apparent tranquility, or we undergo a thorough review to discover if something is wrong and then receive adequate treatment.

Something similar happens with management systems; once we overcome the bitter pill of nonconformities communication, it is time to get back to work to correct the deviations and improve the system. The superficial auditor will leave a good taste in our mouth at the closing meeting, but the problems will remain hidden in the organization when he walks out the door.

The auditor's work comprises several activities, some of which involve a good dose of human interaction with the personnel of the audited organization at various levels: managers, intermediate technicians, and plant operators. This task requires exercising specific relationship skills so that the process goes smoothly and without unnecessary discomfort and tension. The auditor must know how to adopt the right tone, show respect, and recognize the effects of the audit on the organization without detracting from the examination's effectiveness. Thus, being friendly but firm in his determination is key to achieving the objectives of the audit. This appropriate tone often relies on avoiding the most extreme attitudes; for example, not be arrogant, but not so shy that it makes it challenging to guide the audit effectively; neither be so manipulable that they take auditors wherever they want, nor so rigid as not to know how to adapt to the circumstances that may arise, nor be a charlatan or so quiet that it hinders communication with the auditees, nor be excessively distrustful or a naïve who accepts everything at first sight. These are just some examples; the reader will likely find other characteristics to apply this principle (Figure 7.5).

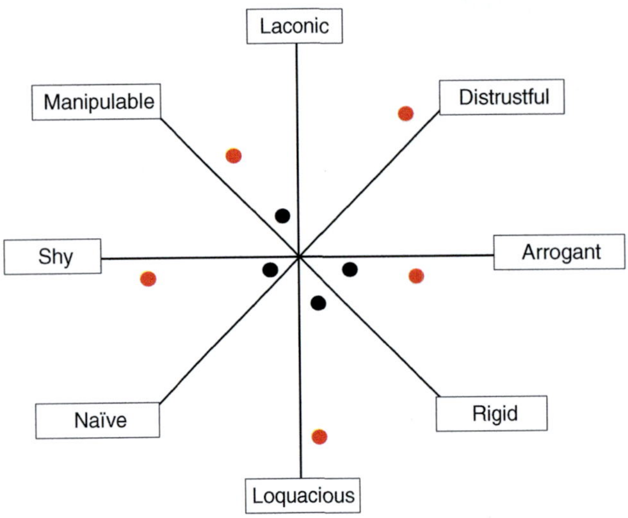

Figure 7.5 The proper attitude of the auditor is often to avoid extremes.

The only quality where there is no risk of going too far is assertiveness. Assertiveness is the quality of being self-assured and confident without being aggressive to defend a right point of view or a relevant statement. It is the attitude that someone adopts to stand up for their rights while respecting those of others. It is like an intermediate state between aggressiveness and passivity, a point of balance in which one's ideas and convictions are expressed and defended without harming others.

7.5 Management Skills

The audit process requires that auditors have specific knowledge about audit principles and methods and exercise certain management skills for the different moments or phases of the process. These capabilities include:

- Plan and organize the work of the team.
- Try to meet the program.
- Know how to interact with the work team.
- Set priorities and focus on relevant aspects.
- Gather information through different sources.
- Conducting interviews with audited personnel.
- Use working documents properly.
- Take notes of findings and evidence.
- Confirm that audit evidence is sufficient to support audit findings and conclusions.
- Prepare the audit report.
- Know how to communicate effectively orally and in writing.

Throughout his professional career, the auditor should improve the performance in each of these skills, especially in those he feels at the start of their career less qualified, to achieve a level of competence sufficient in most of the abovementioned ones to develop the whole process effectively.

References

International Organization for Standardization (2018). *ISO 19011: 2018: Guidelines for Auditing Management Systems*. International Organization for Standardization. Geneva.

8

Audit in Official Control

In the European Union, hazard analysis and critical control point (HACCP) is mandatory for all food business operators involved in producing, processing, and distributing food at any stage after primary production. This requirement was incorporated into community legislation in 1993 with Directive 93/43/EEC on food hygiene. Article 3 of this directive included five points corresponding to the HACCP system but not to the same principles and terms currently used.

In addition, the directive had to be transposed into the national law of each member state to be applicable, which left countries some room for maneuvering on how to achieve the required result.

Regulation 852/2004 on the hygiene package replaces Directive 93/43, strengthening the obligation for food operators in the post-primary production stages to create, apply, and maintain procedures based on HACCP principles, but already incorporating the same terminology and the same seven principles recognized by the Codex Alimentarius.

The change in the food safety model that begins with Directive 93/43/EEC and that the hygiene package consolidates is that food operators, not the health administration, are responsible for ensuring their products' safety and compliance with legal requirements.

The role of official control is to verify that food business operators comply with food law requirements at all production, processing, and distribution stages. Compliance with Regulation 852/2004 requires companies to create, implement, and maintain a self-control system. For that reason, it is the task of the company to design its own HACCP plan according to the characteristics of its production system and its organizational structure in the way it deems most appropriate.

The development of the hygiene package from Regulation 178/2002 led to the publication of two types of rules: some aimed at food operators, such as Regulation 852/2004 on food hygiene and Regulation 853/2004, laying down specific rules for food of animal origin; and others that apply to the competent authorities of the member states such as Regulation 882/2004 on official controls and Regulation 854/2004 on the organization of official controls on products of animal origin intended for human consumption. The latter two were repealed and replaced in 2017 by Regulation 2017/625 on official controls and in 2019 by Regulation 2019/627 on official controls on products of animal origin intended for human consumption, respectively.

If, under the provisions of Regulation 852/2004, the responsibility of the food operator is to implement a self-control system, the role of the health administration is determined by Regulation 2017/625. The function of official control concerning HACCP systems of food businesses is not to give formal approval or certification of conformity of the system nor to participate in the development of the HACCP system of operators. Its task is to evaluate whether it meets the established requirements and whether its operation is adequate to guarantee safe food production. To this end, properly trained official control officers should conduct regular audits of self-control systems per the provisions of the operators' self-control control programs set up by the competent authorities.

HACCP System Auditing for Food Safety: Principles and Techniques, First Edition. Luis Couto Lorenzo.

There may be a temptation to instruct the food operator on designing his self-control manual among official control agents, especially those who have been inspecting the same food operators for many years. In principle, this is not good practice because the control officers are assuming a responsibility that belongs exclusively to the operator; they will be becoming both judge and party, applying a subjective criterion, which will prevent them from acting with the impartiality and independence required in official control. In addition, it is not uncommon for another official control agent to act and do so with different criteria, which will give an image of arbitrariness. A different thing would be to guide the operators in interpreting the HACCP system's legal requirements or to facilitate searching for helpful information to develop or document the self-control system. There is no problem with that, as long as one knows how to differentiate the role played by each one.

Small and/or less developed enterprises do not always have the human resources and knowledge to formulate and implement an effective HACCP plan. In such cases, competent authorities can play a valuable role in providing guidance and information to food operators to facilitate the implementation of the HACCP system, through the publication and dissemination of good hygiene practice guides or sectoral HACCP guides, often with the involvement of stakeholders and representatives of the sectors concerned. These HACCP guides, provided that experts have prepared them, are a handy tool for companies to design and implement HACCP systems and contain valid information on the most appropriate measures and practices for each food sector.

Regulation 2017/625 lays down general principles for the performance of official controls performed in order to:

- prevent, eliminate, or reduce human or animal health risks to acceptable levels; and
- ensure the application of good practices in the food sector and the protection of consumer interests, including food labeling and other forms of consumer information.

In order to achieve these objectives, official control officers have to carry out different activities, methods, or techniques, depending on the nature and purpose of the control in each case. Article 14 of Regulation 2017/625 lists these activities as follows:

- an examination of the controls that operators have put in place and of the results obtained;
- an inspection of
  - equipment, means of transport, premises, and other places under their control and their surroundings;
  - animals and goods, including semi-finished goods, raw materials, ingredients, processing aids, and other products used for the preparation and production of goods or for feeding or treating animals;
  - cleaning and maintenance products and processes;
  - traceability, labeling, presentation, advertising, and relevant packaging materials, including materials intended to come into contact with food;
- controls on the hygiene conditions in the operators' premises;
- an assessment of procedures on good manufacturing practices, good hygiene practices, good farming practices, and procedures based on the principles of HACCP;
- an examination of documents, traceability records, and other records that may be relevant to the assessment of compliance;
- interviews with operators and their staff;
- the verification of measurements taken by the operator and other test results;
- sampling, analysis, diagnosis, and tests;
- audits of operators;
- any other activity required to identify cases of noncompliance.

8.1 Audit vs Inspection

Regulation 2017/625 describes audit and inspection as different official control methods or techniques. However, some people do not see the point in establishing differences between both tasks or do not know well where to put the boundaries between audit and inspection. For this reason, although in practice, we can perceive them as activities that overlap to some extent, it may be good to know the characteristics and differences between both concepts for didactic purposes.

Regulation 2017/625 does not include a definition of inspection, as did the previous Regulation 882/2004, so we can use the one contained in ISO 9000: 2015 Quality Management System – Fundamentals and Vocabulary:

Inspection: determination of conformity with specified requirements.

Among the specified requirements, there are also legal ones. Therefore, the inspection in the official control implies the verification of legal requirements and involves aspects such as:

- Facilities, equipment, and machinery.
- Means of transport.
- Raw materials, ingredients, and technological adjuvants.
- Semi-finished products.
- Animals.
- Materials and objects in contact with food.
- Products and processes of cleaning and maintenance, and protection against pests.
- Traceability.
- Labeling, presentation, and advertising of food.

This definition of inspection is quite different from the audit one we have studied previously. In the following figure, we will describe some of the characteristics that help differentiate both activities (Figure 8.1).

Audit	Inspection
Usually announced	By surprise
Verify self-imposed requirements	Verify legal requirements
Conformity with the procedures	Compliance with legal requirements
Verification of the implementation of the documented management system	More focused on facilities, practices, and records
More planning Systematics	Less planning Sometimes it can be a check of one or a few aspects
More formal implementation and communication	Less formal implementation and communication
Deadlines for corrective action are set by the auditee	Deadlines for correcting deficiencies are set by the inspector
Communication of the result by means of the audit report	The result of the inspection is communicated in the inspection record

Figure 8.1 Differences between inspection and audit.

Figure 8.1 emphasizes the differences between the two activities to explain better what characterizes each concept. In practice, there may be cases where it is easy to distinguish them and situations where it is more challenging to delimitate the two tasks. For example, if we compare the inspection carried out in a fish shop, more focused on checking the conditions of exposure of the fish, its degree of freshness, and labeling, or the inspection carried out at a border inspection point of the goods received, consisting of a documentary control, identity, and physical of the products, with an audit of the self-control system in a medium-sized company, following a process that goes through certain phases: communication, document review, plant audit, etc., until the final report is issued, the differences between these activities become evident. However, when we talk about inspections in which we do some planning and use checklists to inspect the different aspects more systematically, it is no longer so easy to distinguish the boundaries between an audit and an inspection.

Furthermore, some differences do not occur on all occasions; for example, although the common thing until now was to communicate the audit beforehand, it begins to extend more and more conducting audits without prior notice in order to verify better certain types of evidence related to the actual conditions in which companies operate on a day-to-day basis, which are not easy to assess when the organization makes arrangements before the arrival of auditors.

Inspection and audit are not opposing activities; they share many aspects in common; in both, we do similar things: we examine facilities, observe practices, review records, talk to the organization's staff, etc. However, there are also differences in the objectives; the method used; the way of reporting and communicating the results; and the consequences of the controls carried out in each case. In this regard, it is convenient to consider them as differentiated activities for different purposes: verifying the functioning and effectiveness of food safety management systems, in the case of audit; and verifying the hygienic conditions of establishments, handling practices, and conditions of food products, in the case of inspection (Figure 8.2).

As we can see, it is hard to set clear limits between inspection and audit; the purpose of control ultimately makes the difference. The auditor or inspector will make greater use of audit or inspection techniques depending on the circumstances and the nature of the control on each occasion. In any case, these common aspects make it possible that the expertise gained with one activity can also be valid for the other one. For example, minimal planning of the items to be examined can improve the effectiveness of inspections in many cases. Reciprocally, the experience gained by inspectors in the exercise of their profession is a merit to be taken into account for their qualification as a food safety auditor since, among other reasons, they know well the food sector, they are used to working under pressure, keeping the information obtained confidential, making judgments based on objective facts, and dealing with the staff of food companies.

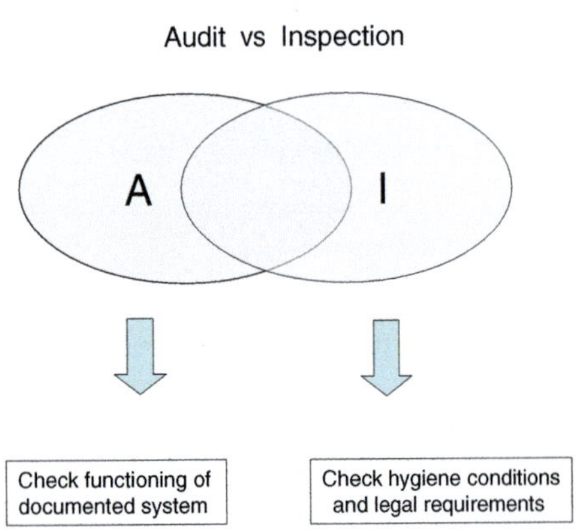

Figure 8.2 Auditing and inspecting have different goals, but share many common aspects.

Regulation 2017/625, as before Regulation 854/2004, also differentiates inspection and audit functions concerning specific rules for products of animal origin. Thus, in its article 18 on the measures to be taken by the competent authorities with products of animal origin intended for human consumption, it refers to audits of good hygiene practices and procedures based on the HACCP system, as opposed to inspection functions where it includes activities such as antemortem inspection, postmortem inspection, animal health and welfare, handling specified risk materials, presence of residues and contaminants, and hygiene of meat production.

Although, as we saw, inspection and audit share some characteristics, they have different purposes, and therefore, they are control units not to do simultaneously. That would be impractical because both their approach and the pace of the process are different. In addition, the frequency established for each may differ; inspections are usually more frequent than audits.

One issue to be addressed when developing official control programs is who is the most suitable personnel to conduct audits of food operators. Can the public health inspector audit the self-control systems of the establishments he routinely inspects? Several factors can be considered in answering this question, including the availability of human resources, but ideally, we must look for other options. In order to ensure compliance with the independence requirement, someone other than the regular inspector of an establishment should be the one who audits its self-control system. In this respect, the familiarity generally created between the inspector and the staff of the establishment he visits regularly can be avoided; otherwise, this familiarity can lead to the appearance of biases and lack of perspective in front of the audited system and also cause a "familiarity bias," i.e. the loss of the ability to perceive certain flaws because they are habitual or after seeing them repeatedly in the establishment. From the side of the audited organization, there may also be reasons to appreciate a greater degree of independence when it is not the usual inspector who performs the audit.

The problem of lack of independence is particularly evident in the case of veterinary slaughterhouse inspectors, who permanently carry out their official control tasks in the establishments assigned to them. In this case, there is direct responsibility for slaughterhouse veterinary inspectors over the control of hygienic conditions and the supervision of hygienic and manufacturing practices, among other tasks; for this reason, if we want to have a greater degree of independence and effectiveness, we must choose auditors who do not maintain that strong relationship with the activities to be audited.

One way to overcome this problem is by rotating or exchanging inspectors between the different areas. This system can be easily included in the control program of the operators' self-control when establishing the criteria for selecting the audit team members to guarantee the impartiality and objectivity of the actions as much as possible.

In managing these auditor exchange systems, one point that should be clear is that we do not question the ability of inspectors to carry out audits; we apply measures to better comply with the requirement of independence. Due to his knowledge and experience in the food sector, an official inspector can readily acquire the competence to conduct audits efficiently. However, doing them in establishments other than those where they habitually carry out their official control tasks will always be better.

A recurring question often raised by people new to auditing is what the official control officer should do if, during an audit, he observes or finds something that could constitute an infringement or pose a serious and immediate risk to public health. The answer is that he should stop the audit and act as an official control agent due to the circumstances that have arisen. ISO 19011:2018 observes this type of situation in its point 6.4.4, "Communication during the audit," laying down that when the evidence collected during the audit suggests an immediate and significant risk to the auditee, it should be communicated without delay to the auditee and, if appropriate, to the audit client. In the case of official control, the audit program should provide for the possibility of such circumstances arising and identify the procedure to follow, the channels of communication, and the responsibilities of each party.

References

Europe, European Parliament and Council (2004). Regulation (EC) No 852/2004 of the European Parliament and of the Council of 29 April 2004 on the hygiene of foodstuffs. *Official Journal of the European Union* 139 (1): 1–54.

Europe, European Parliament and Council (2017). Regulation (EU) 2017/625 of the European Parliament and of the Council of 15 March 2017 on official controls and other official activities performed to ensure the application of food and feed law, rules on animal health and welfare, plant health and plant protection products (Official Controls Regulation). *Official Journal of the European Union* 1–142.

International Organization for Standardization (2018). *ISO 19011: 2018: Guidelines for Auditing Management Systems*. International Organization for Standardization. Geneva.

9

The Audit Process

When audit is defined as a systematic examination, what is meant is that it is a process consisting of a sequence of predetermined phases, so that certain tasks or activities have to be performed in each of these phases. This arrangement of audit activities into defined steps can easily be understood as the result of a logical development based on common sense and many years of experience in conducting audits, be they financial, quality, or other management system audits. In other words, if things are done in a certain way, it is because experience has confirmed the best audit practices to make the process as effective as possible.

The international reference standard for conducting management system audits is ISO 19011:2018, which devotes several chapters to providing guidance on planning and performing the activities that make up the audit process. Since the publication of its first version in 2002 – replacing the previous ISO 10011:1990 – ISO 19011 has undergone several changes to adapt it to the new circumstances that have arisen in the management systems arena and the needs of all parties involved in auditing. Thus, one of the changes introduced in previous versions was the extension of the scope of the audit of quality and environmental management systems to the audit of all management systems, regardless of the discipline for which the system is intended. Compared with the previous 2011 edition, changes in the 2018 version include:

- the addition of the risk-based approach to the auditing principles;
- extended guidance on managing an audit program, including audit program risk;
- expanded guidance on conducting an audit, particularly the section on audit planning; and
- extended general competence requirements for auditors.

The extension of the scope of management system auditing to new disciplines has also led to the need to provide more generic guidance, applicable to a wider range of potential users, including auditors, organizations implementing management systems, small and medium-sized organizations, and organizations required to perform management system audits for contractual or regulatory reasons.

9.1 Stages of the Audit Process

In essence, the audit is merely a process of information gathering for the purpose of determining the conformance of a system with established requirements. Therefore, in general terms and for didactic purposes, the information-gathering process can be divided into four main steps (Figure 9.1):

HACCP System Auditing for Food Safety: Principles and Techniques, First Edition. Luis Couto Lorenzo.

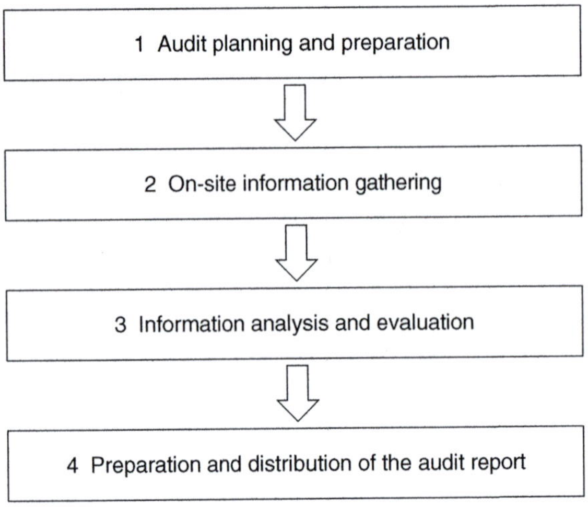

Figure 9.1 Four main steps of the audit process.

The audit is a paradigmatic case where failure to properly plan the activities to be carried out can clearly affect the outcome; that is, fail to prepare, prepare to fail. Therefore, it cannot be stressed enough that planning and preparing for the audit is essential to ensure that the process runs smoothly and effectively. Auditors should prepare for the audit so that when they arrive at the auditee's premises they know what they are going to do, what kind of information they are going to look for, where they are going to look for it, who they are going to talk to, and so on. When an auditor knows what to look for and how to look for it, the whole process becomes more efficient and the auditor conveys an image of professionalism and diligence in the eyes of the auditee from the very first moment.

Once the lead auditor has been appointed to carry out one or more audits planned for a specific purpose, the preparation tasks fall under the full responsibility of the lead auditor. At this stage, prior to each audit, it is necessary for the lead auditor to gather all the information and documentation he/she deems appropriate to adequately prepare the team's activities during the on-site audit phase, such as the size of the organization, the food sector it belongs to, the products it manufactures, the specific legislation it applies, the technology it uses, and any other information that may be helpful, taking into account the objectives and scope defined by the client.

The next phase of gathering information on site is the most visible part of the audit, especially for the auditee, as it involves more interaction between the team members and people in the audited organization. It is also at this stage that the auditor's professionalism and ability to manage the whole process during their stay at the audited organization's premises is most put to the test. Therefore, during their stay at the auditee's premises, the audit team members must use all their relationship and communication skills to maintain a friendly and respectful tone with the auditees, being sensitive to the impact of the audit on their normal activities, but at the same time guiding the process with a steady hand in order to cover all aspects that are necessary for the achievement of the audit objectives.

After gathering information at the auditee's premises, the audit team should meet to analyze and evaluate all observations and findings made by each of its members in order to make a collective and objective judgment as to which of the nonconformities or findings found should be included in the audit report. Under the coordination of the lead auditor, who is ultimately responsible for deciding what to include in the audit report, the team members should present and discuss their audit notes containing their findings and observations to ensure that they are accurate and adequately supported and to identify those that are not well documented or where further evidence needs to be obtained. If this review is carried out while the team is still on the premises of the audited organization, it is always possible to revisit certain areas to resolve any questions or issues that remain unclear.

In this review phase, prior to the closing meeting, the aim is to achieve as much consensus as possible among the team members, trying to get an overall view of the information gathered, which is always more balanced and less subjective than each individual contribution. The audit team should check whether the findings and observations are supported by sufficient objective evidence, whether they are consistent with the reference criteria used, and whether they are within the scope defined for the audit, bearing in mind that it is always better to question them now than to have to retract a nonconformity presented to the auditee because of a lack of evidence to support it or because the noncompliant criterion has not been identified.

The task of drafting the report is, as mentioned above, the responsibility of the lead auditor. Based on the work documents and the notes provided by the team members, the lead auditor should draft the audit report with the intention of providing as accurate, faithful, and objective record as possible

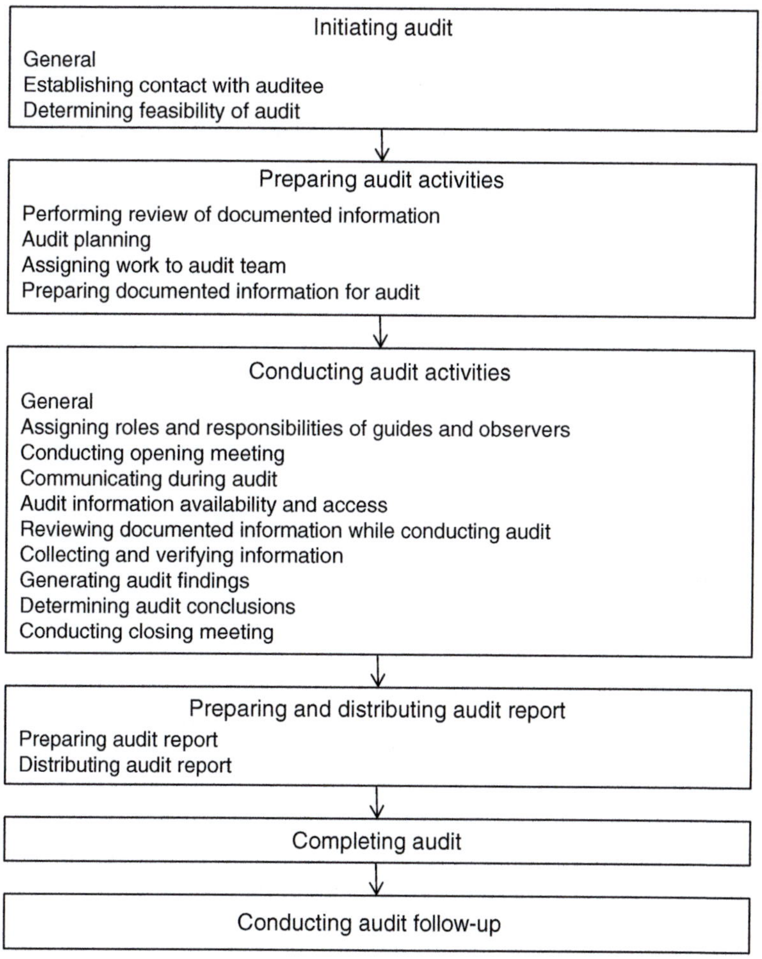

Figure 9.2 Overview of auditing activities.

of the results of the audit and the context in which the audit was carried out. The audit report is the most important and long-lasting document in the whole process, and it is therefore necessary that the information it contains is correct and contains accurate data so that anyone reading it at a later date can have a clear picture of the process and its final outcome.

When the lead auditor has completed the draft report, it is usual to pass it on to the team members to review and check for any errors or parts that need to be changed; in any case, once the opinions and contributions of all members have been considered, the final decision is made by the lead auditor (Figure 9.2).

9.2 Initiating Audit

The audit team is responsible for carrying out the audit, but the process has already started before it is even known who the auditors will be. The start of the audit is determined by the client, who is the person or organization requesting an audit and who actually initiates the whole process.

Therefore, the first thing to know when preparing for an audit is what the client wants, why they are requesting an audit, what information they want to obtain, and why they need it. Are they doing it for commercial reasons, because a customer is asking for it, or because they want to approve a supplier, or do they want to know how their management system is working to see if there are any failures or dysfunctions that need to be corrected?

Once the client is clear about the objectives of the audit, it is normal for the client to contact the audit organization or auditors, whether in-house or outsourced, and other interested parties as appropriate. They will then jointly determine the details of the audit(s) that the client wishes to carry out, such as:

- Scope: functions, activities, and areas to be audited.
- **Type of audit**: internal, external, system, product, etc.
- Audit criteria.
- Number of audits.
- Locations.
- Timing.
- Resources.
- Selection of auditors.

9.3 Selection of Auditors

The person responsible for managing the audit program should appoint the lead auditor. The lead auditor, once selected, is then responsible for selecting the other members of the audit team, a decision that may be taken alone or jointly with the program managers and/or the client.

In any case, whoever is involved in the selection process of the auditors, the most important thing is to ensure that the team, adding up the competencies and skills of each of its members, has the overall competence necessary to achieve the defined objectives, that they are independent of the activity or area being audited, and that they are free from bias and conflicts of interest.

In internal audits, and especially in small organizations, the principle of independence from the audited area is always more difficult to achieve, but the organization should make every effort to minimize this risk and promote as much objectivity as possible.

In addition to these considerations, other factors need to be taken into account when deciding on the size and composition of the audit team, such as:

- The type of audit.
- The complexity of the audited organization.
- The food sector of the organization.
- The technology of the company's processes.
- The audit methods selected.
- The legal, contractual, and other requirements that may affect the organization.
- The ability of the audit team members to interact effectively with the audited organization's personnel and to work together.

9.4 Team Size

ISO 19011:2018 defines the audit team as follows:

Audit team: One or more persons conducting an audit, supported if needed by technical experts. And adds two notes:

Note 1: One auditor of the audit team is appointed as the audit team leader.
Note 2: The audit team can include auditors-in-training.

This means that an audit can be carried out by one person; it is not always necessary to have several auditors. In any case, this option is more appropriate for simpler audits, in small companies with low complexity in their processes, and where a competent auditor with experience in the food sector of the organization to be audited is available.

However, quite a few authors recommend that teams should consist of two or more auditors. One of the reasons for this is that the competencies and skills that each auditor brings to the table complement each other, making the competence of the team as a whole more appropriate. For more complex and specialized audits, it is rare for a single person to have all the required competencies, knowledge, or skills. This means that personnel should be selected to ensure that the overall competence of the team is sufficient to achieve the audit objectives, taking into account the scope and criteria to be applied. In cases where the area to be audited involves specific processes or technology, the competence of the auditors may not be sufficient, and qualified experts in that area should be included as part of the team.

Another beneficial aspect of teamwork is being able to count on the support of a colleague, especially in more tense or conflictual situations or when acting under pressure from the audited organization. Not only is this comforting from an emotional point of view, but it is also important to defend the audit team's actions and criteria for achieving the audit objectives in the face of any reluctance expressed by the audited organization's managers.

On-site audit is a very demanding process by itself, requiring almost constant concentration to interact with people and review documents, trying to cover all the aspects included in the scope, and then processing all the information obtained to meet the audit objectives. In this context, it can be said that four eyes are better than two and that working in pairs makes the process more manageable for the auditors and more efficient, as it is more likely to avoid failures in evidence gathering or issues that are left unchecked when the task is shared between two people.

Working in pairs also allows for a division of labor and role differentiation, so that while one auditor is busy interviewing staff or checking documents, the other may be better able to observe the staff's operations or practices more carefully, picking up information or details that might otherwise have been overlooked.

Last but not least, teamwork provides a more objective overview than the individual contributions of each auditor. When findings and observations are pooled, it is always easier to neutralize more subjective positions, analyze whether there is a sufficient basis for documenting nonconformities, and see whether the criteria have been well applied or whether the audit objectives have been achieved, before presenting the results to the auditee.

However, these positive aspects of teamwork can be undermined if there is a lack of understanding between team members or if mistakes are made by some team members, such as making inappropriate comments, expressing different opinions in the presence of the auditee, lack of mutual support, not following the audit plan, getting lost in irrelevant issues, talking too much, or adopting a consultant approach rather than an auditor approach.

In addition to the auditors, the audit team may include other members such as technical experts and trainee auditors, as well as observers agreed by the client or the audited organization, who are not part of the team and cannot interfere with the audit activities.

9.5 Establishing Contact with Auditee

The initial contact with the auditee is a task that falls under the responsibility of the lead auditor, who should contact the organization to be audited using the means he considers most appropriate: telephone, email, or other tools, and identify the contact person designated to communicate with the auditors.

The purpose of this initial contact is to inform the organization that an audit of their organization is planned to take place on a forthcoming date in accordance with the previously agreed program, informing the operator of the objectives, scope, criteria, legal requirements, methods, schedule, and composition of the audit team.

Announced or surprise audits? The most common practice, at least so far, is for audits to be announced to the audited organization in advance. However, unannounced audits are becoming more common, and in fact, some certification schemes require an unannounced audit every specified period. Whether mandatory or chosen by the operator, unannounced audits do serve to demonstrate an additional commitment to food safety. It is of even greater interest in the food sector because of the importance of both good handling or manufacturing practices and the maintenance of clean and tidy premises in the food safety management system. Therefore, if the audit is unannounced, the hygiene and health conditions found in the plant will always be a truer reflection of the usual reality of the company than if what we see is a situation prepared for the day when they know the auditors are coming.

At the initial contact, the lead auditor should set the audit dates according to the agenda or availability indicated by the organization to be audited. The initial contact with the auditee is also used to request information relevant to the planning and preparation of the audit, such as working hours or shifts, the need for protective equipment, or requirements for access to the site. In this way, the lead auditor uses this initial contact to request the company's management system documentation. Often, this will not be all of the system documentation, nor will it be necessary because it is too voluminous or not in a format that can be shared, but at least some parts of the hazard analysis critical control point (HACCP) manual, such as the flow diagrams, the control chart showing the critical control points (CCPs) that have been identified, or any other documentation that gives an idea of the size and type and number of products that the operator produces.

Another aspect that needs to be clarified during this initial contact with the auditee is the audit methods and audit communication channels, as well as the availability of technologies in case the auditors wish to conduct an audit or parts of it remotely, e.g. to collect information, conduct interviews, or observe handling or manufacturing practices with the help of a remote guide who may use streaming device or augmented reality technologies to facilitate interaction. Remote auditing has grown significantly in recent years and can have both advantages and disadvantages; the main disadvantage is probably that not being on site limits the auditors' ability to observe and check things for themselves, but this can be offset by improved efficiency, time savings, and the ease of increasing the sample size by receiving data telematically. Therefore, if used in the right way and for the right purpose, remote auditing can be of great benefit in the future.

9.6 Determining the Feasibility of the Audit

Before proceeding with the sequence of audit activities, it should be determined whether it is feasible to carry out the audit in order to achieve the stated objectives.

Occasionally, when contacting the audited organization, circumstances may arise that make it necessary to consider whether the process should be continued or postponed to a more appropriate time. This may be due to the auditee's inability to attend the audit, e.g. because of the absence of those responsible, coincidence with other audits scheduled for the same dates, works or refurbishment of the premises, changes in the activity or ownership of the business or even lack of cooperation, or the system documentation being not sufficiently developed.

In other cases, the reasons for the lack of feasibility may originate from the audit program management or the auditors, e.g. insufficient information to prepare the program and plan the audits, lack of human or material resources, or lack of time to complete the audit program on time.

It is not uncommon that, when contacting the auditee to set audit dates, it may be necessary to reschedule the audit to dates when the auditee can attend the audit, or there may be other circumstances affecting the auditors that make it necessary to reschedule the audit days within a reasonable period of time.

9.7 Preparation of Audit Activities

9.7.1 Initial document review

The purpose of this initial document review is to obtain sufficient information about the organization to be audited in order to better prepare the on-site audit activities.

Based on the documentation provided by the audited organization, the initial review can be carried out either in the auditor's office, conveniently without the need to travel and without time constraints, or at the auditee's premises. However, although the first option has its advantages, it may not be possible to obtain a copy of the required documents, either because they are not in a suitable format or because of the auditee's reluctance, and in that case, there is no alternative but to carry out the review on the auditee's premises.

In addition to the basic documentation of the HACCP system, there are other documents of the management system that can also be useful in this phase of preparation of the activities, such as the process map, the list of documents in force (if the system is subject to document control), the organization chart of the company, or even, depending on the preferences of each auditor, some type of records, such as the reports of previous audits or customer complaints. However, records are usually left for the on-site audit phase.

In addition, the preparation of the audit requires other information on the characteristics of the organization to be audited, which is essential to get an idea of the type of company and the workload and duration that the whole process may entail, such as the following:

- Size and turnover.
- Number of employees.
- Number and type of products manufactured.
- Processes it carries out.
- Scope of its activities within the supply chain.
- Number of sites or locations.
- Timetable and shifts.
- Technology used.
- Legislation and specific regulations that apply to it.

9.7.2 Preparation of the Audit Plan

Based on the information gathered about the organization to be audited, the lead auditor should prepare the audit plan, which describes the planned activities to be carried out at the organization's premises. Preparation of the audit plan is the responsibility of the lead auditor, but depending on what has been agreed in the audit program, the client may also have the right to review and approve the plan before it is communicated to the auditee. On the other hand, it is normal to involve the other members of the team in the preparation of the plan so that they can provide their input on what may be the most appropriate timing.

Before moving on, it is important to clarify what an audit plan is and how it differs from an audit program, as this is an issue that is sometimes confused or misused. According to ISO 19011, the definitions of audit plan and audit program are as follows

Audit plan: description of the activities and arrangements for an audit.
Audit program: arrangements for a set of one or more audits planned for a specific time frame and directed towards a specific purpose.

As can be seen, the audit plan describes the agenda or schedule of activities to be performed in a single audit and is prepared by the lead auditor, whereas the audit program, which has a broader scope, refers to the stated objectives, resources, and follow-up of a series of audits planned over time and is prepared by the audit program management.

The audit plan is used to better manage resources and to allocate time as efficiently as necessary to cover all areas included in the scope. The audit plan is helpful for the audited company to know in advance what activities will be carried out on its premises, what areas will be involved, and how long the whole process will take, so that it can make the necessary preparations to assist the audit team without any unforeseen problems or conflicts, for example, to ensure that all persons responsible for the management system are present during the audit and can devote some or all of their time to assisting the audit team.

For the team members, the preparation of the audit plan may serve as an exercise to try to anticipate and foresee problems that may arise on the appointed day and to consider other aspects such as logistical arrangements, resources available on site, work schedules, shifts in the company. The level of detail of the plan may vary according to the type of audit and the scope and complexity of the activities to be carried out, but the plan should be understood as flexible enough to allow changes to be made as necessary as the audit activities unfold.

The typical content of an audit plan is as follows:

- Nature or type of audit.
- Audit objectives.
- Scope, including identification of functional units of the organization and processes to be audited.
- Audit criteria and reference documents.
- Locations, schedules, dates, and expected duration.
- Audit methods.
- Identification of audit team members, their roles, and responsibilities, including the presence of observers, experts, or trainee auditors.
- Schedule of meetings to be held with the auditee.

ISO 19011 adds other elements that may also be included in the audit plan if appropriate, such as:

- Identification of the auditee's representative for the audit.
- Working and reporting language.
- Audit report topics.
- Logistical and communication arrangements.
- Specific measures to address risks to the achievement of the audit objectives.
- Confidentiality and information security issues.
- Follow-up from a previous audit.
- Follow-up activities to the planned audit.
- Coordination with other audit activities in the case of a joint audit (Figure 9.3).

9.7.3 Allocation of Tasks to the Audit Team

In order to optimize the audit process and manage resources efficiently, the lead auditor should allocate tasks to the team members. Taking into account the competence of the team members and in consultation with the team, the lead auditor should assign responsibilities for auditing different elements of the system. This may be based on different areas, processes, production lines, prerequisite programs, or specific sites, as decided on a case-by-case basis.

Although at certain times, such as meetings with representatives of the auditee, the whole team will be present, it is normal for the auditors to work separately during the on-site audit activities, once the allocation of tasks has been agreed. This should not preclude frequent communication between the auditors at various times during the audit to exchange information and views on the progress of the process or to make changes to task assignments if necessary to meet the audit objectives.

Audit Plan

Audited company: Cárnicas Monte Louro

Address: Polígono Agrela-Bens, B-45, A Coruña

Dates: May 15 and 16, 2023

Time scheduled: 09:00 h to 14:00 h

1. **Audit team**

Surname	Name	Position
Taboada Rodríguez	Dolores	Lead auditor
Souto Valcárcel	Manuel	Auditor

2. **Purpose:**
Verification that the food business operator has put in place a self-control system based on the HACCP principles, that it is adequate to control the hazards associated with the activities carried out, and that it is properly maintained.

3. **Scope of the audit:**
All activities and functions related to processed food: HACCP and hygiene requirements.
Period covered: From 2022

4. **Audit criteria**: Hygiene packet regulations: Rº 178/2002, Rº 852/2004, Rº 853/2004 and Rº 2073/2005; Regulations: Rº 1333/2008, Rº 1069/2009, Rº 142/2011 and Rº 1169/2011; CXC 1-1969 General principles offood hygiene
Audit method: Representative on-site sampling

5. **Overall schedule:**

May 15, 2023

1 phase 9:00 h:
Opening meeting
Introduction of the audit team and explanation of the objective and scope of the audit.
System documentation initial review

2 phase 10:00 h to 14:00 h:
Tour of the facilities
Review of documentation
On-site audit activities

May 16, 2023

2 phase 9:00 h to 12:00 h:
Continuation of on-site audit activities

3 phase 12:00 h to 13:00 h:
Audit team meeting

4 phase 13:00 h to 14:00 h:
Closing meeting
Presentation of the audit findings

Figure 9.3 An example audit plan.

9.8 Opening Meeting

According to the audit plan, the opening meeting generally marks the beginning of the visible phase of the audit, although, as we have seen, the process has started earlier, both for the audit team members – with the planning and desk review activities – and for the auditee itself, which is most likely to have made some preparations before the audit team arrived.

The purpose of the opening meeting is to:

- Introduce the members of the audit team to the members of the organization involved in the audit and, reciprocally, introduce the management representatives to the audit team.
- Confirm the scope and objectives of the audit plan.
- Ensure that the planned audit activities can be carried out.
- Clarify any doubts or questions that arise or are raised by the attendees.

The opening meeting should be attended by all members of the audit team and, from the audited organization, those responsible for the management and maintenance of the food safety system. Therefore, if only members of a technical or middle level are present, such as typically the quality manager, technical director, or production manager, among others, and no representative of the management level or top management of the company is present, this may be interpreted by the audit team as a sign of a lack of interest in the outcome of the audit or a lack of management commitment to the implementation and continuous improvement of the system and ultimately to the food safety of its products. This may be confirmed or assessed during the audit, depending on the results obtained.

The level of detail and formality of the opening meeting may vary depending on the nature of the audit and the auditee's familiarity with the audit activities. For internal audits or in small companies, the opening meeting may be limited to a simple communication or explanation of what the process will entail and answering any questions raised by the auditee. In other situations, however, it may be more formal and may even require a register of attendees.

In addition, there are other matters that may also need to be addressed at the opening meeting, such as:

- Access to system documentation and records related to its implementation.
- Clarifying the lines of communication with the audited organization, i.e. who will be the interlocutors or the people who will accompany the audit team.
- Determine the need for, and role of, guides or assistants.
- Determine the role, if any, of other members of the audit team, such as experts, observers, or trainee auditors.
- Availability of working rooms for the audit team to review documentation or meet.
- If necessary, further specify the schedule of on-site activities. This is because some people may only be available at certain times, in which case it is necessary to agree with the auditee the most appropriate time for them to be present to audit the areas in which they work.
- Advise the audit team of relevant health, safety, security, and emergency procedures.
- Confidentiality requirements or clauses.
- Language of the audit.
- Arrangements for stops or breaks, meals, refreshments, etc.

From the very first contact with the auditee's staff, the audit team needs to display an appropriate attitude: friendly and attentive to questions or issues raised by the auditee, but at the same, time confident and taking the leadership role appropriate to them to guide the whole process.

The opening meeting should be conducted by the lead auditor, who is responsible for trying to keep to the agenda and avoid unnecessary deviations or delays. This is not always an easy task, especially when senior management representatives are present who may be more inclined to take control of the meetings in their company. In any case, if the lead auditor observes that the planned schedule is beginning to be unduly delayed, he should point out the need to move on to the next stage of the process. Most references suggest that the opening meeting should not last much longer than 30 minutes. On occasions where less formality is required, such as internal audits, five or ten minutes may be sufficient.

9.9 The Audit Tour

After the opening meeting, it is usual for the audit team, accompanied by one or more guides, to carry out a full tour of the auditee's premises, which may include the outside of the factory. Although this is usually done at this stage, this does not preclude it from being done before the opening meeting, as may be the case with companies that start work very early, in which case the auditors may need to visit the facilities first if they wish to observe certain activities, and the opening meeting may be scheduled for a later time when all those who plan to attend can be present.

The main purpose of this visit is to familiarize the audit team with the layout of the auditee's facilities, but it can also be used to observe activities and collect some data, depending on the audit team's preferences. The duration of this visit will vary: it will obviously depend on the size of the industry, the nature of the processes, and the complexity of the operations, but also on the auditor's willingness to take the time to ask for explanations, observe the processes, or talk to the operators at this point or to leave it for a later visit. In any case, the pace and scope of the visit should be set by the lead auditor to avoid it becoming a typical guided tour of the facilities, which is usually done for clients or customers or for commercial purposes and which has little to do with the objectives of the audit. To provide some guidance, some authors consider that an average of one hour may be sufficient to carry out this tour, although in some companies more time may be required.

In some publications, the verification of the flow diagram is placed in the schedule of activities between the opening meeting and the document review in the plant. In fact, during the initial visit to the premises, the audit team will observe the processes and interview the personnel, thus gaining an idea of the sequence of operations that make up the production process. However, again, depending on the activity of the company and the characteristics of the processes, it is not always possible to verify the flow diagram during the initial walk-through. Thus, depending on the degree of mechanization of the company, manual activities are easier to observe; depending on the length of the production cycle or the fact that there may be activities that are not carried out every day, it may not be possible to verify the entire flow diagram in a single session.

Therefore, although the initial visit will help the audit team to become familiar with the processes, in order to verify the flow diagram at the level of detail required for the audit, it is more appropriate for the team to spend more time on it, to prepare it, and often, due to time constraints, to focus only on the phases of the process that are considered to be the most critical.

9.10 Document Review During the Audit

After the opening meeting and before starting the audit activities at the production site, the audit team normally moves to the room offered by the auditee to carry out the document review of the management system. This document review is a continuation of what may have been done previously, depending on the parts of the system that the auditee's representative has made available to them in advance, with the particularity that they now have access to all the documentation related to the implementation and management of the self-control system.

Continuing with the two-stage audit scheme, the first stage is to verify that the documented system complies with the HACCP principles and requirements, including the legal requirements that may be applicable depending on the food sector and the activity carried out by the audited company. The second objective is to gather information on the activities, processes, and elements of the management system in order to better prepare the audit team for their visit to the company and, in the second stage, to verify that the company applies the procedures of the system as described, i.e. if they do what they say they do.

When auditors are given all the system documentation, which in some companies can be considerable, it is easy for some to feel overwhelmed by the task ahead or, especially in the case of new auditors, to feel that they do not know where to start. It is like a mixture of a writer's blank page syndrome and an artist's stage fright. However, this initial feeling of hesitation is something that, as we will see below, should disappear as soon as the auditor knows what to do, especially if he has been through the same situation more than once. It is all about getting started.

Another difficulty that the auditor has to deal with is the variety of formats in which the HACCP system can be documented: depending on the characteristics of the company (size, activity, number of products, etc.), according to the preferences of the HACCP team, it can be integrated into a quality management system, it can form part of the documentation of a food safety standard or certification scheme adopted by the company, or even – and this is becoming increasingly common – it can be documented electronically. This question, which can be somewhat misleading when the auditor receives the documentation, as if we cannot see the forest for the trees, is not really a problem if we know what to look for. In other words, if the auditor knows the structure of the HACCP plan and its components, it does not matter what format or how they are arranged in the system documentation.

This is the moment to go back to the HACCP basics of the Codex, which defines the elements that a HACCP system must contain and that must be present in one way or another in the company's documentation, whatever format is chosen. Thus, following this scheme, the auditor will start by looking at how the products are described, where the flow diagrams are, how the hazard analysis has been documented, what the preventive measures are for each of the hazards, and so on until all the phases and principles have been completed.

When carrying out the document review, it is normal to go from the general to the specific: beginning with the structure of the system – the table of contents, the parts into which the HACCP plan is divided, and the prerequisites – and then going down to the lower levels of the documentation hierarchy – procedures, work instructions, specifications, etc.

By examining the documentation, the auditor can get an idea of the competence of the HACCP team that developed the HACCP system and the information and input data used to carry out the HACCP study, which is as important as the final result for the evaluation of the self-control system. When reviewing the documentation, the audit team should consider whether the information contained in the documentation meets the established criteria and, in particular, whether it is:

- **Complete**: it covers all aspects required by the HACCP system.
- **Correct**: in accordance with the HACCP system and any applicable legal requirements.

- **Consistent**: with the activities of the company and the associated documentation of the system.
- **Up to date**: the documents are kept up to date and reflect changes in the company.

Although included in this stage of the process, this does not mean that the document review during the audit has to be done all at once. On the contrary, many auditors prefer to alternate periods of document review with fieldwork so that, depending on the parts of the system or the circumstances of the audit, the audit team can combine the time spent in the office reviewing documents and making notes with the time spent gathering information and looking for evidence in the field in the way that is most practical for each auditor to carry out the checks that he considers necessary. In calculating these times, it is expected that at least 50% of the total time should be spent inspecting facilities, observing processes, interviewing personnel, and reviewing records in production areas to avoid reducing the process to a simple desk audit. A number of standards or certification schemes require this split of time between factory and site audits to be taken into account.

One thing to bear in mind from the outset when reviewing documentation is that it is not possible to review all the documents in the system; it is not possible – there is not enough time – but it is not necessary. The audit is essentially a sampling exercise, and it is precisely the competence or skill of the auditor to select the parts he considers most appropriate, to obtain the information he needs to draw reliable conclusions and thus generate the confidence expected by all parties. When dealing with system documentation, especially if it is voluminous or in different parts, the auditor should not hesitate to ask the accompanying person or the person responsible for the auditee to assist in locating the documents to be examined; this can save unnecessary effort and avoid wasting time, which only detracts from the effectiveness of the audit.

In addition to checking that the HACCP plan complies with the Codex principles and methodology and that it contains all the elements required for the documented system, the document review is also used to prepare for the on-site audit. As the auditor becomes more familiar with the management system of the audited organization, he can select the elements or parts of the system to be verified on site. It should be noted that many management system documents are both evidence – the proof that this part is documented – and audit criteria, i.e. the reference that defines how things should be done. Therefore, when reviewing the documentation, whether it is a procedure or a work instruction, the auditor can take notes on how a particular task or process is to be carried out and then check on the ground whether it is actually being carried out in the required way. To illustrate this better, we can use the example of a cleaning plan: the auditor, when reviewing the plan, first checks that the method for cleaning a particular piece of equipment is described, and then, after taking note of some details of this method (frequency, person responsible, products used, etc.), goes to the plant to check that it is cleaned in the described way. In fact, the two parts always work together: first, seeing how things should be done and, second, checking that things are being done as described.

The preparation of the on-site audit process requires the auditor, when carrying out the document review, to start selecting the areas to be visited, the processes to be observed, the people to be interviewed, the records to be requested in each department, the data to be verified, etc. At this point, the auditor's experience or skills are fundamental in order to know, given the time constraints, how to select the elements that can provide the most useful and reliable information, to verify the correct implementation of the HACCP procedures. The planning of the collection of evidence is mainly done at this stage, but this does not prevent it from being revised or refined at a later stage.

9.11 Conducting On-site Audit Activities

This is the main and most characteristic stage of the audit, at least for the staff of the audited organization, where the auditors are engaged in gathering the information they need to verify how the management system operates in practice.

According to the two-stage audit scheme described above, once the desk review has been carried out to assess the conformity of the HACCP manual with the Codex principles and applicable requirements, it is now time to check how the provisions and procedures of the system are implemented, i.e. whether they are carried out as described and to what extent they are being complied with (Figure 9.4).

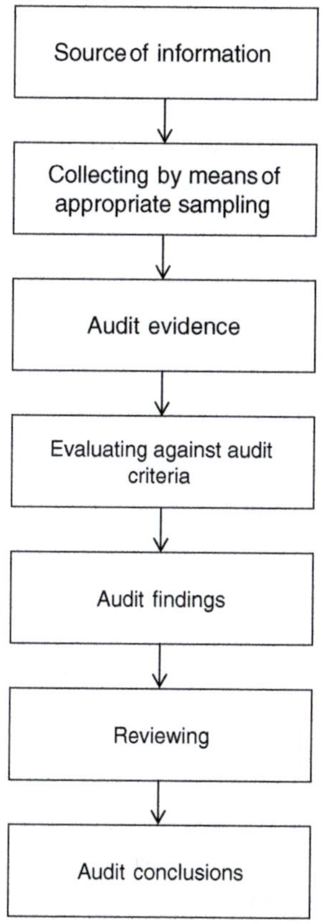

Figure 9.4 Collecting and evaluating information.

It is a common assumption that failures in the HACCP system can be due to poor design of the HACCP plan or poor implementation or maintenance of the system. Both aspects are fundamental and interrelated; they depend on each other. The importance of design has already been discussed in previous chapters, but it is no use having a correctly drafted and valid HACCP manual if it is not implemented in an adequate manner by the company's personnel and maintained over time so that the HACCP system becomes an integral part of the organization's production activities.

The task of the audit team at this stage of the process is to gather, through appropriate sampling, the information necessary to reach reliable conclusions in accordance with the requirements of the auditing principles. The information sought by the auditor should be useful and relevant to the stated objectives and consistent with the audit scope and criteria. Only information that is verifiable can be accepted as valid in the audit, so that any interested party can verify the truth and accuracy of the data or information supporting the auditor's conclusions.

As seen above, the auditor's role is not to present subjective considerations or assumptions to the client or the audited entity, but to present valid conclusions based on the objective evidence gathered during the audit. The subject matter with which auditors work and their leitmotiv ("driving motive" in German) are objective evidence, and it can be said that the whole audit process is geared toward finding ways of obtaining them. The audit process moves on two legs: the established criteria and the search for evidence to contrast with these criteria.

9.12 Sources of Information

In order to carry out information gathering on site, the audit team should consider the strategy that will best enable them to obtain the evidence they need in accordance with the objectives of the audit. Depending on the audit criteria and audit objectives, it is necessary to determine the type of evidence required to support the findings and conclusions of the overall process.

Audit evidence should be:

- **Sufficient**: sufficient and representative to enable the auditor to make a judgment on a reasonable basis.
- **Relevant**: in accordance with the audit objectives and criteria.
- **Reliable**: the auditor should assess the reliability of the source and the quality of the information obtained.
- **Verifiable**: capable of being corroborated by any interested party.
- **Objective**: free from bias and subjectivity.

The information or data collected may be of a qualitative or quantitative nature. Qualitative information refers, for example, to whether or not an activity or action is being carried out as intended. Quantitative information, on the other hand, involves counting or measuring various parameters to determine the extent to which a procedure is being carried out or how effective it is.

The information or objective evidence required can be obtained in different ways. The audit team may use a variety of sources of information, or a combination of sources, depending on the scope and complexity of the audit:

- Observation of the activities, working environment, and conditions in which they are carried out.
- Documents of various types, such as manuals, standards, procedures, instructions, specifications, drawings.
- Physical examination or inspection.
- Interviews with employees and others.
- Analytical or rational.
- Records, such as monitoring records, corrective actions, and audit reports.
- Information from sampling programs: microbiological analysis, chemical analysis, etc.
- Data summaries, analyses, and performance indicators.
- Information from external sources: customer feedback, complaints, claims, product returns, etc.
- Other databases and Internet.

9.12.1 Observation of Activities

Observation of people, activities, and working conditions is a direct source of information. Observation means looking carefully, and this means that the auditor needs to understand what he is seeing and what is going on, in order to compare it with the way it should be done, i.e. with what is specified in the operator's procedures. This is often not an easy task; it depends on the nature and complexity of the processes: it is not the same to observe a manual process as a more mechanized one where we do not even see the product being transformed. It also depends on the auditor's knowledge of the technology and processes in the audited organization.

Observation takes time; it is not a matter of looking around the facilities or at the people working, but of paying attention to understand the process and, if possible, to find out what goes in, what comes out, what stays in, how it is transformed, what controls are in place, etc., to the extent necessary for the purposes of the audit (Figure 9.5).

The auditor must observe the activities from a suitable position, close enough to the process to be able to gather the information he needs, but without disturbing the workers or interfering with their work. There are several reasons for this: firstly, in order to minimize disruption to the company's productive activities and for health and safety reasons, but also so that workers can carry out their tasks naturally, as they normally do, unaffected by the presence of the auditor too close to them.

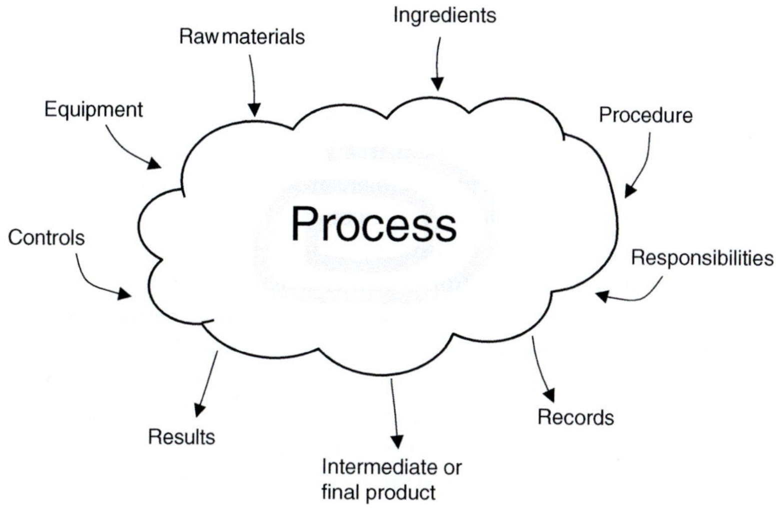

Figure 9.5 Mind map for observing a process.

In order for this source of information to be most effective, the auditor should plan what information he will be looking for, where he will get it, and even at what time if, for example, he wishes to observe practices at different times of the day or during different shifts. He must then understand what he has seen and heard, assess whether it is consistent with the system's procedures, and corroborate this information by other means if he deems it appropriate.

9.12.2 Physical Examination

Obtaining evidence by physical examination is another source of information that the auditor may use. Depending on the characteristics of the process and the nature of the information to be verified, the auditor may measure certain physical values or process parameters or carry out tests as deemed necessary.

Physical inspection may be carried out on product characteristics as well as on facilities and working conditions. To carry out the measurements, the auditor may use his own instruments, respecting the rules for the use and safety of the equipment, or ask the workers to carry out the test with the operator's equipment or instruments.

Physical examinations are a direct means of obtaining information and are therefore one of the most useful and reliable forms of evidence, since they cannot be easily altered or falsified.

9.12.3 Review of Documents and Records

Documents and records are some of the most typical sources of information in auditing. The system documents describe what is to be done, and the record records what has been done. Technically, a record is a type of document whose purpose is to collect information about the outcome of a process. ISO 9000:2015, Quality Management Systems – Fundamentals and Vocabulary, defines it as follows:

Record: document stating results achieved or providing evidence of activities performed.

With appropriate sampling, documents usually provide very useful information for audit purposes; however, not everything can be verified by reviewing documents, and other methods of obtaining evidence should be used.

The auditor may examine different types of documents. Some are of internal origin, such as those forming part of the management system: procedures, work instructions, plans, specifications, etc., or those generated by the organization's productive and commercial activities; and others are of external origin, i.e. those provided by other companies as a result of their professional or commercial relationship with the audited organization, such as customers, suppliers, service providers.

Records may be in paper form or, increasingly, in electronic form. If in electronic form, the auditor should consider the integrity and authenticity of such records and assess the measures taken by the operator to control access, to identify the user(s) authorized to manage such records, or to prevent them from being inadvertently or deliberately altered or tampered with.

Records are a fundamental source of information in auditing, but the quality of the evidence they provide varies, and not all of it is equally reliable. In the critical spirit that characterizes his professional performance, the auditor must assess the reliability of the information to be used as evidence, depending on its origin and nature, as well as the circumstances in which it was obtained. A record containing specific information related to the traceability of a product, which can be verified, or a record of external origin, such as a supplier's delivery note, is usually more reliable than those related to the routine performance of activities consisting of ticking boxes or crosses on a template, which may always be subject to the temptation of being filled in automatically or fraudulently.

For example, it is not uncommon to find certain types of records that appear to have been completed in a single sitting, neatly, with answers that are always perfect or with no incidents recorded, which may lead us to doubt their authenticity or the reliability of the evidence they provide. Auditors should be aware of such practices and, if they suspect their use, it is better to proceed and try to rely on other sources of information to obtain the evidence they are looking for, rather than inferring a deviation from the alleged falsity of the records, which would create an embarrassing situation with the auditee.

A similar situation occurs when the auditor requests a record from the auditee, and following the auditee's absence and some delay in providing the record, the auditor considers that there are signs that such a record has been made after the request. Again, if the auditor suspects that this is the case, the auditor's attitude should not be to infer the existence of a noncompliance directly, but to increase the sampling on the aspect he is examining and, this time, trying to be present at the place where such records should have been kept in order to avoid creating them on the spot.

9.12.4 Interviews

Interviews are another vital source of audit information. Indeed, some authors (David Mills 1993) argue that in their relationship with the organization's staff, auditors rarely stop asking questions while on site. This makes the audit a virtually continuous dialogue, and the information gathered orally is a fundamental part of the audit process. On the other hand, there may be some information that can only be obtained in this way and some that can be obtained orally but needs to be corroborated in some other way.

As we shall see, interviews are an essential part of the management system audit, but to be effective, the auditor needs to know how to conduct them. Conducting a good interview is not an easy task; it requires preparation and practice. There are a number of established guidelines and techniques that the auditor needs to practice if he is to improve his skills in using this source of information.

Before starting to ask questions of the organization's staff as part of the on-site audit process, the auditor must first consider the type of information he wants to obtain and then identify or select the people who can best provide it. Once the auditor has a preliminary idea of the people he wants to interview, while remaining open to changes that may occur, he should also decide on the most appropriate time and place to conduct the interview.

With the information gained from reviewing the management system documentation, the auditor can prepare for the interviews by taking notes or preparing a list of questions in advance. In some cases, questionnaires have been prepared for use at this stage of the audit. However, while these questionnaires or notes prepared by the auditor can be helpful, they should never be followed in such a rigid way that they affect the flow of the interview or prevent the auditor from going off script or adding further questions based on the information provided by the interviewee or if he finds a topic of interest to the audit objectives.

As with any communication process, there are several factors that can influence the effectiveness of interviews, not only the type of question and the way it is asked but also elements of nonverbal communication such as the auditor's attitude, tone of voice, posture, gestures, and also the context in which the interview takes place, including both the physical space and cultural or social differences that the auditor should consider when conducting the interview.

The first moments of the interview are fundamental, as they determine the climate and the dynamics of the whole process; many people may feel uncomfortable or intimidated by the audit process or by the interview itself, and for this reason, the auditor must know how to reduce tensions and create a climate of trust as much as possible.

Another aspect to be taken into account is the role or position that the person to be interviewed occupies within the company; it is not the same to interact with a technical director or a quality manager as it is with an operator who performs a basic function on the production line, which does not mean that the latter is not often the person who can provide the most valuable information to verify certain issues, because they are closest to the production process and know what really happens in the plant, or simply because they are the person responsible for implementing the procedure that the auditor wants to verify. In these cases, the auditor must adapt his language to the technical level of the operator being interviewed, avoiding the use of technical terms and formulating the question in such a way that it can be perfectly understood by that person.

9.12.4.1 Types of Questions

There are different types of questions that the auditor may use during the interview, depending on the purpose or type of information to be obtained:

- **Open-ended questions**: These are questions that cannot be answered with a "yes" or "no" but require further explanation. They provide the most information and can often be asked in the following order:
  - What?
  - Why?
  - When?
  - How?
  - Where?
  - Who?
- **Closed questions**: These are questions that can be answered with "yes" or "no" and can be used by the auditor to confirm certain information.
- **Specific questions**: These are asked to obtain specific information or data.

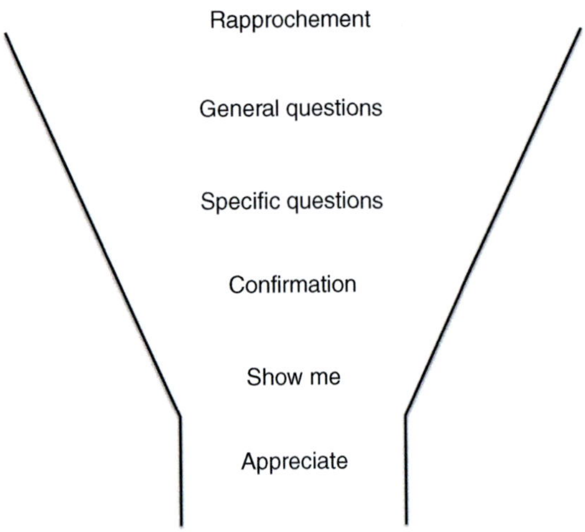

Figure 9.6 Funnel interview phases.

- **Hypothetical questions**: These are very useful to know how to act in certain situations or what kind of actions to take (Figure 9.6): for example, "What would you do if…?" or "If this were to happen…?"

In general, interviews are usually divided into more or less clearly defined phases, although it is not always necessary to follow all the steps, depending on the purpose of the interview or the degree of formality.

The first stage is the rapprochement between the interviewer and the interviewee. The auditor should start by introducing himself and explaining why he wants to ask him some questions. The first questions should be of a general nature and are often intended to encourage the person to talk rather than to obtain specific information. For example, it is very common to start by asking about the person's work; most people are happy to talk about the activities they carry out in their job. Once the auditor feels that the initial nervousness has dissipated and a more conducive atmosphere has been created, he can then introduce questions on more specific topics or specific data in order to obtain the information or evidence he is looking for.

During the interview, the auditor's main task is to listen, to listen actively: to show interest in what the interviewee is saying and try to understand the information he is conveying. Throughout this time, it is important that the auditor adopts the right posture: looking the interviewee in the eye, at an appropriate distance, smiling and nodding to reinforce his explanations when necessary. Questions should be asked one at a time, at an appropriate pace, without overloading the interviewee; this is particularly to be avoided when two auditors go together – if they are not very experienced – and both ask questions, sometimes almost at the same time and on different issues.

The auditor should ask questions in a clear, unambiguous way, in a language that the interviewee understands perfectly, avoiding technicalities, especially when addressing less qualified staff, as mentioned above. The auditor needs to make sure that the interviewee has understood the question clearly and, conversely, that the auditor understands what the interviewee is saying. If the answer is ambiguous or unclear, the auditor should not hesitate to insist on rephrasing, if necessary. It is also important to manage the use of pauses and silences. Silences can be positive or negative. A positive silence is one that the auditor uses after an overly brief answer from the interviewee, gently pressuring the interviewee to continue and complete his answer with more information. A negative silence, on the other hand, is one in which the interviewee may become blocked, not knowing what to say, and which, if prolonged, may cause anxiety or create an undesirable climate of tension, which the auditor should avoid.

Another aspect that the auditor needs to be aware of is the influence of the guide during the interview. Many workers may feel self-conscious or pressured to speak in the presence of a guide, who may often be a manager or in a position of authority over their work. The auditor should be aware of such situations and try to minimize this effect by trying to keep the guide in the background of the interview, for example by positioning relative physical distances between them, and especially by not letting them answer for the interviewee. If this happens, the auditor should quickly, politely but clearly, warn the guide that he wants to hear the answers from the mouth of the person being audited. If this is not done, it is easy to miss the chance to get the evidence he is looking for.

As the interview progresses, it is important that the auditor takes notes of the information and details provided by the interviewee in order to be able to properly document the audit findings; relying on memory is never a good plan as it is easy to forget any data or details needed to document the findings after some time has passed. Seeing the auditor taking notes can cause some discomfort to the interviewee if he thinks that what the auditor is writing down is all the negative aspects of what he is reporting. The auditor should find a way to reassure the interviewee, for example by asking before the interview if he would mind taking some notes, by explaining that it is just a record of data, or by doing it in a more discreet way, without constantly holding the notebook or questionnaire in front of the interviewee.

Once the auditor has covered the points he intended to make in the interview, he can ask closed questions to check that he has understood the information given by the interviewee. These are questions of the type: "So, as I understand it, you do this, don't you?," which the interviewee usually answers with "yes" or "no" and which serves to confirm the data and ensure that there are no misunderstandings. The idea is to follow a cycle in which a request for clarification from the auditee is followed by a question to confirm the information.

Although oral information is very valuable, the auditor must seek to cross-check it in a number of ways before accepting it as the only objective evidence of nonconformity. For this reason, it is essential that the auditor, having listened to the interviewees, attempts to verify the facts for himself. To this end, the auditor should not hesitate to use the "show me" technique. In other words, after listening to the auditee's explanations, he should ask for verification using a statement such as the following: "OK, I think I understand, but now please show me."

"Show me," two words that, for David Mills, "should be engraved on the heart of the true auditor" and without which "no form of auditing would be of any value," for nothing can replace the value of what an auditor has personally seen, heard, read, or witnessed.

At the end of the interview, the auditor should always thank the interviewees for their participation and cooperation.

Becoming a good interviewer requires skill and practice. It is an activity involving a high degree of human interaction that can be influenced by several factors, such as the context, the working environment, and the personality of each interviewee. The auditor needs to know how to adapt to each employee's personality and the circumstances.

It is clear that an audit is not an event that is greeted with great enthusiasm by the receiving organization; on the contrary, it is often viewed with some concern or suspicion because of the findings and the consequences that may follow. In any case, not everyone reacts or cooperates with the audit in the same way; it depends largely on the motivation and commitment to maintain and improve the system and, ultimately, on the maturity of the company's food safety culture. The auditor must be sensitive to the moods and attitudes of individual employees. For example, it is not uncommon to find people who are unhappy in their jobs or who do not feel valued for the work they do, and who are unlikely to be very cooperative with the audit and provide a lot of information about their work. The auditor's questioning strategy should not be the same for this type of person. A different approach may be used, for example with questions such as "What would you do to improve this process?", "How do you think this task could be done more effectively?", or "What would you need to make the process more efficient?", which will certainly motivate the auditee to participate much more, seeing that there is finally a person who is interested in their ideas or their situation.

Another type of employee the auditor may encounter are those who, regardless of their level of satisfaction with their job, want to use the audit to send messages to management about problems they have detected in their department or other departments, so that their superiors are aware of them and can solve them. There is no harm in listening to each other; on the contrary, this information can be very valuable, as long as the auditor then verifies it and the indications given and gets a better idea of what the situation is and what kind of evidence supports it. As noted above, oral information is usually not sufficient evidence and needs to be verified by other means (Figure 9.7). Scott Parsowith (1995) sums it up perfectly in this sentence: "Verify what is said. Seek and ask for facts. Believe the confessions and verify the complaints."

9.12.5 Analytical

Analytical evidence is a type of indirect evidence created by the auditor through rational reasoning and the use of other sources of information. Evidence can be divided into two types: direct or natural evidence and indirect or created evidence.

Natural evidence is the thing or event itself. It does not require any mental effort to understand. It is directly observed. For example, if an auditor observes an operator's improper handling or a structural defect in a facility, the mere observation is evidence.

Indirect evidence is not directly observable. It requires proof or rational argumentation to be apparent. Analytical evidence is of this type and requires the auditor to be able to reach reliable audit conclusions by comparing, calculating, reasoning, or breaking down information into its components. The use of this type of evidence is common and very important to the objectives of the audit. It tests the auditor's ability and professional judgment to investigate and detect errors that would remain hidden to those performing a cursory review. Often, what is essential is not visible to the eye.

Indirect evidence is obtained by evaluating physical, documentary, and oral evidence. The auditor must assess the quality of the evidence obtained in terms of the relevance and reliability of the

Do	Don't
Smile	Speak with arrogance
Look at the interviewee	Argue
Speak plainly and clearly	Asking more than one question at a time
Listen actively	Asking questions that induce a response
Reassure the interviewee	Giving opinions
Be impartial	Be emotional
Wait for the answer	Hesitate to seek further evidence

Figure 9.7 Dos and don'ts in interviews.

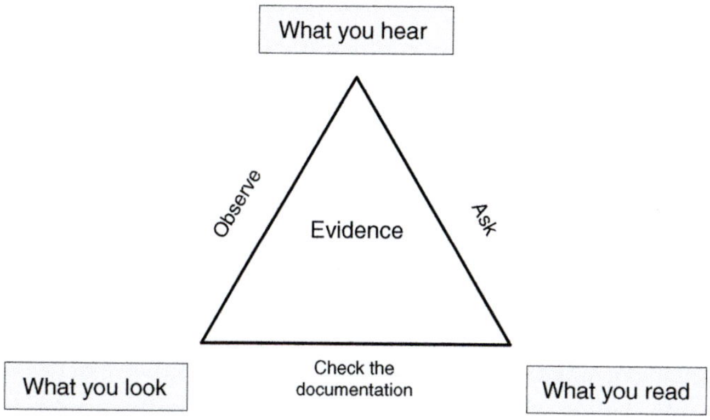

Figure 9.8 Audit triangle evidence.

information on which it is based and check that there is a correlation between the different sources of information, i.e. what is heard, read, and observed must be related and refer to the same situation or facts. Evidence from different sources must be consistent. There should be no discrepancy between what is found in one source and what is found in another (Figure 9.8).

More assurance is ordinarily obtained from consistent audit evidence obtained from different sources or of a different nature than from items of audit evidence considered individually (International Standard on Auditing 500, 2010a).

When reviewing records and when listening to statements, the auditor cannot take at face value all the information he receives, but must consider the degree of reliance he places on it according to his professional judgment. In the case of oral information, sometimes because of the type of response or the way in which it is given, the auditor may perceive signs that suggest that what is being said does not quite correspond to reality, that it is incomplete, or that there are aspects that are not being mentioned. It should be borne in mind that auditing is about detecting deviations in procedures, which often account for no more than 5% or 10% of the time (Ambrose 2013). This is the object of the audit, not the other 90% or 95% where everything is working as expected. For this reason, Patrick Ambrose advises to be on the lookout during interviews for what he calls "telltale words": telltale words, indicating that there is more to the story than is being told, referring to expressions such as:

- "Usually, we…"
- "Most of the time…"
- "Sometimes…"
- "I guess…"

The auditor, when hearing such expressions, should continue to ask questions in a more direct way or probe further to see exactly what kind of situation is being alluded to.

In the field of auditing financial and management systems, certain generalizations are made about the reliability of different types of evidence, which are widely accepted by auditors, albeit with exceptions and taking into account the circumstances of each case. These generalizations, which are perfectly applicable to quality or food safety audits, are as follows:

- Evidence from external sources is more reliable than evidence from internal sources.
- Evidence obtained directly by the auditor (e.g. by observation or inspection) is more reliable than evidence obtained indirectly or by inference.
- Documentary evidence is more reliable than oral evidence.

- Evidence from original documents is more reliable than evidence from photocopies, facsimiles, or electronic documents, whose reliability depends on controls over their preparation and preservation.

9.13 Sampling in Auditing

Sampling is inherent in auditing; each time the auditor requests a record, selects the people to be interviewed, or decides on the products or activities to be examined, he selects a particular item from the total population. Audit sampling can be defined as the use of procedures to select an item or unit from the total population of items or units. Normally, in an audit, it is neither practical nor efficient to examine all available information when records are numerous or costly to obtain because they are too widely dispersed. However, in some cases, it may be possible for the auditor to examine 100% of the items if they are small in number and important or high risk. For example, in a company where there are three metal detectors on the production lines, the auditor may request the records of the test runs of all three metal detectors because of the small population and the risk to the safety of the products.

In management system audits, the use of statistical sampling plans is not common, but sampling is based on the auditor's judgment; however, knowledge of statistical sampling procedures and methods may be required to assess the validity of the auditee's sampling plans, e.g. those used for batch acceptance.

The objective of sampling is to provide sufficient and appropriate information to enable the auditor to draw reliable conclusions on the audit objectives. All sampling involves risk. Sampling risk is defined as the risk that the auditor's conclusion based on a sample may differ from that which would be reached by applying the same audit procedures to the entire population, because the samples selected are not representative of the population from which they were drawn. Sampling risk can lead to two types of erroneous conclusions:

- Concluding that procedures and controls are better implemented than they are, i.e. that nonconformities exist but are not detected.
- Concluding that procedures and controls are not being applied correctly when they are being applied correctly, i.e. concluding that nonconformities exist while the system is functioning reasonably well.

Sampling requires consideration of both the quantity and quality of information collected. Sufficiency and adequacy of evidence are linked. Sufficiency is a quantitative measure of audit evidence; the amount of evidence collected is sufficient when the auditor believes, based on his experience and professional judgment, that it is sufficient to draw reasonable conclusions and satisfy stakeholders of the validity of his conclusions. Adequacy is a qualitative measure of the evidence, i.e. its relevance and reliability to support the auditor's conclusions. The amount of evidence required depends on the auditor's judgment of the risks of an erroneous conclusion and the quality of the evidence obtained; the higher the quality, the lower the amount required.

Factors that the auditor may consider to increase the number or quality of units sampled include the following:

- The greater the importance or significance of the audit finding, the greater the sample size or quality should be.
- The greater the risk of reaching an erroneous conclusion, the greater the sample size or quality should be.

- The higher the level of assurance the auditor wishes to obtain before reaching a conclusion on a finding, the higher the sample size or quality should be.
- The cost (in time or money) of obtaining more evidence; at some point, this cost may outweigh the perceived improvement in the auditor's level of confidence in his conclusions.
- If the auditor has doubts about compliance or the extent of the problem he has identified, he should increase the number of samples or obtain higher-quality samples.

There are generally two types of sampling used in audits:

- Judgment-based sampling.
- Statistical sampling.

9.13.1 Judgment-based Sampling

Judgment-based sampling relies on the auditor's knowledge, skills, and experience. In using this method, the auditor selects samples based on what he believes will provide the best information to meet the audit objectives. Often, these samples are drawn from areas or items that are associated with a higher risk to the entity or from items where the auditor's experience indicates that problems are likely to occur.

The positive side of this type of sampling is that it allows the auditor to focus his efforts on certain areas and perform a more thorough analysis to identify problems or failures that might be missed by a more superficial examination. This is very useful in identifying weaknesses in the system and areas for improvement. The downside of this type of sampling is that not all parts or elements of the system receive the same level of attention from the auditors; there may be areas that are historically audited the most and others that are less thoroughly examined and where deviations are less likely to be detected. It is an unwritten law of auditing that "if you look for it, you will find it." As a result, the audit picture is often somewhat asymmetric, unbalanced, and focused more on finding errors than on assessing the functioning of the system as a whole. In addition, judgment-based sampling lacks statistical validity and therefore leads to greater uncertainty in drawing conclusions about the population as a whole.

In any case, in judgment-based sampling, it is important that the auditor always keeps the following rules in mind:

- The sample is chosen by the auditor. It is not the one they give him, nor the one they find first.
- The sample is selected at random.
- It is selected on a global basis. This means that the largest possible universe or population must be taken into account. This means that the auditor first has to get an idea of what the globality is and from there carry out the sampling. To give an example, if an auditor wants to check that suppliers are approved and asks for a list of suppliers, it is clear that all those on the list given are approved suppliers, but the point is that there may be suppliers of raw materials, products, or services that are not on the list and are therefore not yet approved. For this reason, the auditor must first get an idea of the globality of the products or services supplied to the operator and from there conduct the checks he deems appropriate.

9.13.2 Statistical Sampling

Statistical sampling is not common in food safety management system audits, but depending on the nature or objectives of the audit, a decision may be made to use a statistical sampling method in some cases. For a sampling method to be considered statistical, it must have the following characteristics:

- Random selection of the sample, i.e. when all elements of the population have the same probability of being selected.
- Application of probability theory to evaluate the results of the sample, including measurement of sampling risk.

Samples can be taken from the whole population, or the population can be divided into subgroups or strata, from which samples are taken separately. In order to make inferences about the whole population from the sample obtained, the population must be homogeneous, in addition to being randomly selected. Inferences can also be made from a single stratum, but then, they may not be valid for the whole population.

The most commonly used statistical sampling techniques in auditing are:

- Simple random sampling.
- Systematic random sampling.
- Stratified random sampling.
- Cluster sampling.

In simple random sampling, all units have an equal chance of being selected, and all units are selected at random, e.g. using random number tables or computer programs.

Systematic random sampling is most often used when the population is very large. The difference between this type and simple sampling is that once the sample size is determined, a sampling interval is defined, for example: every ten units. The first sample is selected at random, with the means we have seen, and from then on, the subsequent samples are taken according to the interval defined, in this example: every 10 units.

In stratified random sampling, the total population is divided into strata or categories with respect to some characteristic of interest, and a certain number of samples are taken from each stratum.

Finally, there is cluster sampling, which, although used in other disciplines, is clearly an exception in the field of auditing. In cluster sampling, the population is naturally divided into groups in which all the variability of the population is assumed to be found, and samples are taken randomly from each of these groups.

The sampling plan should take into account whether the outcomes to be audited are discrete or continuous variables. We can therefore distinguish between two types of sampling:

- Attribute sampling is used when there are only two possible outcomes for each sample: correct/wrong, fit/not fit.
- Variable sampling is used when the sample result can be given in a continuous range, e.g. the number of nonconforming products on a production line and the number of out-of-range or out-of-specification values in monitoring records.

All sampling involves some degree of risk or uncertainty because, by not testing the entire population, there is a risk that the selected sample will have values that differ from the value of the entire population. The level of risk in a sampling plan can be quantified by the alpha (α) and beta (β) significance levels.

The following elements can be distinguished in a sampling plan:

- **Acceptable quality level (AQL)**: the highest defect or failure rate of a supplier's process that is considered acceptable.

- **Rejectable quality level (RQL)**: the highest defect rate or defect rate that the consumer is willing to tolerate in an individual lot.
- **Alpha significance level (α)**: the (manufacturer's risk) probability of rejecting a sample with a level of defects below the AQL, which should therefore be accepted.
- **Beta significance level (β)**: the (consumer risk) probability of accepting a sample with a number of defects above the RQL, which should therefore be rejected.

Once these values have been chosen, the sample size can be calculated using the method described in ISO 2859-4: Sampling Procedures for Inspection by Attribute – Part 4: Procedures for Assessment of Declared Quality Levels, or using Minitab software.

9.13.3 Sampling Techniques

A sample in auditing is any piece of information, such as documents, records, measurements, statements, observations, products, obtained from the various sources discussed above, that supports the evidence and is relevant to the objectives of the audit.

The auditor may sample as widely and as intensively as he considers necessary, depending on the quality and quantity of evidence required to reach reliable audit conclusions.

Within the scope set out in the audit plan, the audit team can select the processes, activities, areas, or departments of the organization to be sampled and decide on the level of detail or intensity to be applied to each. For example, the auditors decide on the breadth of the sample according to the number of production lines or areas of the organization, and then for each of these, they decide how many elements or units they will examine. In this way, the audit becomes a tool of analysis, of dissection, which can be used as widely and intensively as deemed appropriate for the purposes of the audit, taking into account time constraints.

9.13.3.1 Vertical Audit vs Horizontal Audit

Two sampling approaches can be used in audits:

- Horizontal sampling approach.
- Vertical sampling approach.

 Horizontal sampling approach: it is the examination of different samples of a single element of the system and is related to the requirements specified in a single procedure, e.g. records related to the training of the organization's employees or records of incoming raw and auxiliary materials.

 It is used to check in detail the implementation of a procedure or a specific element of the management system in the different departments or sections of the food business.

 Vertical sampling approach: it is the examination of a sample in several elements of the system and against the requirements specified in several procedures. In this case, the auditor may follow the production process of a product and request samples from different elements of the system. For example, when examining the delivery notes, the auditor may check whether the suppliers are approved, whether the specified traceability requirements are met, whether the equipment used for reception is calibrated, and the training records of the operators carrying out these activities. These are samples of different items taken at the same point, related to different procedures or prerequisite programs.

 This type of sampling is very useful as it allows for the quick sampling of different elements that may be related and helps to better manage the time available for the audit.

9.14 Tools to Guide the Audit

The on-site audit process is a very demanding task that requires the auditors to be in an almost constant state of concentration in order to understand the processes being observed, ask questions, listen to the interlocutors, assimilate the information received, evaluate the findings, assess whether there are any nonconformities, and take note of all details, attempting to check all the elements foreseen in the audit plan, systematically and with sufficient and appropriate sampling – all within the time constraints of the audit.

When on site, the auditor needs to know what to do at all times. To this end, as we have seen, the planning of activities by the various departments or units of the entity is essential. But it is also essential for the auditor to be able to lead the process proactively, always indicating where he wants to go and what he wants to see, without getting carried away or distracted by irrelevant issues.

Due to the complexity of the audit process, a number of tools are often used to help the auditor guide the process and ensure that no element is left unchecked. The most commonly used tools include audit trails and questionnaires or checklists.

9.14.1 Audit Strategies

When preparing for the audit, the audit team should consider the strategy or path to be taken during the information-gathering activities at the auditee's premises. This involves determining in advance the order of the areas, departments, and activities to be followed, taking into account the interrelationships between the various elements, so that the information can be gathered in a logical and methodical way, helping the auditor to rationalize the process and make better use of the time available.

The audit team can follow a number of paths, each of which has its advantages and disadvantages and which can also be complementary. The auditor must choose the most appropriate approach in each case, depending on the characteristics of the company, the objectives of the audit, or the auditor's working methods. Among the strategies that can be used to guide the audit are the following:

- Toward the flow of the manufacturing process.
- Against the flow of the manufacturing process.
- Vertical.
- Department method.
- Element method.
- Random.
- Tracking technique.

  In the sense of manufacturing flow: it consists of following the manufacturing stages of a product in chronological order. It makes it possible to see the inputs, outputs, and interfaces of the process. The auditor can use the flow diagram in the HACCP manual or make his own to guide the process.

  In the opposite direction to the production flow: this is the same as above, except that the direction of the path is from the end of the process to the beginning. Outputs are examined before process inputs, which can be more revealing in some cases.

  Vertical: this approach is used to assess how food safety policy and objectives are communicated from higher levels of the organization to lower levels and the extent to which instructions and regulations received by staff are implemented, or conversely, to see how incidents that occur in the plant are communicated and managed by the higher levels.

Department method: the audit is organized according to the different departments of the organization. The auditor focuses on all the operations carried out in each department separately. This makes it possible to check the requirements of several elements of the system in a single department.

Element method: this is a very common method that consists of dividing the task between the different elements that make up the management system. In the case of HACCP, this can be done by auditing the HACCP plan and each of the prerequisite programs of the management system separately. In this way, an element of the system is audited across several departments. For example, if the element of the system to be audited is the cleaning and disinfection plan, the audit will be carried out by visiting the different departments or areas of the company to see how this plan is implemented in each of them.

Random: Also known as the "discovery method," this is a method that does not appear to follow any logical path or pattern. However, this does not mean that there has been no planning; often the auditor, when reviewing documentation, will note in his notebook the elements or issues on which he wishes to obtain more information in the factory, without this relationship corresponding to any of the routes we have seen.

Tracking technique: this technique is widely used, especially for traceability exercises. It consists of selecting a batch of products and following it through all the elements of the management system: raw materials, ingredients, auxiliary materials, controls received, equipment used, records of treatments, tests carried out, etc. It can be applied in both directions: backward from the finished product or from raw materials to finished products.

9.14.2 Checklists

Checklists, also known as aide-mémoire, aids to memory, questionnaires, or memorandum, are one of the most characteristic working documents for auditors.

ISO 19011 indicates that audit team members should collect information relevant to the assigned tasks and, as necessary, prepare working papers to support and record audit evidence. Working papers may include:

- Checklists.
- Sampling plans.
- Forms for recording information such as evidence, findings, and minutes of meetings (Figure 9.9).

A checklist can be defined as a list of items to be considered during the audit that can be used to assess the implementation of the HACCP system, including the hygiene prerequisites, design, implementation, and maintenance of the HACCP plan (FAO-WHO 1998).

Checklists are a very useful tool for auditors to guide the on-site audit process in a logical way, without leaving items or elements unchecked, and to keep in mind the information and details to be verified. However, under no circumstances should they be followed in a rigid manner, limiting the audit activities, the ability to observe, or preventing the order of the questionnaire from being changed or omitted depending on the information obtained during the audit.

Depending on their origin, design, and intended use, checklists can have a more or less formal character. In general, two types of checklists can be distinguished:

- Generic checklists.
- Personal checklists.

HACCP principle	Checklist
Preparation	Who was on the HACCP team?
	What is the likely knowledge level of the individuals?
	Has external expertise been sought when necessary?
	Has the scope been clearly defined?
	Have all activities been included?
	Has the product been properly described?
	Is the flow diagram correct?
Principle 1 Conduct a hazard analysis	How was the hazard analysis conducted?
	Were only significant hazards identified?
	Have all raw materials been included?
	Have all process steps been considered?
	Have the hazards been specifically identified by type/source or have they been generalized?
	What information sources were utilized?
	Have appropriate control measures (CMs) been identified for each hazard?
	Will the CMs control the hazards and how was this validated?
	Are all the CMs in place at the plant level?
Principle 2 Determine the critical control points (CCPs)	How were the CCPs identified?
	Did each significant hazard undergo a systematic consideration?
	How are the hazards that are not controlled by CCPs addressed?
Principle 3 Establish critical limits	How were the critical limits established?
	What validation exists to confirm that the critical limits control the identified hazards?
	Have critical limits been established for each CCP?
	How do they differ from operational limits?
Principle 4 Establish a system to monitor the control of the CCP	Have realistic monitoring schedules been established?
	Do they cover all CCPs?
	What is the status of monitoring equipment?
	Are the CCP log sheets being used at all CCPs?
	Is there any evidence that procedures are not being followed consistently?
	Are monitoring personnel and their deputies properly identified and trained?
Principle 5 Establish the corrective action to be taken when monitoring indicates that a particular CCP is not under control	Have the corrective actions been properly defined such the control is regained?
	What evidence is there to demonstrate that this is being done in the event of a CCP deviation?
	Has corrective action been recorded and how is the effectiveness being verified?
	How has the authority for corrective action been assigned?
	How is nonconforming product controlled and is this clearly recorded?
	Are there clear disposition actions listed?
Principle 6 Validate the HACCP plan and then establish procedures for verification to confirm that the HACCP system is working as intended	Have verification procedures been clearly and appropriately established?
	Have responsibilities for verification procedures been allocated?
	Is there documented evidence of system validation?
	Are verification procedures being carried out effectively?
	Is there a formal system to trigger amendments?
	How are the data from HACCP being used to improve the system?
	Is there a regular review of CCP failure and product dispositions?
	Are prerequisite programs included within the verification procedures?
Principle 7 Establish documentation concerning all procedures and records appropriate to these principles and their application	Does the documentation cover all the HACCP system operation?
	How is the documentation controlled with regard to updates and issues?
	Are the records accessible?
	Are all documents accurate and current?
	Are the HACCP records clearly identified by unique reference numbers?
	What format is being used to document the system?
	How is the change control document managed?

Figure 9.9 FAO - WHO HACCP checklist.

Generic checklists are produced by several types of organizations for use by auditors working under their direction or on their behalf. These checklists should be designed to cover at least the minimum requirements of the HACCP system and legal criteria; they help to standardize the work of auditors. In order to facilitate their use, the generic checklists should be accompanied by a manual or reference guide to assist in the understanding of the requirements and their consistent application.

Personal checklists, on the other hand, are checklists prepared by the individual auditor for personal use and for a single audit. They are notes taken by the auditor when reviewing the auditee's system documentation in preparation for the on-site audit. They serve as a memorandum containing a series of data and details about procedures, records, responsibilities, problems, control points, etc., that the auditor wants to have at hand so that they are not overlooked and can be checked more accurately.

Because of their specific nature, personal checklists can take different forms: a list of questions, a double-entry table, a flowchart, or even a mind map showing the points to be looked at and the information to be remembered. Among the different options, each auditor chooses the one that best suits him or the one that best fits his way of working, always with the intention of making them a real help and not an additional burden to his work.

When developing checklists, they can include a list of questions such as what, how, when, who, where; the place to write down the possible answers: yes, no, applies, does not apply; or a classification of the degree of compliance: excellent, good, medium, and poor. They can also be developed by listing the activities to be checked.

Checklists also serve as a support for the notes that the auditor takes during the information-gathering process to document the evidence, findings, and other details that will be needed for the drafting of observations and nonconformities that may arise.

The two types of checklists are not mutually exclusive; the auditor can use a generic checklist provided by the auditing body, which lists the items and minimum requirements to be checked, and at the same time can prepare his own notes to help him carry out the checks in the organization.

In addition to these two types of checklists, there is another type of checklist that we can call nonspecific because it can be used at any time and for any type of management system audit. This nonspecific type of checklist includes the following three types of checklists:

- Corrective actions.
- Document control.
- Customer requirements.

Apart from the latter, which has more to do with the requirements of quality management systems, the first two apply to the implementation of the HACCP system.

The corrective action checklist is something that is always on the auditor's mind. It means that for every deviation or failure that occurs, those responsible for the system must establish some kind of corrective or preventive action. Therefore, if a nonconformity or adverse event is found during the audit, the auditor will check what action the operator has taken to resolve it and prevent it from happening again. If there is no evidence of the action taken, then there is likely to be a possible failure in the implementation of the system.

Something similar happens with the control of documentation: when confronted with any document that is presented, a procedure, a work instruction, a format, etc., the auditor will almost automatically check whether this document is included in the system, whether it is up to date, or whether it meets the rest of the documentation requirements.

9.14.2.1 Types of Questions in Checklists

One aspect to consider in checklists is the type of questions they contain and how they should be used to gather audit evidence. This is a source of confusion, particularly for novice auditors, and leads to a lack of understanding of how to use checklist questionnaires correctly.

The key point is that the questions contained in the checklists – the generic ones in this case – are questions to be answered by the auditor as a result of the evidence-gathering process, not by the auditee. We are referring to general questions that are related to the objectives and requirements to be checked during the audit, not to the way in which evidence is obtained through interviews, so they are questions that are not posed as such to the auditee.

For example, a checklist might include questions such as:

- Are the products described accurately?
- Is the process flow diagram correct?

It is clear that these are not the questions that the auditor will ask in his interviews with the auditee to obtain evidence. Rather, they are requirements that the auditor has to check and which he will only be able to answer adequately after searching for the information in the organization and gathering the necessary evidence, often in several ways.

Thus, following the terminology used by Patrick Ambrose, we can distinguish two types of questions in checklists:

- Audit questions.
- Evidence questions.

Audit questions are those asked directly to the auditee and therefore follow the interview technique described above.

Evidence questions are those that the auditor is expected to answer. They are general questions, the wording of which is linked to the requirements or objectives of the audit. They are questions to be answered during the audit, but not questions that the auditor must ask.

Other authors prefer to differentiate questions according to the levels they occupy on a scale:

- Big question.
- Main question.
- First-level sub-question.
- Second-level sub-question.

Starting from the general to the specific, the top level would be the "big question," which refers to the objectives of the audit. This gives rise to the main questions, which are subdivided into first-level sub-questions, and these in turn are subdivided into second-level sub-questions, which are those used by the auditor to obtain information and evidence for the audit. The higher-level questions would be the evidence questions in the classification mentioned by Patrick Ambrose, and the lower-level questions would be the audit questions that the auditor asks the interviewee.

The evidence gathered is compared with the criteria to obtain the findings, from which conclusions are drawn to determine whether the objectives have been met and thus to answer the big question at the start: moving upward – from the particular to the general.

It is generally accepted that checklists have positive aspects for audits, but also that they can cause problems if they are not well designed or if they are used by inexperienced people. Some of the advantages and disadvantages of checklists can be seen in the table below (Figure 9.10).

In summary, checklists are useful to the auditor, provided they are used by trained people and understood for what they are: a tool to guide the audit, not a task to be performed. The purpose of the audit is not to complete the checklist forms.

Pros	Cons
Memory aid	Time-consuming
Provide consistency/uniformity	Can be distracting
Help to systematize work	The order of the questions does not help
More efficient time management	Not specific in some cases
Useful for taking notes	Can be a tick box only
Help keep the audit focused on the requirements	Confusing the novice auditor
Bring more transparency to the process	Auditing is more than just covering checklists

Figure 9.10 Pros and cons of checklists.

9.15 Counter-auditing Techniques

The conduct of the audit process requires the establishment of an appropriate climate of respect and cooperation between the audit team and the auditees. The effective conduct of the audit requires the collaboration of the auditee in a number of ways: granting access to premises, providing requested documentation, reporting honestly on what is asked, and generally providing whatever resources are needed to carry out the audit plan.

However, in a more or less deliberate, almost instinctive way, some companies or their employees may use certain tactics that are intended to impede or mislead the work of the auditor to some extent. These tactics or actions by auditees are known as counter-audit tactics and include practices such as the following:

Time wasting: Whether more or less deliberate, this is a very common practice in audits. Auditees will never be in a hurry; the hurry to follow the plan and try to check everything that is planned is on the side of the auditor. As a result, there can be a constant waste of time, from the opening meeting, which can take longer than it should, to long coffee or refreshment breaks, to interviews where the subject is diverted and conversations are started on topics that are not related to the audit objectives, and so on.

Delays in providing documentation: Related to the previous point, auditees may be late in providing documents or records requested by the auditor, as if they were operating in slow motion. Apart from the loss of time, the auditor must be vigilant to ensure that such documentation is not produced at the time. To this end, as mentioned above, if the auditor notices that this might be happening, he should accompany the person to the place where the record should be filed.

Elusive auditees: Some people who are needed for the continuity of the on-site audit are difficult to locate or seem too busy to attend to the auditors. Tracking them down is certainly an extra burden on the audit team. In these cases, it is necessary to rely on the guide to help locate these people and make them available to the auditors within a reasonable timeframe.

Chatty auditee: The problem with this type of auditee, who tends to talk too much, is not only the waste of time involved but also the extra effort required by the auditor to concentrate on the point being audited or verified. Once the auditor is aware that he is dealing with this type of person, he should conduct the interview in a firm manner, modifying the type of questions asked to elicit shorter and more concrete answers.

Wrong or out-of-date documents: The auditee's intentional or unintentional provision of out-of-date documents or documents that are not what was requested wastes the auditor's time and additional effort. In such cases, the auditor needs to identify such documentation errors as early as possible to avoid confusion and wasted time in reviewing the wrong documentation.

Auditee-driven sampling: It is also common and quite instinctive for the auditee to want to present their management system in the way they prefer, focusing on aspects that they consider to be their strengths. In this way, the auditee will steer the audit through the points that interest them and show the auditors the records and evidence that they have selected. As mentioned above, it is always the auditor's responsibility to select the sample in order to check what he is looking for and to ensure that the sample is representative and unbiased.

Exploiting the emotions of the auditor to compromise their impartiality and professionalism: This may happen either by being excessively hospitable, offering benefits, gifts, or meals, or trying to foster comradeship and complicity with the audited personnel; or, on the other hand, by creating a hostile climate toward the auditor and trying to intimidate him and make him lose his nerve and self-confidence. Both situations should be avoided from the outset by refusing any gifts or benefits offered and by not responding to any provocation from the auditee. It is easier said than done, but there is no other way to manage the audit.

Handing evidence on a silver platter. Bolder and more experienced auditees may hand the auditor evidence on a silver platter, so that once the auditor has found something, his attention will relax and he will not be as vigilant in finding other evidence in that area or department that the auditees are trying to hide. Brazilian auditors call this tactic "Boi de piranha," which is a very revealing expression once you know its origin, because it refers to a bull (usually an old or sick one) that cattle ranchers used to sacrifice when they had to cross a river with piranhas, letting this animal pass first to entertain the piranhas and save the rest. This tactic is based on the eagerness of some auditors to find nonconformities, so that once they find one, they are more inclined to relax their search. And this is a mistake, because the objective of an auditor is not to look for nonconformities, but to check the functioning of the management system, always maintaining the same level of attention, whether nonconformities appear or not (Couto and Bertolino 2023).

The use of such counter-audit tactics is not surprising. The auditee will always try to minimize the impact of the audit (Figure 9.11). Therefore, we cannot expect the auditee to do the auditor's job for him by handing him the evidence on a plate, except for the "Boi de piranha."

The auditor's role is to check that the system is working and to identify any deficiencies or weaknesses so that they can be corrected. The auditee's role is to find as few issues as possible and for the audit report to be favorable. So he will try all sorts of tricks to distract the auditor from his objective. This is the game of auditing. If the auditee hides data or does not tell things as they are, it is not for personal reasons; the auditor should not take it personally, as it is just part of the game. The important thing is that the auditor knows this, accepts the challenge, and learns to enjoy the competition (Ambrose 2013).

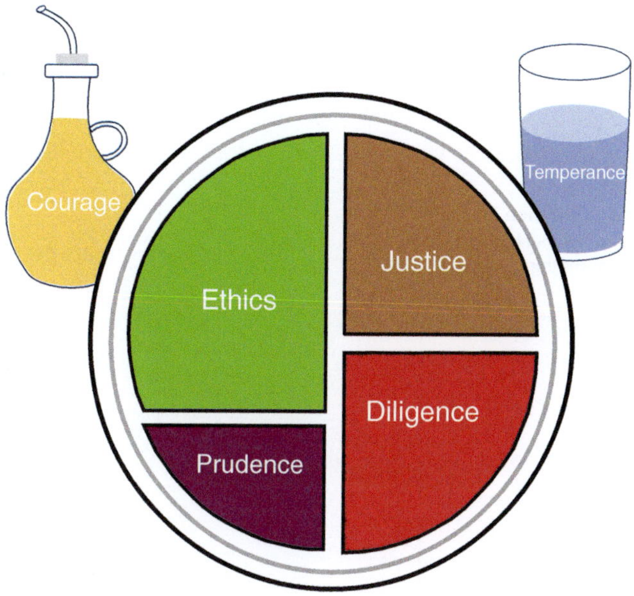

Figure 9.11 Healthy diet of an auditor.

9.16 Generating audit findings

The generation of audit findings is the crucial phase of the audit. It can be said to be the point at which, based on the information gathered and the elements used as criteria, the whole process is catalyzed and the outcome of the audit is determined. ISO 19011 defines audit findings as follows:

Audit findings: results of the evaluation of the collected audit evidence against audit criteria.

Note 1: Audit findings indicate conformity or nonconformity.
Note 2: Audit findings can lead to the identification of risks, opportunities for improvement or recording good practices.
Note 3: In English if the audit criteria are selected from statutory requirements or regulatory requirements, the audit finding is termed compliance or non-compliance.

The generation of audit findings therefore consists of comparing the information gathered or observed, based on facts or objective data, with the set of standards, procedures, requirements, and objectives that make up the audit criteria. Not all information obtained during the audit process serves as evidence, only that which is verifiable and relevant to the audit objectives (Figure 9.12).

As mentioned above, evidence and criteria are the two legs of the audit process. While on site, the auditor will be constantly thinking about how to obtain the information that will serve as evidence to determine whether the audit criteria are being met and whether it is sufficient and appropriate to draw reasonable conclusions (Figure 9.13).

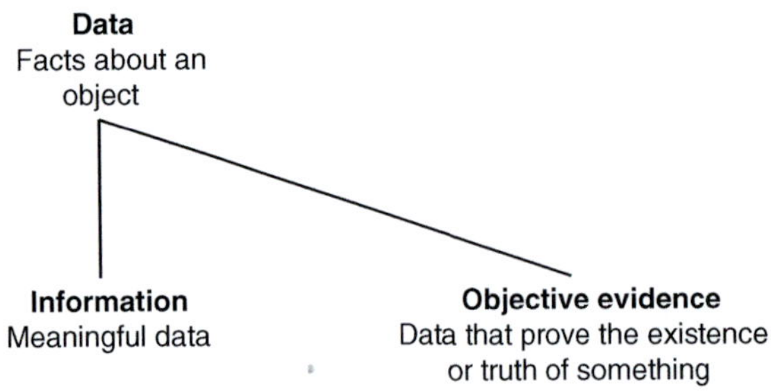

Figure 9.12 Data and objective evidence.

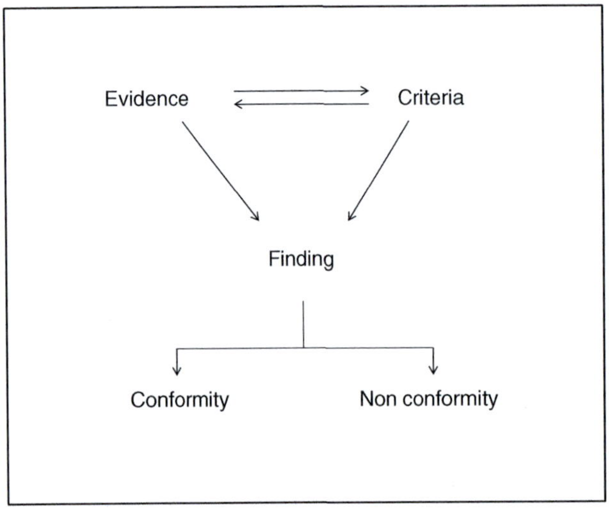

Figure 9.13 Evidence, criteria and findings.

In determining the extent to which the audit criteria are met, the auditor needs to analyze the information gathered in order to make a reasonable judgment. Sometimes, when the information available to the auditor is not sufficient or does not allow a clear conclusion to be drawn, the auditor may need to increase the sample size and investigate further to resolve doubts and make a decision on a sufficient basis.

The auditor's characteristic determination to look for more evidence may, when viewed from the outside, give the impression that he has no other motivation than to find errors or deviations, as if he enjoys writing nonconformities or something similar. However, this is not the reason for the auditor to act in this way. What the auditor really wants is not to make mistakes. If it is clear that an area or department is performing satisfactorily, no auditor will be reluctant to say so. But if there is evidence that something is not working well or an intuition that there may be deviations, then it is their job to find that out, to bring it to light, not to take for granted what is not the case. That would be an inexcusable error and would diminish the effectiveness of the audit and its value in improving the system.

When gathering information on the plant floor, the auditor must remain aware, constantly asking what he is missing, what details are being overlooked, or if there is something he is forgetting to check. Especially when everything seems to be going smoothly, he may not be digging deep enough.

If the auditor discovers something that he considers to be a deviation or nonconformity, the right thing to do is to report it immediately to the auditee or the person accompanying them at the time: to explain what the evidence is and the reasons for it. In this way, the guide or those present can see the situation or facts that are being used as evidence, give their explanations if they wish, and also get an idea of how the audit is progressing. This will avoid unpleasant surprises for the auditee when he thinks everything is going well and suddenly all the nonconformities are presented to him at the closing meeting. If the auditor is not sure whether he has found a nonconformity on site, he should say so at the time, but make it clear that he needs more information to confirm whether there is a nonconformity.

If a nonconformance is identified during the audit and corrected on the spot by the auditee, the circumstance or fact remains an audit finding and does not disappear because it was corrected on the spot. The auditor will then consider how to describe it and how it affects the outcome of the audit.

9.17 Classification of Findings

The evaluation of findings can lead to different results:

- Conformity.
- Opportunity for improvement.
- Observation.
- Nonconformity (Figure 9.14).

The terms conformity and nonconformity are defined in ISO 19011:2018 as follows:

Conformity: fulfillment of a requirement.
Nonconformity: nonfulfillment of a requirement.

A **requirement** is defined as a need or expectation that is stated, generally implied or obligatory.

Observations are discrepancies with the requirements of the system that do not show a failure to meet the requirements but that weaken or may weaken the effectiveness of the system. In general, observations that are not corrected may end up as nonconformities.

An **opportunity for improvement** is a finding where there is compliance but it is determined, based on objective criteria, that there is room for improvement to optimize an activity, task, or process.

Each audit should be based on a defined set of audit objectives, scope, and criteria, which should be consistent with the overall objectives of the audit program. According to ISO 19011:2018, the audit objectives define what is to be accomplished by the individual audit and may include the following:

- determination of the extent of conformity of the management system to be audited, or parts of it, with audit criteria;

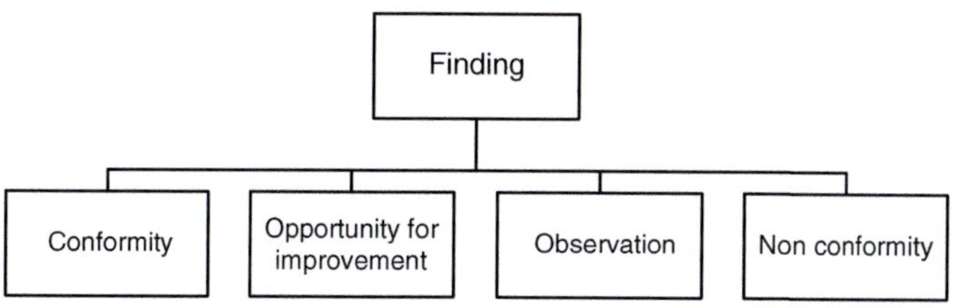

Figure 9.14 Classification of findings.

- evaluation of the capability of the management system to assist the organization in meeting relevant statutory and regulatory requirements to which the organization is committed;
- evaluation of the effectiveness of the management system in meeting its intended results;
- identification of opportunities for potential improvement of the management system;
- evaluation of the suitability and adequacy of the management system with respect to the context and strategic direction of the auditee;
- evaluation of the capability of the management system to establish and achieve objectives; and
- effectively addressing risks and opportunities, in a changing context, including the implementation of the related actions.

The purpose of the audit is therefore, from a positive point of view, to verify the conformity of the system, not to detect failures, let alone to point the finger of blame. However, if nonconformities are found, they should be properly documented and clearly communicated.

Classification of findings is a useful way of reporting the outcome of the audit, giving all stakeholders an overview of both the process and the individual elements tested. It helps the auditee to better understand the extent and seriousness of the various deviations and thus to prioritize corrective actions. And for the auditor who participated in the audit, as well as others who did not participate in the audit but may have an interest in the process, this classification is useful for following up on nonconformities and verifying their closure.

9.18 Elements of a Finding

Before reporting a finding, the auditor needs to consider whether he has all the elements necessary to determine the existence of a finding. A well-documented finding should include the following elements:

- Scale, level of noncompliance or degree of compliance.
- Description or statement of the nature of the finding.
- Evidence gathered to support the finding.
- Criterion or requirement evaluated (Figure 9.15).

If the result of the finding is conformance, the statement can only be made based on the audit criteria against which conformance is demonstrated and the evidence supporting it.

If the result of the finding is a nonconformity, the auditor should consider whether it meets the conditions for being documented as a nonconformity. Two elements are essential in the documentation of a nonconformity: the supporting evidence and the audit criteria. If either is missing, it cannot be documented as a nonconformity.

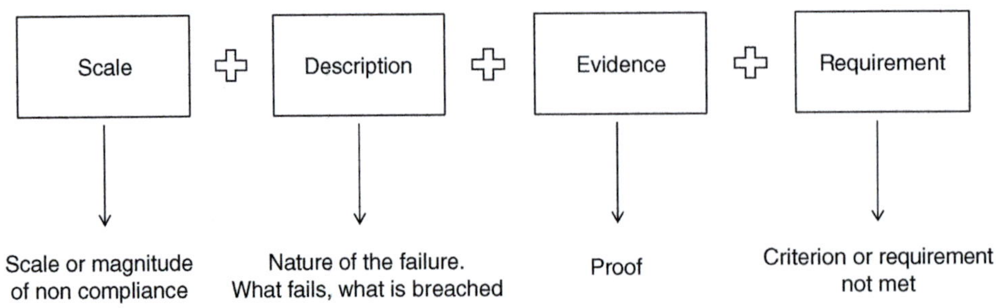

Figure 9.15 Elements of a finding.

If the auditor has an indication that a nonconformity may exist, he should attempt to obtain evidence to support such a statement in order to dispel the doubt. If the auditor can obtain sufficient and appropriate evidence for the criterion being assessed, then a nonconformity exists. However, if the evidence is insufficient or he is unsure of the accuracy of the supporting data, he should not report a nonconformity. The lack of evidence or criterion will often result in the auditor downgrading the observed facts and presenting them as an observation or an opportunity for improvement.

Another element to be included in a nonconformity, provided it can be quantified, is the magnitude or degree to which the nonconformity occurs. In fact, it is common to start the wording of the nonconformity with phrases such as "In some cases," "In six out of ten cases observed," "In general." The reference to magnitude is very useful for both the auditor and the auditee, as it gives an idea of the significance of the nonconformity and also helps to determine the nature of the nonconformity, i.e. whether it is accidental or systematic. When the corrective action is determined and the cause of the nonconformity is analyzed, the treatment of the nonconformity may vary considerably depending on whether the event is incidental or whether it is considered to be systematic.

Depending on the sample size, if the level of nonconformity found is low, the auditor may classify the finding as an observation rather than a nonconformity. In any case, if the failure rate is low, but the deviation has a direct impact on the safety of the food products or concerns an issue of major relevance to the company, it should be classified as a nonconformity.

The description of the nonconformity should be clear, simple, but precise, avoiding circumlocutions or excessively long sentences; normally, a nonconformity is explained in a few words. Anyone reading the report should be able to understand it.

The statement of nonconformity should contain sufficient information to allow an analysis of the cause of the problem. Indeed, for some authors, it is important for the auditor to ask himself, before writing a nonconformity, whether he can identify an effective and feasible action to correct the nonconformity by applying the test: "So, what now?" HACCP must be a useful tool for managing food hazards. And for the factory responsible to sleep soundly. The purpose of the audit is to check that the operator's HACCP system is working properly and is effective and to correct any deficiencies that may undermine the system. Therefore, if nonconformities do not lead to effective corrective action, there is little point in overwhelming the auditee with nonconformities whose resolution is beyond their reach and which ultimately serve only to discourage them and make them feel that the system is nothing more than a burden.

In addition to these elements, the nonconformity report may include other information such as the procedures not followed or the locations where the nonconformity was found. Such details will help those reading the report at a later date to get as accurate a picture as possible of the facts and the nonconformities found (Figure 9.16).

Must be	Must not be
Objective	Subjective
Precise	Wrong
Reproducible	Ambiguous
Relevant	Trivial
Useful for improvement	Not useful for improvement

Figure 9.16 Drafting of nonconformities.

9.19 Classification of Nonconformities

It is only logical to assume that not all nonconformities have the same value or significance within the food safety management system. Nor do all requirements. Some findings are more serious than others, and it is therefore common for auditing bodies to use a classification system for nonconformities. This provides a better understanding of the audit outcome and helps the auditee to prioritize corrective actions.

The classes and the terms used to describe them vary from one organization to another. Classifications may include: critical, significant, and nonsignificant; knockout (KO) and major; major and minor; significant and nonsignificant; systematic and isolated; A, B, and C; etc. Each organization chooses the one it considers most appropriate, and the auditors working under its direction should use the same classification system in order to achieve the greatest possible consistency in their assessments.

Regardless of the system chosen, it is important for the auditor to be clear about the criteria used to classify nonconformities according to their significance. In general, the criteria used for classification are:

- Frequency of the problem, i.e. whether it is isolated or systematic.
- Severity.
- The risk associated with the product.
- Criterion or requirement not met.

For example, in the system used by Brand Reputation Compliance Global Standard (BRCGS), the levels of nonconformity are defined as follows:

- **Critical**: where there is a critical failure to meet food safety or legal requirements.
- **Serious**: where there is a significant failure to meet the requirements of a "statement of intent" or any clause of the standard or where there is a situation which, based on available objective evidence, would raise significant doubt as to the conformity of the product being supplied.
- **Minor**: where a clause has not been fully complied with, but based on objective evidence, the conformity of the product is not in doubt.

On the other hand, International Featured Standards (IFS) defines two types of nonconformities:

- **KO**: When a specific requirement named KO requirement is not implemented. In IFS, there are ten KO requirements that are essential and address key issues that must be implemented by the production site to achieve compliance.
- **Major**: A Major nonconformity can be issued to any regular requirement (which is not defined as a KO requirement) when there is a substantial failure to meet the requirements of the standard, which includes but is not limited to food safety and/or legal requirements of the production and/or destination countries, or when a process is out of control that might have an impact on food safety (Figure 9.17).

Findings and nonconformities are an essential part of the audit report; therefore, when the report is given to the stakeholders at the end of the audit, they should be provided with a document containing all the information related to the statement of nonconformities and the data on which the nonconformities are based. In addition to the report, another way of recording nonconformities is through the use of nonconformity notes. Nonconformity notes are documents that may follow a predefined format with a number of fields or sections to record all information and details related to

<table>
<tr><th colspan="6">Nonconformity note</th></tr>
<tr><td colspan="3">NCN No:</td><td colspan="3" rowspan="2">Audited organization</td></tr>
<tr><td colspan="3">Area</td></tr>
<tr><td colspan="3">Localization</td><td colspan="3" rowspan="2">Reference</td></tr>
<tr><td colspan="3">Date</td></tr>
<tr><td colspan="6">Nonconformity</td></tr>
<tr><td colspan="6"></td></tr>
<tr><td colspan="2">Auditee</td><td colspan="2">Auditor</td><td colspan="2">☐Major ☐Minor</td></tr>
<tr><td colspan="6">Corrective action</td></tr>
<tr><td colspan="6"></td></tr>
<tr><td colspan="2">Auditee</td><td colspan="2">Auditor</td><td colspan="2">Date</td></tr>
<tr><td colspan="2">Date of completion</td><td colspan="2">Follow-up</td><td colspan="2">Observations</td></tr>
<tr><td colspan="6">Auditor verification</td></tr>
</table>

Figure 9.17 Nonconformity note (NCN).

the nonconformity statement. Although the number and layout of the sections will vary according to the preferences of each body, there are generally four parts:

- **Header:**
  - Company name.
  - Department, area, or function audited.
  - Nonconformity number.
  - Reference standard.
  - Date.
- **Top section**: Description of nonconformity. To be completed by the auditor.
- **Middle section**: Description of corrective action: date, deadline for completion, and signature of person responsible. Completed by the auditee.
- **Bottom section**: Reserved for the follow-up of the corrective action and its verification by the auditor to close the corrective action.

Although widely used in quality management system audits, nonconformity notes are not mandatory. The advantage of nonconformity notes is that they make it possible to follow the stages through which each nonconformity passes, from its declaration to its closure: when the corrective action taken by the company is verified by the auditor as effective.

9.20 Drafting of Nonconformities

As we have just seen, nonconformities – unless there are none at all – are probably the most important record of the audit; not only are they the result of all the work of the audit team during the previous phases, but they will determine how the post-audit follow-up will be. Therefore, auditors should pay particular attention to their drafting, applying auditing principles and describing the type of noncompliance, the evidence on which it is based, and its magnitude accurately. When properly drafted, whether the auditee likes or dislikes what they report, they will not hesitate to accept the nonconformities, provided they are reasonable, are well founded, and can be useful for improvement, helping to locate system problems and failures, identify their causes, and establish appropriate corrective actions.

Some examples of nonconformities are given below, first the wording of the nonconformity and then the areas for improvement:

Example 9.1
"Some operators in the processing room should be trained to ensure proper knowledge and implementation of the HACCP system."

Comment

This wording has several shortcomings; the elements to be included in a nonconformity are not well specified: nature, evidence, and extent of the nonconformity. In addition, the auditor indicates what the corrective action should be, when this is a function not of the auditor but of the audited organization.

Example 9.2
"The equipment maintenance plan is not properly carried out. No maintenance records are provided."

Comment

This example does not clearly describe the nature of the noncompliance. We do not know whether the maintenance plan is carried out according to their procedures or whether the plan is carried out but there is a deviation in the way it is recorded. Both are nonconformities, but the nature of the nonconformity is different and so are the corrective actions. The evidence and extent of the variance is also not included.

The wording could be as follows:

> "Contrary to what is stated in the plan for the maintenance of installations and equipment (PG-09), the monthly record templates (F2-PG-09) are not covered with the frequency defined in the procedure. Of the monthly records requested for the first half of 2023, the responsible person only provided the one for the month of February".

Example 9.3
"The traceability plan is not adequately followed; some crates of cheese are found without the label identifying the production batch."

Comment

In this case, the nature of the nonconformity is correctly described and the evidence on which it is based is provided, but the magnitude of the nonconformity needs to be specified.

This nonconformity could be worded as follows:

> "10 plastic crates of cheese were found in the ripening room without the production batch label, which is not in compliance with the traceability plan (PG-11)".

Example 9.4
"Sampling is not carried out correctly. Three verification analyses of the beef line are missing."

Comment

This nonconformity does not describe which procedure or requirement is not complied with. The extent of the nonconformity is not clear.

It could be worded as follows:

> "The verification procedure of the L+D plan (IT-04) is not correctly applied: of the four quarterly analyses required, corresponding to the beef line for the year 2023, only the one for the second quarter is provided".

Example 9.5
"Some products are not correctly labelled."

Comment

This nonconformity refers to compliance with a legal requirement for food labeling. However, no evidence is provided, and the extent of the problem is not mentioned. In fact, it is unclear whether there is a nonconformity at all, since, in the absence of specificity, these products may be at a pre-labeling stage.

This nonconformity could be worded as follows:

"In some cases, it has been observed that there are products with deficiencies in their labelling, examples: on the label of salchichón, the additives E 252 and E 621, contained in the preparation used in its manufacture, are not included in the list of ingredients; on the label of fresh chorizo, the information on the allergens: lactose and soya, contained in the preparation used in its manufacture, is not included." As can be seen from this new wording of the noncompliance, there is a systematic failure because the text on the labels used for these two products does not correctly indicate the ingredients.

Example 9.6

"There is no evidence of corrective action in the monitoring of CCPs."

Comment

The noncompliance does not describe the evidence, nor the location, nor does it give an idea of the extent of the noncompliance.

It could be worded as follows:

> "Contrary to what is stated in the HACCP manual, it is noted that corrective actions were not taken following the occurrence of failures in the monitoring of the pasteurization CCP; for two weeks in March 2023 the pasteurizer recording system was out of order, without evidence that this occurrence was reflected and the corrective actions described were taken."

Example 9.7

"Reception registers are not correctly covered."

Comment

This nonconformity does not describe the evidence and the extent of the nonconformity.

It could be worded as follows:

> "In most cases it was found that the milk tank reception records (F-15-01-02) do not contain all the data required in the record; the reception temperature and the pH and acidity test values of the milk received are not recorded. For example: milk reception records for the months of October, November and December 2023."

Example 9.8

"The temperature of the product does not comply with the HACCP manual. An operator takes measurements at the exit of the freezing tunnel with values of −15 °C, −14 °C and −16 °C."

Comment

The nature of the deviation is quite clear, but the criterion should be better specified. It could be worded as follows:

> "Contrary to operational procedure P.08.04, the values of the frozen tuna product at the exit of the tunnel are lower than the value of −18 C set in this procedure, e.g. values of −15 °C, −14 °C and −16 °C measured by the line operator on 15.11.23."

Example 9.9
"During the visit to the intestines processing room, two open metal containers were observed, which are not airtight and without a label indicating the category of by-product."

Comment

The evidence is clear, but not the nature of the failure, nor the criterion. And not much information is provided on the extent of the deviation to be able to determine its severity. A possible wording could be:

> There is no evidence of effective monitoring of hygienic conditions in the gut processing room, on 16.11.23 two open, unsealed and unlabeled containers are observed, with no such incidence reflected in the monitoring records or corrective action taken by the monitor.

Example 9.10
"In no case are there any documented shelf life studies, nor is there any reference to the packaging material in the product data sheets."

Comment

There are two quite clear pieces of evidence, but referring to different criteria, not mentioned in the wording. The absence of documented shelf life studies would be a nonconformity with regard to the validation of the system and therefore with HACCP principle 6. On the other hand, the absence of information on packaging materials is a deficiency in the product description. The tasks to be carried out by the food business operator to correct these deviations are of a very different nature and therefore, if the audit is to be useful in improving the system, the nonconformities need to be better described in terms of their nature and their location within the HACCP system.

A possible wording would be:

> "There is no evidence of shelf life validation for the product(s) (name the products)."

"In the technical data sheets of the products (name products) the section on packaging materials has not been completed."

9.21 Preparing Audit Conclusions

Prior to the closing meeting, the audit team should meet to review the findings and information gathered during the audit and prepare the conclusions to be presented to the auditee.

This sharing should serve to minimize the subjectivity of the auditors and personal bias in the assessment of the facts observed and should seek to achieve, through consensus among all members of the audit team, a contrasting and more objective view than the individual contributions of each of the auditors.

The purpose of this meeting is to check the accuracy of the data and to analyze the findings to see whether they are correctly documented, i.e. whether they are supported by sufficient and appropriate evidence, and to identify the audit criterion that has not been met. In this way, this prior review acts as a filter to prevent deficient or unsubstantiated findings or nonconformities from arriving at the closing meeting that are difficult to defend against objections or challenges that may be raised by the auditee's representatives. Therefore, it is always better for the audit team to identify them first and, if necessary, correct or withdraw them, than to have to back down in the face of the auditee's demand for nonconformity. This will always be a disadvantage for the audit team.

9.22 Closing Meeting

At the end of the on-site audit phase, it is customary to hold a closing meeting between the audit team and representatives of the audited organization. The purpose of this meeting is to present the findings and conclusions of the audit in a way that can be understood and accepted by the auditee.

ISO 19011 includes this meeting as a characteristic activity of the audit, which is recommended because it is the best way for the audit team to communicate the outcome of the audit directly to the auditee, who in turn can ask for clarification, express their views, or make objections to the findings presented. If this meeting does not take place or if its content is inadequate or inappropriate, misunderstandings may arise and the auditee may receive the final report containing data or issues that have not been properly understood or communicated to the auditee and for which the auditee has not had the opportunity to seek clarification or, where appropriate, raise objections.

At the closing meeting, at least the same persons should be present as at the opening meeting. However, this does not preclude the possibility that, in view of the findings, other persons may attend the closing meeting, not only from the top management but also from the heads of the departments affected by the audit findings. In any case, it is important that the people who have the responsibility and authority to deal with the nonconformities are present.

As with the initial meeting, the lead auditor should maintain control of the meeting. Although the other team members are present and may intervene at certain points, the role of the spokesperson is that of the lead auditor, and it is important to avoid any expression of disagreement among the team members, as this would seriously undermine the credibility of the team in the eyes of the auditee. Any differences of opinion or interpretation that may exist between the auditors should have been resolved in the previous phase of reviewing the findings and preparing the conclusions. At this stage, it is important to demonstrate unity and support the conclusions communicated by the lead auditor in a consistent manner.

If the auditee feels that the findings are not reasonable or are not sufficiently justified, either by the nature of the evidence on which they are based or by the criteria applied, it is likely that the auditee will raise the tone of its interventions. For this reason, it is again essential that the lead auditor demonstrates good interpersonal and communication skills to try to keep the meeting on track: maintaining a polite tone and avoiding unnecessary discussion or heated exchanges.

It is also important for the auditors to be able to highlight the positive aspects and strengths found during the audit, recognizing the efforts made by the organization to maintain and improve the management system. This provides a more balanced view and prevents the meeting from focusing solely on the negative aspects, with the resulting impact on the morale of the auditees.

When communicating the results, auditors should be frank, not afraid to give bad news, but without giving the impression that they are pleased to report the nonconformities found. In any case, the findings presented at the final meeting should not come as a surprise to the audited organization, as it is normal for the guides and observers accompanying the auditors to be kept informed of the progress of the audit and the findings throughout the process.

During the meeting, it is normal for the auditee to express doubts, raise objections, or even try to challenge the validity of the findings. This is expected and should be taken for granted by team members. Therefore, in an appropriate manner and without arrogance, the auditors should listen carefully to the reasons or assertions that the auditee may express and seek to clarify doubts and resolve disagreements as far as possible, but not in a way that would amount to capitulation on the part of the audit team or result in an unjustified downgrading of the audit findings. Provided that the auditors have done their work well and reached consistent conclusions based on correctly identified objective evidence and accepted criteria, there is no reason to withdraw findings or downgrade nonconformities. However, if this condition is not met, the audit team can only withdraw with the consequent loss of credibility of its performance.

Only the provision of new evidence by the auditee would be grounds for withdrawing a nonconformity at the closing meeting, accepting the evidence provided, and apologizing for the misunderstanding. This may be the case when certain types of documents are requested during the on-site audit, which are not made available to the auditors at the time, and only at the end, at the closing meeting, are they able to provide them, which is considered acceptable and allows the nonconformity or observation to be withdrawn.

9.23 Audit Report

After completion of the on-site activities, the team leader should prepare the audit report. Although there is a possibility of presenting it at the closing meeting, the report is normally prepared later, after leaving the site, within the timeframe agreed with the audit client or set out in the program.

While there is no longer the time pressure to write the report that characterizes the whole audit, it is not advisable to delay it too long; firstly, if it is done early, the auditors will still be fresh with the facts and circumstances surrounding the audit and will be better able to capture them in the report; secondly, when the auditee has to take corrective action on the nonconformities reported at the closing meeting, the subliminal message sent by the delayed report is that these may not be as important and urgent in the end as their severity might have suggested.

Based on the working papers and notes collected by all team members, the lead auditor should prepare and sign the audit report. Although the ultimate responsibility rests with the lead auditor, this does not mean that the report should not be prepared in a consensual manner. Therefore, once the draft has been prepared, it is recommended that it be reviewed by the auditors to see if there are any shortcomings, aspects that are difficult to understand, or points that need to be modified.

The report is the most important record of the audit; it is the final outcome of the audit and the document that will be available for use or consultation by the client, the auditee, and other interested parties. It is essential, accordingly, that particular care is taken in its preparation; a poorly prepared or inadequate report can ruin all the work done or significantly reduce the effectiveness of the audit.

The audit report should provide a complete, accurate, concise, and clear record of the audit. In addition, the report should fulfill the following functions:

- Provide information on the degree of conformity of the management system of the audited organization.
- Determine, i.e. be the cause of, corrective or improvement actions to be taken by the auditee.
- Guide the follow-up of the audit.
- Consultation document for other stakeholders.

The content and format of the audit report can vary considerably from one organization to another, depending on the type of audit and the objectives set. However, there is a minimum content and certain information that should always be included. ISO 19011 indicates that the following should be included or referenced:

- the objectives of the audit;
- the scope of the audit, in particular the identification of the entities of the organization and the functional units or processes that will be audited;
- identification of the audit client;
- identification of the audit team and the audit participants from the audited entity;
- the dates and places where the audit activities were performed;
- the audit criteria;
- the audit findings and supporting evidence;
- the audit conclusions; and
- a statement of the extent to which the audit criteria have been met.

This content can be divided into different parts according to the organization's preferences, such as introduction, summary, list of nonconformities, observations, and conclusions.

Other information that may be included in the report includes the audit plan, areas within the scope that were not covered, opportunities for improvement, unresolved divergent opinions between the audit team and the auditee, a statement about the confidential nature of the data, a comment or clause about the uncertainty associated with the audit sampling, identified good practices or strengths, and the distribution list for the report.

The language used in the report should be clear, precise, but simple and free of unnecessary rhetoric. The presentation should be objective, presenting the facts fairly and avoiding personal interpretations by the auditors. Phrases containing subjective impressions such as "in my opinion," "in my judgement," "it seems that"; or imperatives "the food operator must" are not part of the scope and rationale of audits. It should always be remembered that nonconformities are raised against the audit criteria, not against the auditor's opinion. Although it is sometimes stated in the audit program or agreed with the client that recommendations are to be included in the audit report, it is not the auditor's role to propose solutions or indicate corrective actions; this is a task for the audited organization when establishing the action plan.

The audit report should include all data and details deemed necessary to describe the nature of the nonconformities or observations and to facilitate analysis of the causes leading to them, but should never include the names of individuals so that they can be associated with the nonconformities or observations described. Audits are carried out on the system, not on people; their main objective is to detect failures and deviations in order to improve the management system, not to find the culprits or those responsible for nonconformities in order to reproach them. This would have very negative consequences, not only for the individuals concerned but also for the usefulness of the audits in the future.

Once the report has been finalized, it should be reviewed and approved by a person responsible for the management of the audit program or by the audit body's nominee, as appropriate and in accordance with the audit program procedures.

The audit report is confidential, and therefore, except as required by law, neither the audit team nor any other person involved in the management of the audit program should disclose its contents or the contents of any other document or information obtained during the audit without the expressed consent of the client and, where appropriate, the auditee.

9.24 Completing Audit

According to ISO 19011, an audit is completed when all activities included in the audit plan have been completed or as agreed with the client. In some cases, unexpected situations may arise that prevent the audit plan from being completed, for example, problems in obtaining documentation or if the evidence gathered indicates that the objectives cannot be achieved, which may lead to an agreement to suspend or terminate the audit earlier than planned.

9.25 Audit Follow-up Activities

If the audit findings indicate the need for corrections or the initiation of corrective actions or opportunities for improvement, audit follow-up activities may be identified depending on what is stated in the audit program or agreed with the client.

The responsibility for initiating such actions rests with the auditee, which shall take the following steps for each departure identified during the audit:

1. Identify the causes that led to the nonconformities and their consequences and determine the extent of the problem identified.
2. Determine corrective actions and, where appropriate, immediate remedial actions. Corrective actions should always be aimed at eliminating the cause(s) of nonconformity, not at justifying or resolving the findings on which the nonconformity is based.
3. Indicate the expected timeframe for resolution.
4. Once corrective action has been taken, verify its effectiveness in resolving the cause(s) of nonconformity.

Where appropriate, the auditee should inform the follow-up auditor of the actions it intends to take, e.g. by providing the action plan to resolve the nonconformities within the expected timeframe.

After the specified or agreed period has expired, the auditor may verify whether the corrective action has been properly implemented and whether it has been effective in eliminating the cause of the nonconformity, in which case the auditor may close the nonconformity. This verification may be included in a subsequent audit.

9.26 Managing an Audit Program

This is the last section of the book, but chronologically it should be the first in the part on audit methodology, because before any individual audit is carried out, the program that defines what is to be achieved needs to be established.

Auditors do not normally work alone, but within an audit program that defines the scope, objectives, and available resources, under the direction of an organization to which they report and which is responsible for coordinating the audit teams and monitoring the program.

Chapter 5 of ISO 19011:2018 is intended to provide guidance on the development and management of an audit program, including the setting of program objectives and the coordination of audit activities.

The development and management of an audit program is the responsibility of an organization, or the part of an organization that is capable of program management. The person responsible for program management does not necessarily need to be part of the audit group, but should have the audit knowledge and experience to perform the role effectively, as well as competence in areas such as management system standards, the auditee's activities, products, and processes, legal requirements, and other applicable requirements.

The audit program should gather sufficient information to be able to carry out audits effectively within the specified timeframe. It should include:

- Objectives for the audit program and individual audits.
- Risks and opportunities associated with the audit program and actions to address them.
- Scope, number, type, and timing of audits.
- Program procedures.
- Audit criteria.
- Audit methods.
- Selection of audit teams.
- Identification of resources required.
- Relevant documented information.

The management of an audit program described in ISO 19011 is based on the structure of a continuous improvement cycle. That is, it follows the four phases of the plan–do–check–act (PDCA) cycle. The "plan" phase defines the objectives, scope, and resources required. In the "do" phase, the objectives for each individual audit are defined in more detail, the criteria and methods to be used are determined, the auditors are selected, and the responsibilities are assigned for each of the audits.

In line with the continuous improvement approach, once the audits foreseen in the program have been carried out, the activities need to be followed up to assess whether the expected results have been achieved, to check whether the objectives have been met, and also to identify shortcomings and opportunities for improvement in the audit program.

An important part of the review and monitoring of the audit program is the evaluation of the performance of the audit team members. As noted above, the effectiveness of an audit program depends to a large extent on the professional competence of the auditors involved in its implementation. For this reason, it is important that a monitoring system is put in place to determine whether the auditors are performing to the professional standard expected of them and to evaluate their performance on an ongoing basis.

ISO 19011 identifies some of the possible methods for evaluating auditors, such as:

- Review of audit reports and other documentation.
- Feedback from the client or auditee.
- Observation of audit activities, e.g. audits in the presence of a witness.
- Interviews with team leaders or auditors.
- Testing.
- Technical meetings of groups of auditors to share results, procedures, and other issues.

Assessment of team performance should be used to identify training needs and to promote the auditor's continuing professional development. In addition to training, other activities such as attending meetings, seminars, conferences, and mentoring contribute to the maintenance and continuous improvement of auditor competence.

In conducting the review, the person responsible for managing the audit program should consider:

- Results and trends observed in the monitoring of the program.
- Compliance with program procedures.

- Changes in stakeholder needs and expectations.
- Program records.
- New or alternative audit techniques.
- Effectiveness of measures to mitigate risks identified in the program.
- Confidentiality and information security issues.

All of this information should be considered as input to the program review to complete the cycle of continuous improvement.

References

Ambrose, P.W. (2013). *ISO 9001:2008: Process Auditing Checklist*. Amazon.com.

Couto, M. and Bertolino, M.T. (2023). *Contra-Auditoria em Sistemas de Gestão e outros contratempos*. Niterói: RJ. GSI.

BRCGS (2022). *Global Standard Food Safety Issue 9*. London: BRCGS.

International Auditing and Assurance Standards Board (2010a). *ISA 500 Audit Evidence*. www.ifac.org

International Featured Standards (2023). *IFS Food Version 8*. Berlin.

International Organization for Standardization (2018). *ISO 19011:2018: Guidelines for Auditing Management Systems*. International Organization for Standardization. Geneva.

International Organization for Standardization (ISO) (2015). *ISO 9000:2015: Quality Management Systems – Fundamentals and Vocabulary*.

FAO and WHO (1998). *Guidance on Regulatory Assessment of HACCP*. Geneva: World Health Organization.

Mills, D. (1993). *Quality Auditing*. Dordrecht: Springer.

Scott Parsowith, B. (1995). *Fundamentals of Quality Auditing*. ASQC/Quality Press.

Case Study 1

Slaughterhouse for Rabbits, with an Adjoining Cutting Plant

Note: Due to the length of this complete hazard analysis and critical control point (HACCP) plan, this case study only includes the parts that are considered most important such as product description, hazard analysis, assessment of their degree of significance, and the management of critical control points (CCPs) and operational prerequisite programs (OPRPs).

1 HACCP Team

Managing director
Production manager
Veterinary
Quality technicians (2 people)
Quality manager
Maintenance responsible

2 Scope

Slaughtering, dismembering, and cutting of chilled or frozen rabbits in carcasses and in parts (such as thigs, loins, shoulders, livers) packed under vacuum in plastic bags or in modified atmosphere in plastic trays with plastic film or standard packed in plastic boxes or plastic bags.

3 Description of Product

3.1 Raw Material: Rabbit Meat

3.1.1 General

Rabbit meat from other industries: the edible part of the muscles of rabbits slaughtered under hygienic conditions in establishments of other industries approved by the competent authorities and approved by the establishment. The meat must be clean, sound, and properly prepared and

HACCP System Auditing for Food Safety: Principles and Techniques, First Edition. Luis Couto Lorenzo.

must include skeletal muscle with fat, parts of bones, tendons, nerves, and blood vessels, which normally accompany the muscle tissue and which are not separated from it during the process of meat separation.

3.1.2 Origin

Rabbit meat must come from approved establishments that meet the health requirements laid down by the countries of the European Union. Provided that these requirements are met, there are no restrictions as to the origin of the meat, as long as this origin is adequately identified and does not cause confusion or misunderstanding in the operations to be carried out in the industry.

3.1.3 Characteristics or Requirements to be Met by the Product

3.1.3.1 General Information on the Meat

- Whole carcasses and cuts presented in a good state of freshness, with no obvious signs of chemical or biological deterioration;
- well-developed muscle mass, particularly at the level of the breast and muscles, with no obvious signs of emaciation;
- absence of partial amputations of the limbs;
- absence of fractures in the bony structure of the carcass;
- absence of malignant tumors;
- absence of fairly extensive or disseminated hemorrhages or clots;
- absence of abscesses, tumors, blood or serous infiltrations, or other significant pathological lesions that may contaminate or give an unpleasant appearance;
- absence of fecal matter or other foreign matter;
- color: color characteristic of the species; pale color with a low index or intensity of red in muscles such as the longissimus dorsi, or slightly more intense in the thighs;
- appearance: meat with characteristic texture; no problems with pale, soft, and exudative muscles. No problems with deep red and hard carcasses.
- presence of a health mark on the product and/or packaging.

3.1.3.2 Microbiological and Physicochemical Characteristics In addition to the applicable microbiological requirements laid down in existing legislation, in particular Regulation (EC) No 2073/2005 of 15 November 2005 on microbiological criteria for foodstuffs, as amended by Regulation (EC) No 1441/2007 of 5 December 2007, the products covered by the management system shall comply with the following microbiological and physicochemical parameters (Tables 1–3):

Table 1 Microbiological criteria at the end of the production process.

Parameter	Criterion	Tolerance
Total mesophilic aerobic content	$\leq 1 \times 10^5$ cfu/g	$\leq 1 \times 10^6$ cfu/g
Total Enterobacteriaceae	$\leq 1 \times 10^3$ cfu/g	
Salmonella	Absent in 25 g	
Staphylococcus aureus	$\leq 1 \times 10^2$ cfu/g	
Escherichia coli	$\leq 1 \times 10^2$ cfu/g	$\leq 5 \times 10^2$ cfu/g
Listeria monocytogenes	Absent in 25 g	

Table 2 Microbiological end-of-life criteria.

Parameter	Criterion	Tolerance
Salmonella	Absent in 25 g	

Table 3 Physicochemical criteria and other specifications for the final product.

Parameter	Minimum	Maximum
Temperature (°C) – fresh product	0 °C	+4 °C
pH	5.6	6.6
Caliber of carcass	0,600 kg	3,000 kg
Residues of antibiotics in meat and other compounds (heavy metals, polychlorinated biphenyls (PCBs), dioxins, melamine, mercury, chlorates)	Maximum residue levels (MRLs) established by current legislation	
Allergens (including traces): cereals containing gluten (rye, barley, oats, spelt, kamut, and cereal products); crustaceans and crustacean products; eggs and egg products; fish and fish products; peanuts and peanut products; soya and soya products; milk and milk products (including lactose); nuts (almonds, hazelnuts, walnuts, cashew nuts, groundnuts, Brazil nuts, peanuts, peanut and peanut products, soya and soya products); milk and milk products (including lactose); nuts (almonds, hazelnuts, walnuts, cashew nuts, cashew kernels, Brazil nuts, Brazil nuts); walnuts, Brazil nuts, pistachios, macadamia and Queensland nuts and walnut products; celery and celery-based products; mustard and mustard-based products; sesame seeds and sesame-based products; sulfur dioxide and sulfites in concentrations greater than 10 mg/kg but not exceeding 10 mg/l expressed and SO_2	Absent	
Other (including traces): beef; pork; poultry; shellfish; gelatin; fructose; maize; cocoa; yeast and yeast products; phenylalanine glutamate; benzoic acid (E210 to E213); P. hydroxybenzoate (E214 to E219); tartrazine (E102); BHA/BHT (E320 to E321); sunset yellow FCF (E110); yellow (E110); amaranth (E123); cochineal red (E124); cinnamon; vanilla; legumes; coriander	Absent	
Irradiated substances	Absent	
GMO (same <1%)	Absent	

3.1.3.3 Other Requirements

- Full labeling in accordance with current legislation (Regulation (EC) 1169/2011).
- Fresh product: up to three days from the date of slaughter on arrival at the industry (tolerance up to five days).

3.2 Description of the Intended Use

FRESH RABBIT MEAT (in any presentation)

1) **Intended use**

   Consumption after production, as it is a raw product. It must be fully cooked.

   Due to its microbiological, physicochemical, and nutritional properties, it is a product that can be consumed by all sectors of the population, including high-risk groups (children, the elderly, the immunosuppressed, pregnant women, etc.) and is generally recommended as an integral part of the diet of people with cardiovascular problems and hypercholesterolemia.

2) **Target population**
   - "Children: eat well cooked. This group is particularly sensitive to the possible presence of broken bones or meat fragments, as well as any other foreign objects that could cause broken teeth, choking, or digestive problems.
   - The elderly: consume well cooked. As in the previous point, this is a risk group that is particularly sensitive to the possible presence of broken bones or bone fragments in the meat, as well as any other type of foreign object that could cause broken teeth, choking, or digestive problems.
   - Immunocompromised and pregnant women: consume well cooked. This group is particularly sensitive to the presence of pathogens such as *Listeria*, *Salmonella*, *E. coli*, or *Campylobacter*, among others. Special care must be taken when handling the products to avoid contamination of the food".

3) **Allergies or intolerances**

   No intolerance or allergy to rabbit meat has been described or documented in the medical literature.

4) **Handling, storage, and preservation instructions or recommendations**

   Taking into account the risk groups and the importance of the correct preparation of the product, with the aim of eliminating any type of pathogenic microbial contamination, it is advisable to implement measures to disseminate the correct practices for the preparation, conservation, and storage of the product, through the label, advertising, website, etc., with the aim of avoiding possible food intoxications or toxi-infections.
   - **Conservation**: Keep refrigerated between 0 and 4 °C. Minimize the time that the product can be above 4 °C (e.g. during transport from the point of purchase to the refrigerator).
   - **Handling**: Once out of the refrigerator, cook as soon as possible. Wash hands frequently during cooking. Never eat raw. Cook thoroughly before eating.
   - **Storage**: Do not store in the refrigerator in contact with products that can be eaten raw.

4 Flow Diagram

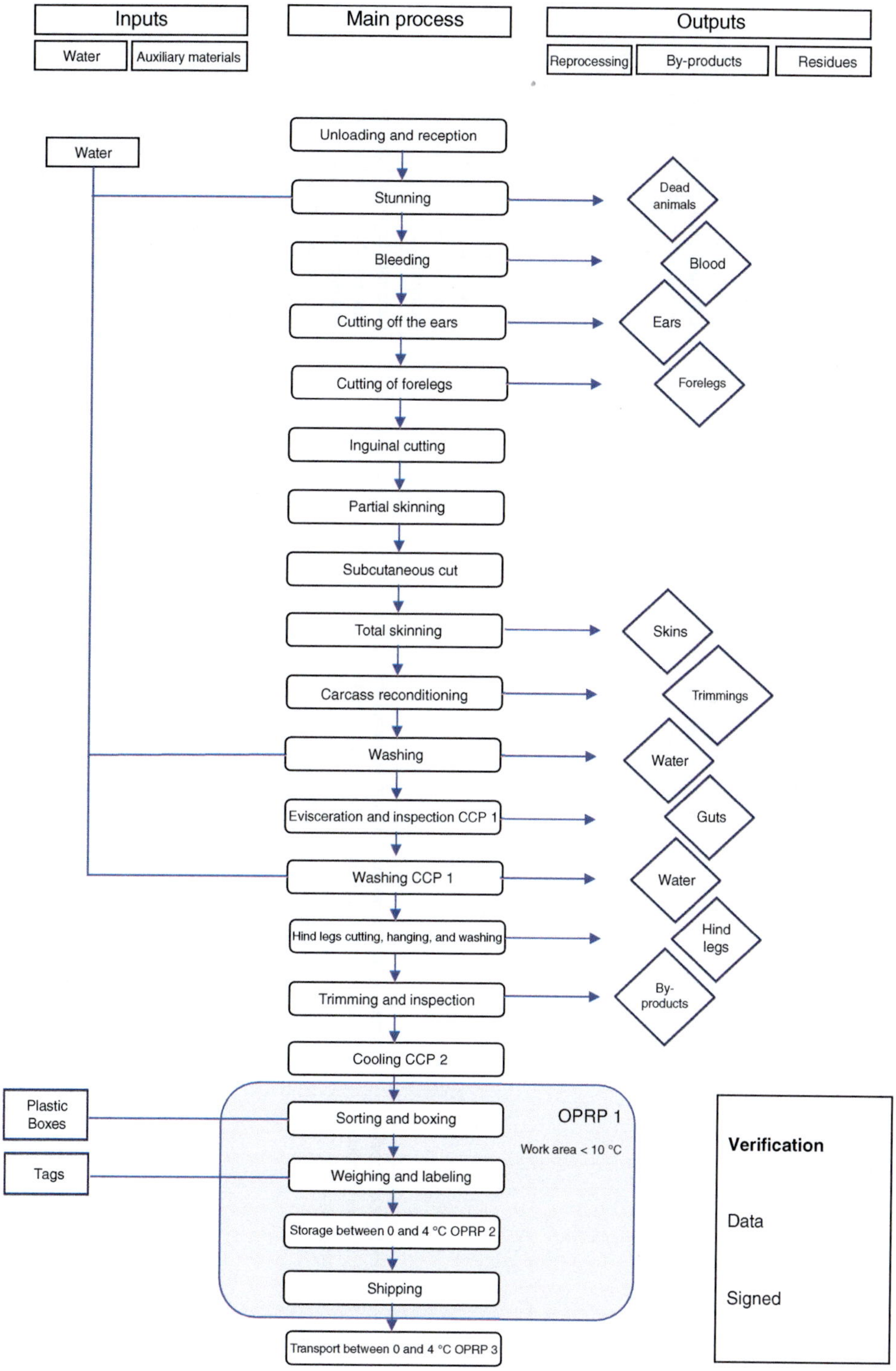

5 Hazard Control Chart

N°	Step	Hazard	Cause	Control measure	Risk analysis			
					Likelihood	Severity	Risk value	Justification
1	Unloading and reception	Biological. Possible zoonoses	The animals present zoonotic diseases or infections, especially bacterial (myxomatosis, hemorrhagic virus, tularemia)	Control of suppliers: periodic visits/audits to suppliers to know their performance, handling conditions, etc.; inspection antemortem; increase precautions during the process if symptoms or evidence of possible zoonosis is detected	2	2	4	There is a low probability of occurrence. Although it is always possible to find occasional cases in some batches, no complete batches affected by possible zoonosis have been recorded in recent years. The current control systems implemented in farms have minimized the possibilities of the appearance of zoonotic agents with the possibility of posing a hazard to food safety or even safety at work. The severity is very moderate, in case of occurrence, because affected animals are subjected to an antemortem inspection and postmortem inspection in which they could be detected and separated in case of incidents.
		Biological. Possible parasites (cysticercus and coccidia)	Animals present parasitic diseases due to inadequate handling conditions on the farm. Affected carcasses or parts of carcasses enter the food chain because they are not correctly identified and disposed of	Control of suppliers: periodic visits/audits to suppliers to know their performance, handling conditions, etc.; inspection antemortem; increase precautions during the process if symptoms or evidence of possible parasites is detected	1	2	2	Carcasses or parts of carcasses (especially livers) with possible parasitosis are easily identifiable and therefore easily segregated and eliminated. No cases of complaints or nonconformities have been recorded in recent years associated with this hazard. If any carcass or part of carcass were to enter the food supply chain and be ingested, it could cause some harm to the person, although this is very remote, since the parasites would be eliminated during the cooking of the product
		Biological. Microbiological and pathogenic contamination	The use of compressed air by the operators in the stunning area to get rid of excess hair on the gloves	Control of compressor equipment. General maintenance procedure for compressor installations	1	1	1	The probability is very low because the possible contamination of the gloves through the air will only affect the carcasses prior to dressing. The skin of the carcasses will be removed later, preventing contamination of the product

	Unloading and reception	Chemical. Residues of antibiotics or other medicinal substances (including coccidiostats)	The animals have been medicated, and the withdrawal periods have been insufficient or they have been medicated with unauthorized substances. The animals have incorporated other undesirable substances into their metabolism (by exposure, contact, direct administration, etc.) that may enter the food supply chain	Control of suppliers: periodic visits/audits to the suppliers to know about their performance, handling conditions, etc. Three-plate inhibitor testing of production batches on carcasses upon leaving the slaughterhouse	2	3	6	The probability of occurrence is low, since over the last ten years only two positives have been detected in this type of substances. There is no bibliographic evidence or records in other industries of a higher incidence. If the product reaches the market with residues of this type of substances, they would not have to cause a serious health problem for people, but in a systematic way, they could accumulate in human tissues and cause bacterial resistance to drugs, accumulation of heavy metals, etc., which in the long term could cause a public health problem (in this area, work is done on plants to reduce the use of antibiotics in rabbit farms, and some measures have already been implemented)
		Chemical. Residues of contaminants and toxic substances: Cd, Pb, dioxins, PCBs, melamine, mercury, chlorates, etc.	The animals have been either exposed to environments with the presence of these contaminants or intentionally contaminated, or as a more possible but unlikely cause, the animals have been contaminated through feed	Control of suppliers: periodic visits/audits to suppliers to know their performance, handling conditions, etc. Request analysis or declaration or absence of dioxins and PCBs from the feed manufacturers. Carrying out analyses to rule out the presence of contaminants according to the annual analytical plan	1	3	3	The probability of occurrence is low, since over the last five years only two positives have been detected in this type of substances. There is no bibliographic evidence or records in other industries of a higher incidence. If the product reaches the market with residues of this type of substances, they would not have to cause a serious health problem for people, but in a systematic way, they could accumulate in human tissues and cause bacterial resistance to drugs, accumulation of heavy metals, etc., which in the long term could cause a public health problem (in this area, work is done on plants to reduce the use of antibiotics in rabbit farms, and some measures have already been implemented)
2	Stunning	Biological. Introduction of pathogens into the food chain	Animals that die during transport or during waiting time are introduced into the production line. If the reason for the death of the animals is unknown, pathogens (bacterial, parasitic, or viral) may be introduced into the food supply chain	Good handling practices: dead animals during transport or while waiting must be separated and segregated as a category 2 by-product and treated as such Hygienic removal of by-products	1	5	5	The probability of occurrence of this hazard is remote as dead animals during transport are easily identifiable, separated, and removed The highest severity was considered in case of occurrence of the hazard as the agent that caused the death of the animal is not known. The principle of prudence leads us to place ourselves in the most unfavorable situation

(*Continued*)

N°	Step	Hazard	Cause	Control measure	Risk analysis			
					Likelihood	Severity	Risk value	Justification
3	Bleeding	Biological. Pathogenic and spoilage microbiological contamination	Contamination from cutting: poorly disinfected knife	Good hygiene practices Rotation of knives and hygienic storage of knives not in use Cleaning and disinfection of knives Sterilization at $T > 82\,^{\circ}\mathrm{C}$ or equivalent method	2	2	4	Low probability of occurrence: no cases were recorded in the industry, and no reference literature was found to report incidences associated with this possibility Moderate severity: the product should be consumed cooked (indicated on the label)
		Physical. Physical contamination by metals from knives	Knife blades that break or lose part of their cutting edge	Preoperational control of knives and metal elements: identification, integrity, cleanliness, etc. Good handling practices for workers: on-site inspection of cutters and metal elements Visual inspection of carcasses at later stages of the process	2	2	4	Low probability of occurrence: no cases were recorded over the last 10 yr Moderate severity: according to the literature consulted, the size of the metallic elements to be in the margins of causing harm to the consumer and the impossibility of detection means that hazardous sizes can be easily detected during visual inspections even by the ultimate consumer
4	Cutting off the ears	Biological. Pathogenic and spoilage microbiological contamination	Contamination from cutting: poorly disinfected knife	Good Handler Practices Rotation of knives and hygienic storage of knives not in use Cleaning and disinfection of knives Sterilization at $T > 82\,^{\circ}\mathrm{C}$ or equivalent method	2	2	4	It is considered that with Good Handling Practices (and control and monitoring of these), the potential contamination for this reason is low The severity is considered low as the product must be consumed cooked and possible contaminants introduced at this stage are sensitive to high temperatures
		Biological. Pathogenic and spoilage microbiological contamination	Contamination by the handler: the handler touches the meat directly with contaminated gloves after the ears have been removed	Good Handling Practices: the handler shall wash hands regularly Good Handling Practices: the handler shall not touch the meat after removal of the ears directly with his hands (even with gloves)	2	2	4	It is considered that with Good Handling Practices (and control and monitoring of these), the potential contamination for this reason is low The severity is considered low as the product must be consumed cooked and possible contaminants introduced at this stage are sensitive to high temperatures

	Cutting off the ears	Biological. Pathogenic and spoilage microbiological contamination	Cross-contamination: contamination through by-products (ears) that may not be removed hygienically	Good handling practices: by-products shall be removed in real time as they are produced Good hygienic practices: handlers shall avoid contact with accumulated by-products. If handling is necessary, hands should be washed and disinfected before returning to handling tasks	2	2	4	It is considered that with good handling practices (and the control and monitoring of these), the potential contamination for this reason is considered to be low The severity is considered to be low as the product must be consumed cooked and the possible contaminants introduced at this stage are sensitive to high temperatures Although other more hazardous contaminants could be introduced if contact occurs (*Salmonella*, *Campylobacter*, *Listeria*, etc.), the correct handling of the by-products and the following of Good Handling Practices make this very unlikely. Moreover, in later stages this hazard can be controlled or reduced by washing or cooling
		Physical. Physical contamination by metal contamination from knives	Knife blades that break or lose part of their cutting edge	Preoperational control of knives and metal elements: identification, integrity, cleanliness, etc. Good Handling Practices for workers: on-site inspection of cutters and metal elements Visual inspection of carcasses at later stages of the process	2	2	4	Low probability of occurrence: no cases were recorded over the last 10 yr Moderate severity: according to the literature consulted, the size of the metallic elements to be in the margins of causing harm to the consumer and the impossibility of detection means that hazardous sizes can be easily detected during visual inspections even by the ultimate consumer
5	Cutting of forelegs	Biological. Pathogenic and spoilage microbiological contamination	Cross-contamination: contamination through by-products (feet) that may not be removed hygienically	Good Handling Practices: by-products shall be removed in real time as they are produced Good handling Practices: handlers shall avoid contact with accumulated by-products. If handling is necessary, hands should be washed and disinfected before returning to handling tasks	1	2	2	It is considered that with Good Handling Practices (and the control and monitoring of these), the potential contamination for this reason is considered to be low The severity is considered low as the product has to be consumed cooked and possible contaminants introduced at this stage are sensitive to high temperatures By-products are removed automatically and in real time so that at no time do they come into direct contact with the carcasses and there is no possibility of cross-contamination

(*Continued*)

N°	Step	Hazard	Cause	Control measure	Risk analysis			
					Likelihood	Severity	Risk value	Justification
	Cutting of forelegs	Physical. Physical metal contamination from the cutoff wheel	Cutting elements or cutoff wheel filings	Preoperational control of cutting disks and metallic elements: identification, integrity, cleanliness, etc. Good Handling Practices for workers: on-site inspection of cutting elements and metal elements Visual inspection of carcasses at later stages of the process	2	2	4	Low probability of occurrence: no cases were recorded over the last 10 yr Moderate severity: according to the literature consulted, the size of the metallic elements is marginal to cause harm to the consumer, and the impossibility of detection means that hazardous sizes can be easily detected during visual inspections even by the ultimate consumer
6	Inguinal cutting	Biological. Pathogenic and spoilage microbiological contamination	Contamination from cutting: poorly disinfected knife	Good Handling Practices Rotation of knives and hygienic storage of knives not in use Cleaning and disinfection of knives Sterilization at $T > 82\,^{\circ}\mathrm{C}$ or equivalent method	2	2	4	It is considered that with Good Handling Practices (and control and monitoring of these), the potential contamination for this reason is low The severity is considered low as the product must be consumed cooked and possible contaminants introduced at this stage are sensitive to high temperatures
		Biological. Pathogenic and spoilage microbiological contamination	Contamination by the handler: the handler touches the meat directly with contaminated gloves or after touching the skin of the rabbit	Good Handling practices The handler should avoid as much as possible touching the carcass directly after touching the skin of the rabbit	2	2	4	It is considered that with Good Handling Practices (and control and monitoring of these), the potential contamination from this source is considered to be low The severity is considered low as the product has to be consumed cooked and possible contaminants introduced at this stage are sensitive to high temperatures

	Inguinal cutting	Physical. Physical contamination by metal contamination from knives	Knife blades that break or lose part of their cutting edge	Preoperational control of knives and metal elements: identification, integrity, cleanliness, etc. Good Handling Practices for workers: on-site inspection of cutters and metal elements Visual inspection of carcasses at later stages of the process	2	2	4	Low probability of occurrence: no cases were recorded over the last 10 yr Moderate severity: according to the literature consulted, the size of the metallic elements is marginal to cause harm to the consumer, and the impossibility of detection means that hazardous sizes can be easily detected during visual inspections even by the ultimate consumer
7	Partial skinning	Biological. Pathogenic and spoilage microbiological contamination	Contamination by the handler: the handler touches the meat directly with contaminated gloves or after touching the skin of the rabbit	Good Handling Practices The handler should avoid as much as possible touching the carcass directly after touching the skin of the rabbit	2	2	4	It is considered that with Good Handling Practices (and control and monitoring of these), the potential contamination from this source is considered to be low The severity is considered low as the product has to be consumed cooked and possible contaminants introduced at this stage are sensitive to high temperatures
		Biological. Pathogenic and spoilage microbiological contamination	Condensation: microbiological contamination through condensate droplets that could accumulate on the ceiling and fall on the product	Activation of exhaust fans when necessary depending on the condensation that accumulates on the slaughterhouse roofs Preoperational and operational checks on condensate accumulation. In case of condensation, activation of extractors Periodic removal of condensate during downtime if the operation of the extractors is not sufficient. This condensate removal on the roof must be carried out without product in the slaughtering line	2	2	4	The probability has been considered low as there are several subsequent stages in which the carcasses are washed, so any drops of condensate that might fall on the carcass carrying contamination would be eliminated in subsequent operations It is also considered low, as the exposure time of an individual carcass to a specific position where condensation could accumulate is very short, so the probability is also low The severity is considered low as the need to cook the product would eliminate any microbiological contamination that could be introduced by condensation droplets It is also considered low as we have not been able to demonstrate that condensate droplets add a significantly different contaminant load or significantly increase the contaminant load that occurs in slaughterhouse operations

(*Continued*)

N°	Step	Hazard	Cause	Control measure	Risk analysis			
					Likelihood	Severity	Risk value	Justification
	Partial skinning			Microbiological control of condensate droplets to rule out the persistence of pathogens Washing of the carcass at various points in the slaughterhouse during the process, thus reducing the impact of point contamination caused by a drop of condensate that could have fallen on the carcass				
8	Subcutaneous cut	Biological. Pathogenic and spoilage microbiological contamination	Contamination from cutting: poorly disinfected knife	Good Handling Practices Rotation of knives and hygienic storage of knives not in use Cleaning and disinfection of knives Sterilization at $T > 82\,^{\circ}\mathrm{C}$ or equivalent method	2	2	4	It is considered that with Good Handling Practices (and control and monitoring of these), the potential contamination for this reason is low The severity is considered low as the product must be consumed cooked and possible contaminants introduced at this stage are sensitive to high temperatures
		Biological. Pathogenic and spoilage microbiological contamination	Contamination by the handler: the handler touches the meat directly with contaminated gloves or after touching the skin of the rabbit	Good Handling Practices The handler should avoid as much as possible touching the carcass directly after touching the skin of the rabbit	2	2	4	It is considered that with Good Handling practices (and control and monitoring thereof), the potential contamination from this source is considered to be low The severity is considered low as the product has to be consumed cooked and potential contaminants introduced at this stage are sensitive to high temperatures

Subcutaneous cut	Biological. Pathogenic and spoilage microbiological contamination	Condensation: microbiological contamination through condensate droplets that could accumulate on the ceiling and fall on the product	Activation of exhaust fans when necessary depending on the condensation that accumulates on the slaughterhouse roofs Preoperational and operational checks on condensate accumulation. In case of condensation, activation of extractors Periodic removal of condensate during downtime if the operation of the extractors is not sufficient. This condensate removal on the roof must be carried out without product in the slaughtering line Microbiological control of condensate droplets to rule out the persistence of pathogens Washing of the carcass at various points in the slaughterhouse during the process, thus reducing the impact of point contamination caused by a drop of condensate that could have fallen on the carcass	2	2	4	The probability has been considered low as there are several subsequent stages in which the carcasses are washed, so any drops of condensate that might fall on the carcass carrying contamination would be eliminated in subsequent operations It is also considered low, as the exposure time of an individual carcass to a specific position where condensation could accumulate is very short, so the probability is also low The severity is considered low as the need to cook the product would eliminate any microbiological contamination that could be introduced by condensation droplets It is also considered low as we have not been able to demonstrate that condensate droplets add a significantly different contaminant load or significantly increase the contaminant load that occurs in slaughterhouse operations
	Physical. Physical contamination by metal contamination from knives	Knife blades that break or lose part of their cutting edge	Preoperational control of knives and metal elements: identification, integrity, cleanliness, etc. Good Handling Practices for workers: on-site inspection of cutters and metal elements Visual inspection of carcasses at later stages of the process	2	2	4	Low probability of occurrence: no cases were recorded over the last 10 yr Moderate severity: according to the literature consulted, the size of the metallic elements is in the margins of causing harm to the consumer, and the impossibility of detection means that hazardous sizes can be easily detected during visual inspections even by the ultimate consumer

(Continued)

N°	Step	Hazard	Cause	Control measure	Risk analysis			
					Likelihood	Severity	Risk value	Justification
9	Total skinning	Biological. Pathogenic and spoilage microbiological contamination	Contamination through parts of the equipment that come into contact with the skinless part of the carcass	Cleaning and disinfection plan Correct sanitization of equipment Cleaning and disinfection products authorized for food industries Equipment maintenance plan (well-maintained equipment causes fewer problems and possible contamination than if it is not)	1	2	2	The likelihood of the occurrence of a hazard of this type is considered low because the equipment is sanitized daily in accordance with a cleaning and disinfection plan The severity is considered low because even if hazards were introduced at this stage, the need for complete cooking of the product would eliminate any existing pathogens (they would never be introduced at high levels at this stage) that are sensitive to high temperatures In addition, at later stages this hazard is controlled and reduced: washing or cooling
		Biological. Pathogenic and spoilage microbiological contamination	Cross-contamination: contamination through by-products (skins) that may not be removed hygienically	Good Handling Practices: by-products shall be removed in real time as they are produced Good Handling Practices: handlers shall avoid contact with accumulated by-products. If handling is necessary, hands should be washed and disinfected before returning to handling tasks	2	2	4	It is considered that with Good Handling Practices (and the control and monitoring of these), the potential contamination for this reason is considered to be low The severity is considered low as the product has to be consumed cooked and possible contaminants introduced at this stage are sensitive to high temperatures By-products are removed automatically and in real time so that at no time do they come into direct contact with the carcasses and there is no possibility of cross-contamination
		Biological. Pathogenic and spoilage microbiological contamination	Condensation: microbiological contamination through condensate droplets that could accumulate on the ceiling and fall on the product	Activation of exhaust fans when necessary depending on the condensation that accumulates on the slaughterhouse roofs Preoperational and operational checks on condensate accumulation. In case of condensation, activation of extractors	2	1	2	The probability has been considered low as there are several subsequent stages in which the carcasses are washed, so any drops of condensate that might fall on the carcass carrying contamination would be eliminated in subsequent operations It is also considered low, as the exposure time of an individual carcass to a specific position where condensation could accumulate is very short, so the probability is also low

Total skinning		Periodic removal of condensate during downtime if the operation of the extractors is not sufficient. This condensate removal on the roof must be carried out without product in the slaughtering line Microbiological control of condensate droplets to rule out the persistence of pathogens Washing of the carcass at various points in the slaughterhouse during the process, thus reducing the impact of point contamination caused by a drop of condensate that could have fallen on the carcass				The severity is considered low as the need to cook the product would eliminate any microbiological contamination that could be introduced by condensation droplets It is also considered low as we have not been able to demonstrate that condensate droplets add a significantly different contaminant load or significantly increase the contaminant load that occurs in slaughterhouse operations
Chemical. Grease and lubricant residues	Contamination of carcass parts that come into contact with the machinery if the machinery has not been properly lubricated	Maintenance plan Good equipment lubrication practices Approved lubrication products suitable for the food industry	2	1	2	The likelihood of introducing a hazard at this stage is considered low as there is a maintenance plan that specifies the conditions under which work equipment must be lubricated All lubricants and/or greases used are food grade and are registered and authorized for food use There are subsequent stages (rinsing) in which any residues introduced would be removed The severity was considered to be very low due to the small amount of the hazard introduced into the food in case it was introduced

(*Continued*)

N°	Step	Hazard	Cause	Control measure	Risk analysis			
					Likelihood	Severity	Risk value	Justification
10	Carcass reconditioning	Biological. Pathogenic and spoilage microbiological contamination	Contamination from cutting: poorly disinfected knife	Good Handling Practices Rotation of knives and hygienic storage of knives not in use Cleaning and disinfection of knives Sterilization at $T > 82\,^{\circ}\mathrm{C}$ or equivalent method	2	2	4	It is considered that with Good Handling Practices (and control and monitoring of these), the potential contamination for this reason is low The severity is considered low as the product must be consumed cooked and possible contaminants introduced at this stage are sensitive to high temperatures
		Biological. Pathogenic and spoilage microbiological contamination	Contamination by the handler: the handler touches the meat directly with contaminated gloves or after touching the skin of the rabbit or other parts that could be contaminated	Good Handling Practices The handler should avoid as much as possible touching the carcass directly after touching the skin of the rabbit or other contaminated parts	2	2	4	It is considered that with Good Handling Practices (and control and monitoring of these), the potential contamination from this source is considered to be low The severity is considered low as the product has to be consumed cooked and possible contaminants introduced at this stage are sensitive to high temperatures
		Biological. Pathogenic and spoilage microbiological contamination	Cross-contamination: contamination through by-products (leftovers) that may not be removed hygienically	Good Handling Practices: by-products shall be removed in real time as they are produced Good Handling Practices: handlers shall avoid contact with accumulated by-products. If handling is necessary, hands should be washed and disinfected before returning to handling tasks	2	2	4	It is considered that with Good Handling Practices (and the control and monitoring of these), the potential contamination for this reason is considered to be low The severity is considered to be low as the product must be consumed cooked and possible contaminants introduced at this stage are sensitive to high temperatures By-products are handled hygienically and are removed only at times when there is no possibility of cross-contamination with food

Carcass reconditioning	Biological. Pathogenic and spoilage microbiological contamination	Condensation: microbiological contamination through condensate droplets that could accumulate on the ceiling and fall on the product	Activation of exhaust fans when necessary depending on the condensation that accumulates on the slaughterhouse roofs Preoperational and operational checks on condensate accumulation. In case of condensation, activation of extractors Periodic removal of condensate during downtime if the operation of the extractors is not sufficient. This condensate removal on the roof must be carried out without product in the slaughtering line Microbiological control of condensate droplets to rule out the persistence of pathogens Washing of the carcass at various points in the slaughterhouse during the process, thus reducing the impact of point contamination caused by a drop of condensate that could have fallen on the carcass	2	2	4	The probability has been considered low as there are several subsequent stages in which the carcasses are washed, so any drops of condensate that might fall on the carcass carrying contamination would be eliminated in subsequent operations It is also considered low, as the exposure time of an individual carcass to a specific position where condensation could accumulate is very short, so the probability is also low The severity is considered low as the need to cook the product would eliminate any microbiological contamination that could be introduced by condensation droplets It is also considered low as we have not been able to demonstrate that condensate droplets add a significantly different contaminant load or significantly increase the contaminant load that occurs in slaughterhouse operations

(Continued)

N°	Step	Hazard	Cause	Control measure	Risk analysis			
					Likelihood	Severity	Risk value	Justification
11	Washing	Biological. Pathogenic and spoilage microbiological contamination	The shower or diffusers are not in proper working order and maintenance; the water pressure is insufficient; etc. All these causes prevent the elimination or reduction of possible contaminant load introduced in previous stages	Maintenance plan. Correct Maintenance of shower equipment and diffusers	2	2	4	The probability was considered low as there are sufficient control measures (water supply, maintenance) to ensure the proper functioning of this stage Since this stage serves to control or reduce hazards produced in previous stages, the principle of prudence was used so as not to consider the probability as very low The severity is considered to be low as the product must be cooked before consumption so that potential hazards introduced at this stage would be eliminated
12	Evisceration and inspection	Biological. Pathogenic and spoilage microbiological contamination	Contamination from cutting: poorly disinfected knife	Good Handling Practices Rotation of knives and hygienic storage of knives not in use Cleaning and disinfection of knives Sterilization at $T > 82\,^{\circ}\mathrm{C}$ or equivalent method	2	2	4	It is considered that with Good Handling Practices (and control and monitoring of these), the potential contamination for this reason is low The severity is considered low as the product must be consumed cooked and possible contaminants introduced at this stage are sensitive to high temperatures
		Biological. Pathogenic and spoilage microbiological contamination	Contamination by ruptured intestinal package	Good Handling Practices: the handler should avoid touching contaminated areas as far as possible. If they do, they should wash and disinfect their hands and utensils Operator training: operators must know how to carry out evisceration tasks to minimize possible breakage of the intestinal package	4	2	8	The probability of occurrence is high. This has been verified after analysis of the processes There are certain Good Handling Practices that minimize the probability of occurrence of these hazards. Control and monitoring of these practices means that this probability is not considered to be very high There are subsequent stages (washing and cooling) where this hazard is controlled or reduced

			Monitoring and quality control				It is considered as low severity because subsequent steps will prevent the development and growth of the introduced potential hazards to unsuitable levels The product must be cooked before consumption so potential hazards introduced at this stage would be eliminated or drastically controlled as they are sensitive to high temperatures
	Biological. Pathogenic and spoilage microbiological contamination	Contamination by the handler: the handler touches the meat directly with contaminated gloves. Contamination of gloves may occur after possible breakage of the intestinal package	Good Handling Practices The handler should avoid as much as possible touching the carcass directly after touching the skin of the rabbit or other contaminated parts	2	2	4	It is considered that with Good Handling Practices (and control and monitoring of these), the potential contamination from this source is considered to be low There are subsequent steps (washing, cooling) that can control or reduce this hazard to acceptable levels The severity is considered low as the product has to be consumed cooked and the possible contaminants introduced at this stage are sensitive to high temperatures
Evisceration and inspection	Biological. Biological contamination: parasites, microbiological agents	Carcasses or parts of carcasses with evidence (or symptoms) of the presence of parasites (coccidia, cysticerci), microbiological agents (*Pasteurella*, *Francisella tularensis*), etc., were not removed	Postmortem inspection Training of operators to carry out ancillary tasks for postmortem inspection Segregation and hygienic removal of by-products Good Handling Practices: the handler should avoid touching contaminated or adjacent areas as far as possible. If they do so, they must wash and disinfect hands and utensils	3	3	9	It has become evident that there are very often carcasses that have to be completely removed or some of their parts because they show symptoms or evidence of pathogens: cysticerci, coccidia, *Pasteurella* This stage has been established to control these hazards at the same time as evisceration. The rhythm of the processes and the operators present allows the necessary segregation of those carcasses or parts affected The official veterinary service also performs control tasks at this stage There are later stages where potential hazards not identified at this stage could be identified (packaging)

(*Continued*)

N°	Step	Hazard	Cause	Control measure	Risk analysis			
					Likelihood	Severity	Risk value	Justification
								For all the above reasons, the likelihood was considered low It is considered to be of moderate severity because in case of reaching the final consumer pathogens could be introduced into the food chain which in many temperature-sensitive cases (the product must be consumed cooked) in high concentrations could produce heat-resistant toxins
	Evisceration and inspection	Biological. Pathogenic and spoilage microbiological contamination	Cross-contamination: contamination through by-products (offal) that may not be removed hygienically	Good Handling Practices: by-products shall be removed in real time as they are produced Good Handling Practices: handlers shall avoid contact with accumulated by-products. If handling is necessary, hands should be washed and disinfected before returning to handling tasks	2	2	4	It is considered that with Good Handling Practices (and the control and monitoring of these), the potential contamination for this reason is considered to be low The severity is considered to be low as the product must be consumed cooked and possible contaminants introduced at this stage are sensitive to high temperatures By-products are handled hygienically and are removed only at times when there is no possibility of cross-contamination with food
		Biological. Pathogenic and spoilage microbiological contamination	Condensation: microbiological contamination through condensate droplets that could accumulate on the ceiling and fall on the product	Activation of exhaust fans when necessary depending on the condensation that accumulates on the slaughterhouse roofs Preoperational and operational checks on condensate accumulation. In case of condensation, activation of extractors	2	2	4	The probability has been considered low as there are several subsequent stages in which the carcasses are washed, so that any drops of condensate that might fall on the carcass and carry contamination would be eliminated in subsequent operations It has also been considered low, as the exposure time of an individual carcass to a specific position where condensation could accumulate is very short, so the probability is also low

		Periodic removal of condensate during downtime if the operation of the extractors is not sufficient. This condensate removal on the roof must be carried out without product in the slaughtering line Microbiological control of condensate droplets to rule out the persistence of pathogens. Washing of the carcass at various points in the slaughterhouse during the process, thus reducing the impact of point contamination caused by a drop of condensate that could have fallen on the carcass				The severity is considered low as the need to cook the product would eliminate any microbiological contamination that could be introduced by condensation droplets It is also considered low as we have not been able to demonstrate that condensate droplets add a significantly different contaminant load or significantly increase the contaminant load that occurs in slaughterhouse operations
Chemical. Allergens: Allergen contamination	Cross-contamination occurs through handlers and food: handlers contaminate food with food they carry or through work clothes	Training of operators Good Handling Practices: rules for staff eating habits as well as work clothes, breaks, etc. Training of operators	1	4	4	It is considered a remote probability because no cases have been recorded historically Monitoring and surveillance of Good Handling Practices as well as training and sensitization of staff reduces the possibility of contamination by this hazard It is considered as serious because although it is well documented that even traces of an allergen can cause very severe anaphylactic reactions in people sensitive to certain types of allergens, all carcasses will be washed before entering the chilling tunnel
Physical. Personal items inside carcasses	Operators could insert personal items (rings, earrings, watch bands, band-aids, etc.) into the carcass cavity	Good Handling Practices: no person carrying personal items such as jewelry, rings, earrings, chains, watches is allowed to enter the handling areas	1	3	3	It is considered a very low probability as there have been no reported incidences of this hazard in recent years Following Good Handling Practices reduces the chances of contamination with this hazard to very low levels

(Continued)

N°	Step	Hazard	Cause	Control measure	Risk analysis			
					Likelihood	Severity	Risk value	Justification
	Evisceration and inspection			If plasters are to be used, they must be bright blue in color				In addition, there are later stages where the product can be visually inspected and possible physical hazards can be detected The severity was considered to be moderate as ingestion of foreign objects could cause broken teeth, tears in the digestive tract, intestinal problems, etc., of a moderate nature
13	Washing CCP 1	Biological. Pathogenic and spoilage microbiological contamination	Potential contaminant load introduced at other stages due to improperly performed washing tasks is not eliminated or reduced	Good Handling Practices: the operator must know how to carry out the carcass washing operation Training of operators	3	3	9	This is considered a moderate probability as it is possible that at this stage not all carcasses with fecal spillage will be identified and washed given the rate of production There are later stages where these carcasses may be identified (packaging) The severity was considered moderate as potential hazards introduced at earlier stages and not controlled or reduced at this stage could potentially cause harm to the health of consumers It was not considered as serious or very serious as there are later stages where the hazard is controlled (cooling or packaging) and because the product must be cooked before consumption and therefore the possible bacteria eliminated as they are mostly temperature sensitive

14 Hinds legs cutting, hanging, and washing	Biological. Pathogenic and spoilage microbiological contamination	Contamination by operators: operators can contaminate the carcass through their gloves if they do not respect good handling and hygiene practices	The handler shall avoid touching contaminated or possibly contaminated parts of the carcass (feet with skin) and subsequently noncontaminated parts of the carcass The handler shall wash hands regularly Operator training: the handler must know how to carry out the operation in a satisfactory and hygienic manner	2	2	4	It is considered that with Good Handling Practices (and control and monitoring of these), the potential contamination from this source is considered to be low Subsequent steps (cooling) help to control the potential hazard introduced The severity is considered low as the product has to be consumed cooked and possible contaminants introduced at this stage are sensitive to high temperatures
	Biological. Pathogenic and spoilage microbiological contamination	Contamination through parts of the equipment that come into contact with the skinless part of the carcass	Cleaning and disinfection plan Correct sanitization of equipment Cleaning and disinfection products authorized for food industries Equipment maintenance plan (well-maintained equipment causes fewer problems and possible contamination than if it is not)	1	1	1	The probability of introducing a hazard at this stage is considered low, as there is a cleaning and disinfection plan that specifies how to carry out the cleaning and disinfection tasks as well as the verification of their effectiveness and the presence of possible traces of cleaning products The contact time of the product with the equipment is very short, so the chances of contamination are reduced The contact area between the food and the equipment is very limited and reduced There are subsequent stages (cooling) in which the possible remains introduced would be controlled The severity was considered to be very low due to the small amount of the hazard introduced into the food in case it was introduced

(*Continued*)

N°	Step	Hazard	Cause	Control measure	Risk analysis			
					Likelihood	Severity	Risk value	Justification
	Hinds legs cutting, hanging, and washing	Biological. Pathogenic and spoilage microbiological contamination	Contamination from cutting: poorly disinfected knife	Good Handling Practices Rotation of knives and hygienic storage of knives not in use Cleaning and disinfection of knives Sterilization at $T > 82\,^{\circ}\mathrm{C}$ or equivalent method	2	2	4	It is considered that with Good Handling practices (and control and monitoring of these), the potential contamination for this reason is low The severity is considered low as the product must be consumed cooked and possible contaminants introduced at this stage are sensitive to high temperatures
		Biological. Pathogenic and spoilage microbiological contamination	Cross-contamination: contamination through by-products (feet) that may not be removed hygienically	Good Handling Practices: by-products shall be segregated and stored hygienically. They should be removed as soon as possible from food handling areas Good Handling Practices: handlers shall avoid contact with accumulated by-products. If handling is necessary, hands should be washed and disinfected before returning to handling tasks	2	2	4	It is considered that with Good Handling Practices (and the control and monitoring of these), the potential contamination for this reason is considered to be low The severity is considered to be low as the product must be consumed cooked and possible contaminants introduced at this stage are sensitive to high temperatures By-products are handled hygienically and are removed only at times when there is no possibility of cross-contamination with food
		Biological. Pathogenic and spoilage microbiological contamination	Condensation: microbiological contamination through condensate droplets that could accumulate on the ceiling and fall on the product	Activation of exhaust fans when necessary depending on the condensation that accumulates on the slaughterhouse roofs Preoperational and operational checks on condensate accumulation. In case of condensation, activation of extractors	2	2	4	The probability has been considered low as there are several subsequent stages in which the carcasses are washed, so any drops of condensate that might fall on the carcass carrying contamination would be eliminated in subsequent operations It is also considered low, as the exposure time of an individual carcass to a specific position where condensation could accumulate is very short, so the probability is also low

			Periodic removal of condensate during downtime if the operation of the extractors is not sufficient. This condensate removal on the roof must be carried out without product in the slaughter line Washing of the carcass at various points in the slaughterhouse during the process, thus reducing the impact of point contamination caused by a drop of condensate that might have fallen on the carcass				The severity is considered low as the need to cook the product would eliminate any microbiological contamination that could be introduced by condensation droplets It is also considered low as we have not been able to demonstrate that condensate droplets add a significantly different contaminant load or significantly increase the contaminant load that occurs in slaughterhouse operations
Hinds legs cutting, hanging, and washing	Physical. Physical metal contamination from the cutoff wheel	Cutting elements or cutoff wheel filings	Preoperational control of cutting discs and metallic elements: identification, integrity, cleanliness, etc. Good Handling Practices for workers: on-site inspection of cutting elements and metal elements Visual inspection of carcasses at later stages of the process	2	2	4	Low probability of occurrence: no cases were recorded over the last 10 yr Moderate severity: according to the literature consulted, the size of the metallic elements to be in the margins of causing harm to the consumer and the impossibility of detection means that hazardous sizes can be easily detected during visual inspections even by the ultimate consumer
	Physical. Physical contamination by metal contamination from knives	Knife blades that break or lose part of their cutting edge	Preoperational control of knives and metal elements: identification, integrity, cleanliness, etc. Good Handling Practices for workers: on-site inspection of cutters and metal elements Visual inspection of carcasses at later stages of the process	2	2	4	Low probability of occurrence: no cases were recorded over the last 10 yr Moderate severity: according to the literature consulted, the size of the metallic elements is in the margins of causing harm to the consumer, and the impossibility of detection means that hazardous sizes can be easily detected during visual inspections even by the ultimate consumer

(*Continued*)

N°	Step	Hazard	Cause	Control measure	Risk analysis			
					Likelihood	Severity	Risk value	Justification
		Physical. Loose bones	Improper cut on the rear leg trimmer causing splinters in the tibia	Proper maintenance and sharpening of the rear leg trimmer blade Visual inspection of carcasses at later stages of the process	3	4	12	Moderate probability. Based on the criteria defined in the AR Instructions and Decisions tab, it can be considered moderate if the maintenance plan is not being properly executed The severity has been considered serious as a loose bone can be ingested by risk groups and cause damage that can lead to hospitalization, broken teeth, etc.
		Chemical. Waste grease and lubricants	Contamination of carcass parts that come into contact with the machinery if the machinery has not been properly lubricated	Maintenance plan Good equipment lubrication practices Approved lubrication products suitable for the food industry	2	1	2	The likelihood of introducing a hazard at this stage is considered low as there is a maintenance plan that specifies the conditions under which work equipment must be lubricated All lubricants and/or greases used are food grade and are registered and authorized for food use If there is contamination at this stage, the probability of detection at later stages is very high and the products affected would be treated as reprocesses or by-products The severity was considered low because of the small amount of the hazard introduced into the food if it was introduced
15	Trimming and inspection	Biological. Pathogenic and spoilage microbiological contamination	Contamination by operators: operators can contaminate the carcass through their gloves if they do not respect good handling and hygiene practices	The handler shall avoid touching contaminated or possibly contaminated parts of the carcass (feet with skin) and subsequently noncontaminated parts of the carcass The handler shall wash hands regularly Operator training: the handler must know how to carry out the operation in a satisfactory and hygienic manner	2	2	4	It is considered that with Good Handling Practices (and control and monitoring of these), the potential contamination from this source is considered to be low Subsequent steps (cooling) help to control the potential hazard introduced The severity is considered low as the product has to be consumed cooked and possible contaminants introduced at this stage are sensitive to high temperatures

Trimming and inspection	Biological. Pathogenic and spoilage microbiological contamination	Contamination from cutting: poorly disinfected knife	Good Handling Practices Rotation of knives and hygienic storage of knives not in use Cleaning and disinfection of knives Sterilization at $T > 82\,^{\circ}\mathrm{C}$ or equivalent method	2	2	4	It is considered that with Good Handling Practices (and control and monitoring of these), the potential contamination for this reason is low The severity is considered low as the product must be consumed cooked and possible contaminants introduced at this stage are sensitive to high temperatures
	Biological. Pathogenic and spoilage microbiological contamination	Condensation: microbiological contamination through condensate droplets that could accumulate on the ceiling and fall on the product	Activation of exhaust fans when necessary depending on the condensation that accumulates on the slaughterhouse roofs Preoperational and operational checks on condensate accumulation. In case of condensation, activation of extractors Periodic removal of condensate during downtime if the operation of the extractors is not sufficient. This condensate removal on the roof must be carried out without product in the slaughtering line Microbiological control of condensate droplets to rule out the persistence of pathogens Washing of the carcass at various points in the slaughterhouse during the process, thus reducing the impact of point contamination caused by a drop of condensate that could have fallen on the carcass	2	2	4	The probability has been considered low as there are several subsequent stages in which the carcasses are washed, so any drops of condensate that might fall on the carcass carrying contamination would be eliminated in subsequent operations It is also considered low, as the exposure time of an individual carcass to a specific position where condensation could accumulate is very short, so the probability is also low The severity is considered low as the need to cook the product would eliminate any microbiological contamination that could be introduced by condensation droplets It is also considered low as we have not been able to demonstrate that condensate droplets add a significantly different contaminant load or significantly increase the contaminant load that occurs in slaughterhouse operations

(*Continued*)

N°	Step	Hazard	Cause	Control measure	Risk analysis			
					Likelihood	Severity	Risk value	Justification
	Trimming and inspection	Physical. Physical contamination by metal contamination from knives	Knife blades that break or lose part of their cutting edge	Preoperational control of knives and metal elements: identification, integrity, cleanliness, etc. Good Handling Practices for workers: on-site inspection of cutters and metal elements Visual inspection of carcasses at later stages of the process	2	2	4	Low probability of occurrence: no cases were recorded over the last 10 yr Moderate severity: according to the literature consulted, the size of the metallic elements is in the margins of causing harm to the consumer, and the impossibility of detection means that hazardous sizes can be easily detected during visual inspections even by the ultimate consumer.
		Chemical. Allergens: Allergen contamination	Cross-contamination occurs through handlers and food: handlers contaminate food with food they carry or through work clothes	Training of operators Good Handling Practices: rules for staff eating habits as well as work clothes, breaks, etc. Training of operators	1	4	4	It is considered a remote probability because no cases have been recorded historically Monitoring and surveillance of Good Handling Practices as well as training and sensitization of staff reduces the possibility of contamination by this hazard It is considered as serious because although it is well documented that even traces of an allergen can cause very severe anaphylactic reactions in people sensitive to certain types of allergens, all carcasses will be washed before entering the airing tunnel
16	Cooling	Biological. Pathogenic and spoilage microbiological contamination	Growth and survival of those bacteria that may have been introduced in previous stages and are present. This stage is specifically designed to prevent the growth and/or survival of microbiota	Maintenance plan Correct functioning of cooling equipment Immediate cooling after slaughter and evisceration operations in the slaughterhouse Sufficient temperature and cooling time. Carcass temperature at tunnel exit $< +4\,°C$	3	4	12	This is considered a moderate probability, based on the principle of prudence, as this step has been incorporated specifically for the control of hazards produced in previous steps (avoidance of microbiological growth). Failure at this stage could result in the initial contaminant load increasing to unacceptable levels There are no subsequent steps specifically established to reduce or control the hazard

							The potential impact on health is considered serious because although the product must be cooked and most of the bacteria are sensitive to high temperatures and therefore eliminated, if the operation at this stage is not correct, the development of toxins of certain bacteria (*Staphylococcus aureus*) or even the growth of other less temperature-sensitive bacteria (*Listeria*) could occur
Cooling	Biological. Pathogenic and spoilage microbiological contamination	Condensation: microbiological contamination through condensate droplets that could accumulate on the ceiling and fall on the product	Activation of exhaust fans when necessary depending on the condensation that accumulates on the slaughterhouse roofs Preoperational and operational checks on condensate accumulation. In case of condensation, activation of extractors Periodic removal of condensate during downtime if the operation of the extractors is not sufficient. This condensate removal on the roof must be carried out without product in the slaughtering line Microbiological control of condensate droplets to rule out the persistence of pathogens Washing of the carcass at various points in the slaughterhouse during the process, thus reducing the impact of point contamination caused by a drop of condensate that could have fallen on the carcass	3	2	6	The probability has been considered moderate since at this stage the product will not be subjected to further washing and the exposure time to possible condensation is longer. However, the turbulence regime inside the tunnel prevents the formation of condensation on the ceiling and walls The gravity is considered low as the need to cook the product would eliminate any microbiological contamination that could be introduced by condensate droplets It is also considered low as we have not been able to demonstrate that condensate droplets add a significantly different contaminant load or significantly increase the contaminant load that occurs in the slaughterhouse operations

(*Continued*)

N°	Step	Hazard	Cause	Control measure	Risk analysis			
					Likelihood	Severity	Risk value	Justification
	Cooling	Chemical. Chemical contamination through the materials used	Contamination through grease or lubricants from the cooling chain by dripping onto the carcass during the cooling period	Maintenance plan Good equipment lubrication practices Approved lubrication products suitable for the food industry	2	1	2	The likelihood of introducing a hazard at this stage is considered low as there is a maintenance plan that specifies the conditions under which work equipment must be lubricated All lubricants and/or greases used are food grade and are registered and authorized for food use If there is contamination at this stage, the probability of detection at later stages is very high and the products affected would be treated as reprocesses or by-products The severity was considered low because of the small amount of the hazard introduced into the food if it was introduced
17	Sorting and boxing	Biological. Pathogenic and spoilage microbiological contamination	Contamination by operators: operators can contaminate the carcass through their gloves if they do not respect good handling and hygiene practices	Good Handling Practices Regular hand washing and whenever necessary after touching potentially contaminated areas	2	3	6	The introduction of such a hazard is considered to be low due to the fact that Good Handling Practices are maintained (and monitored) Subsequent steps (refrigerated storage) would control the potential hazard: e.g. by preventing the multiplication of *Staphylococcus aureus* and thus the formation of toxins It is considered of moderate severity as the potential hazards introduced at this stage would be eliminated during cooking of the product at high temperatures
		Biological. Pathogenic and spoilage microbiological contamination	Contamination from cutting: poorly disinfected knife	Good Handling Practices Rotation of knives and hygienic storage of knives not in use Cleaning and disinfection of knives Sterilization at $T > 82\,^{\circ}C$ or equivalent method	2	2	4	It is considered that with good handling practices (and control and monitoring of these), the potential contamination for this reason is low The severity is considered low as the product must be consumed cooked and possible contaminants introduced at this stage are sensitive to high temperatures

Sorting and boxing	Biological. Pathogenic and spoilage microbiological contamination	Cross-contamination through by-products that could be generated at this stage or through products identified as reprocesses	Good Handling Practices: by-products shall be segregated and stored hygienically. They should be removed as soon as possible from food handling areas Good hygienic practices: handlers shall avoid contact with accumulated by-products. If handling is necessary, hands should be washed and disinfected before returning to handling tasks	2	2	4	It is considered that with Good Handling Practices (and the control and monitoring of these), the potential contamination for this reason is considered to be low The severity is considered to be low as the product must be consumed cooked and possible contaminants introduced at this stage are sensitive to high temperatures By-products are handled hygienically and are removed only at times when there is no possibility of cross-contamination with food
	Biological. Pathogenic and spoilage microbiological contamination	Contamination through ancillary materials that could be in poor cleaning and disinfection conditions and add contaminant load to the products	Good Handling Practices: materials are hygienically used (no contact with soil, by-products, etc.) Review of auxiliary materials before use for direct food contact Good storage practices: materials not directly exposed to the environment, dust, dirt, etc.	1	3	3	The probability of occurrence of this hazard is considered to be very low as the auxiliary materials are stored properly protected from the outside environment, dust, dirt, etc. Before use, operators check that the auxiliary material is in optimum hygienic conditions and discard them if they do not meet these conditions It is considered a moderate seriousness because although the final product must be consumed cooked and a large part of the microbiological contamination will be eliminated with the high temperatures, bacteria such as *Listeria*, which is more resistant to high temperatures, could have been introduced and cause some damage to people's health
	Chemical. Contamination through packaging materials	If the auxiliary materials used for packaging are made from non-food-grade materials, this could lead to migration or even direct contamination of the products	Control of suppliers and auxiliary materials Auxiliary materials are made of material suitable for food use Certificates of suitability, migration, etc.	1	1	1	The contact time of the mark with the carcass is also very limited as it is a highly perishable food that must be consumed in a short period of time not sufficient for the passage of chemical contaminants in unacceptable quantities Control measures are in place to ensure that adequate auxiliary materials are used If the hazard were to occur, its severity would be very low as the contaminated area would be very localized and the contaminant load would be very low

(*Continued*)

N°	Step	Hazard	Cause	Control measure	Risk analysis			
					Likelihood	Severity	Risk value	Justification
	Sorting and boxing	Physical Residues: plastics from tags	Incorrectly fitted or defective tags can result in broken tags remaining in the carcass or even incorrectly fitted tags remaining in the carcass	Supplier control plan: materials suitable for the intended use: resistance Maintenance plan: correct functioning of the marking equipment Monitoring and quality control at subsequent stages	1	4	4	The probability is considered to be very low because if any markings were incorrectly affixed, they would be easily identifiable and removable at later stages If any markings were to leave the industry and contaminate the food it is easily identifiable visually by the consumer It was considered as serious because if it was finally consumed by a risk group such as small children it could cause serious problems in the digestive tract, choking, etc.
		Physical. Personal items inside carcasses	Operators could insert personal items (rings, earrings, watch bands, band-aids, etc.) into the carcass cavity	Good Handling Practices: no person carrying personal items such as jewelry, rings, earrings, chains, watches is allowed to enter the handling areas If plasters are to be used, they must be bright blue in color	1	3	3	It is considered a very low probability as there have been no reported incidences of this hazard in recent years Following good handling practices reduces the likelihood of contamination with this hazard to very low levels It is very likely that in the event of such contamination occurring at the final stage where the consumer proceeds to cook the product, the object causing the contamination will be detected The severity was considered to be moderate as ingestion of foreign objects could cause moderate tooth breakage, tears in the digestive tract, intestinal problems, etc.
		Chemical. Chemical contamination through the materials used	If the materials used in the tag are not suitable for food use, they could cause migration or even direct contamination of the carcass	Control of suppliers and auxiliary materials Labels or tags shall be made of material suitable for use with foodstuffs Certificates of suitability, migration, etc.	1	1	1	The probability is considered remote as the surface area of contact of the marking with the carcass is very restricted to a specific area and within that area a very small surface area The time of contact of the tag with the carcass is also very limited as it is a very perishable food that must be consumed in a short period of time which is not sufficient for the passage of chemical contaminants in unacceptable quantities

							Control measures are in place to ensure that adequate auxiliary materials are used If the hazard were to occur, its severity would be very low as the contaminated area and the contaminant load would be very low
Sorting and boxing	Chemical. Allergens: allergen contamination	Cross-contamination occurs through handlers and food: handlers contaminate food with food they carry or through work clothes	Training of operators Good Handling Practices: rules for staff eating habits as well as work clothes, breaks, etc. Training of operators	1	5	5	It is considered a remote probability because no cases have been recorded historically Monitoring and surveillance of Good Handling Practices as well as staff training and awareness reduces the possibility of contamination by this hazard It is considered as very serious because it is well documented that even traces of an allergen can cause very severe anaphylactic reactions in people sensitive to certain types of allergens
	Physical. Presence of sharp metal parts	During packaging operations, some cuts are made that introduce sharp elements through broken knives (especially the tips), cutter blades, wire cut-resistant gloves, or others	Training of operators: monitoring of operations during the preparation of trays Control of the state of conservation and rotation of knives. Identification of knives Monitoring at subsequent stages	1	5	5	The probability has been considered as low as no cases have been recorded over the last few years. In addition, a bulk product will go through subsequent checks where these pieces of metal are easily detectable The control measures implemented to date have proven to be effective in view of the absence of recorded incidents In addition, there are a series of preventive measures in place that reduce the probability of this hazard appearing in a product available to the end consumer The severity is considered as very serious because in certain risk groups, elderly or children, it could cause very serious harm

(Continued)

N°	Step	Hazard	Cause	Control measure	Risk analysis			
					Likelihood	Severity	Risk value	Justification
18	Weighing and labeling	Biological. Pathogenic and spoilage microbiological contamination	Long waiting times in the packaging room before storage in the cooling chambers at temperatures between 0 and +4 °C could lead to a growth of the initial contamination load of the product	Good Handling Practices Minimum waiting times before storing the product in cold storage: when a pallet is finished, it must be immediately transferred and stored in cold storage	2	4	8	The probability of occurrence of the hazard was considered low as the temperature of the premises where the final product is handled is controlled In addition, there are always personnel available to handle the pallets as they are completed It is considered to be serious because if there is an increase in the contamination load, toxins from certain bacteria (e.g. *Staphylococcus aureus*) could be produced Although the product must be consumed cooked and this would eliminate the possible hazard introduced by temperature-sensitive bacteria (*Salmonella*, *E. coli*, *Campylobacter*, etc.), toxins produced by other families would not be eliminated and could cause damage to the health of consumers
		Biological. Pathogenic and spoilage microbiological contamination	Cross-contamination due to floors that are not properly dried or with the presence of puddles that, after being walked on, cause dirt to be transferred from one area to another	Correct execution of the cleaning and disinfection plan. Adequate training of the cleaning personnel of the facilities Preoperational checks of the facilities. In the event of noncompliance, correction of noncompliance before the start of the working day	2	2	4	Although the likelihood of puddles is moderate, the probability of cross-contamination is considered low as the products (food) should not touch the ground directly at any time (always packed and protected from the ground) and even the boxes containing the product should not touch the ground directly. This reduces the likelihood of contamination and falls within the scope of possible cross-contamination through packaging materials The severity is also considered to be low as we have to take into account that the products are to be consumed cooked so that any possible microbiological contamination should be eliminated during proper cooking

	Weighing and labeling	Physical. Wood scraps	Wooden pallets in an inadequate state of preservation so that they are splintered	Adequate control over the pallets used in packaging. Pallets intact and in a good state of cleanliness and conservation and that do not shed splinters or debris Immediate elimination of any wood that may come off the pallets Protection, if possible, of the carcasses inside plastic bags during packaging	2	3	6	The probability was considered low as at this stage working with wooden pallets is not allowed except for customer requirements. In this case, the use of pallets is controlled. In addition, the carcasses arrive already packed (in boxes and protected with paper) at this stage The severity was considered to be moderate as the ingestion of a piece of wood could cause choking problems or cuts requiring hospitalization. Even so, a piece of wood of a size likely to cause harm should be of a size such that it is easily visible and identifiable
19	Storage between 0 and 4 °C	Biological. Pathogenic and spoilage microbiological contamination	If the storage temperature is not adequate, excessive and inappropriate growth of the initial contaminant load present in the product could occur. This step exists to control this hazard	Maintenance plan for installations and equipment: correct maintenance of installations. Alarm system to warn in case of equipment failure Immediate control of the cold store temperature: real-time temperature display screens Daily control and recording of storage temperatures in the cold stores Quality control and monitoring of processes in situ	2	4	8	The probability of occurrence of the hazard was considered low as there are appropriate control systems in place to keep the product properly refrigerated (including alarms and telephone alerts to responsible personnel) It is considered as serious as an increase in the contaminant load could lead to the production of toxins from certain bacteria (e.g. *Staphylococcus aureus*) Although the product must be consumed cooked and this would eliminate the possible hazard introduced by temperature-sensitive bacteria (*Salmonella*, *E. coli*, *Campylobacter*, etc.), toxins produced by other families would not be eliminated and could cause damage to the health of consumers

(*Continued*)

N°	Step	Hazard	Cause	Control measure	Risk analysis			
					Likelihood	Severity	Risk value	Justification
20	Shipping	Biological. Pathogenic and spoilage microbiological contamination	Long waiting and/or loading times at the shipping dock before loading onto trucks at temperatures between 0 and +4 °C could lead to a growth of the initial contamination load of the product	Good Handling Practices: minimum loading times Controlled temperature conditions at the shipping dock Check temperature of vehicles prior to loading Review of the hygienic conditions of the vehicles before loading	2	4	8	The probability of occurrence of the hazard was considered low since the temperature of the premises where the final product is handled (including the shipping dock) is controlled In addition, there are always personnel present during the loading of the lorries, so that if any circumstance were to occur that prevented immediate loading, the product could be stored again in the refrigeration cold store This is considered a serious problem, as an increase in the contamination load could lead to the production of toxins from certain bacteria (e.g. *Staphylococcus aureus*) Although the product must be consumed cooked and this would eliminate the possible hazard introduced by temperature-sensitive bacteria (*Salmonella*, *E. coli*, *Campylobacter*, etc.), toxins produced by other families would not be eliminated and could potentially cause damage to the health of consumers
		Physical. Wood scraps	Wooden pallets in an inadequate state of preservation so that they are splintered	Adequate control over the pallets used in packaging. Pallets intact and in a good state of cleanliness and conservation and that do not release splinters or debris Immediate elimination of any wood that may come off the pallets Protection, if possible, of the carcasses inside plastic bags during packaging	2	3	6	The probability was considered low as at this stage working with wooden pallets is not allowed except for customer requirements. In this case, the use of pallets is controlled. In addition, the carcasses arrive already packed (in boxes and protected with paper) at this stage The severity was considered to be moderate as the ingestion of a piece of wood could cause choking problems or cuts requiring hospitalization. Even so, a piece of wood of a size likely to cause harm should be of a size such that it is easily visible and identifiable

21	Transport between 0 and 4 °C	Biological. Pathogenic and spoilage microbiological contamination	Inappropriate transport temperature may lead to excessive and inappropriate growth of the initial contaminant load in the product	Transport procedure Temperature recording during transport. Incident control Equipment maintenance plan: periodical revisions of the transport vehicles' refrigeration equipment	3	3	9	The probability was considered low as the transport vehicles have the necessary mechanisms for the maintenance, control, and recording of the temperature during the time the products remain inside There are procedures and control measures in place to reduce the probability of occurrence of this type of hazard It is considered as serious as an increase in the contaminant load could lead to the production of toxins from certain bacteria (e.g. *Staphylococcus aureus*) Although the product must be consumed cooked and this would eliminate the possible hazard introduced by temperature-sensitive bacteria (*Salmonella*, *E. coli*, *Campylobacter*, etc.), toxins produced by other families would not be eliminated and could cause damage to the health of consumers
		Biological, Pathogenic and spoilage microbiological contamination	It could happen that the product is transported with other meat or food products between which cross-contamination could occur	Transport procedure Review of loading conditions, conditions inside the lorry, other transported materials, etc. Recording of possible incidents	1	3	3	The probability of occurrence of this hazard is low because trucks are checked before loading. Loading of foodstuffs with other nonpermitted products (chemicals, hazardous materials, biological agents, etc.) is not allowed The severity would be moderate in any case, as microbiological cross-contamination would be reduced or even eliminated in case of occurrence, taking into account that the product has to be consumed cooked at high temperatures
		Chemical. Cross-contamination with unsuitable substances	It could happen that the product is transported with other chemicals that could transfer traces or residues to the food	Transport procedure Review of loading conditions, conditions inside the lorry, other transported materials, etc. Recording of possible incidents	1	5	5	The probability of occurrence of this hazard is low because trucks are checked before loading. Food is not allowed to be loaded with other nonpermitted products (chemicals, hazardous materials, biological agents, etc.) The severity could be very high if the cargo is not screened and chemical (or even biological) agents could completely contaminate the food by spillage and cause serious damage to health

CCP No. 1 – Evisceration and postmortem inspection: fecal matter, tumors, parasites, other agents.

Step	Hazard identified	Critical limit	Monitoring		Correction or corrective action	Responsibility and authority	Records
			How	Who			
Washing	Biological contamination by parasites or other zoonotic agents (*Pasteurella*, etc.) as well as evidence of disease or other affections of the meat	Absence of carcasses with signs or evidence of parasites or other pathogens	Examination and inspection of all carcasses at the time of evisceration. Symptoms or evidence of disease or pathogenic agents should be looked for: Coccidia/cysticerci in liver or other viscera Tumors or pus-filled areas Other evidence of disease: strange colorations, atypical odors, liquids inside the tissues, inappropriate conformations, etc.	Personnel authorized and accredited as postmortem inspection assistants are trained to do this inspection and rejection work. This work is also done and verified by the official veterinary service The quality manager (or person delegated by him/her with sufficient knowledge)	If during the operations carcasses with possible contaminants are found, a decision must be taken, which can be: total rejection or partial rejection (the official veterinary service has the final decision and the highest authority) If total rejection is necessary, the carcass must be disposed of in the category 2 by-products container available in the evisceration room If it is sufficient to partially reject the contaminated part of the carcass, it shall be rejected to the by-product container and the noncontaminated part of the carcass shall continue to hang on the line	The highest authority is that of the official veterinary service to take the appropriate decisions for meat (total rejection, partial rejection) The assistant inspection or production staff may take decisions (always more restrictive than the official veterinarian, if necessary for commercial reasons) on total or partial rejection of carcasses	F.045 Slaughter record form Record of non-conformities, corrections, and corrective actions.

Washing	Biological contamination: remains of feces or dirt on the carcasses due to breakage of the intestinal contents during evisceration or previous stages. This incidence can cause contamination of carcasses with fecal bacteria (*Salmonella*, *E. coli*, *Campylobacter*, etc.) to unacceptable levels	Operational limit: carcasses with feces > 1.5 % (3 units max. at standard working speed for about 5' of overhaul) (max. 3 units at standard operating speed for about 5' of revision) Critical limit: carcasses with feces > 3 % (maximum 7 units at standard working speed for about 5' of overhaul) (maximum 7 units at standard operating speed for about 5' of overhaul) Critical limit: carcasses with feces > 3 % (maximum 7 units at standard operating speed for about 5' of overhaul)	On at least 1 occasion per working shift(*), a statistical check is made for deviations. At a point on the line where full access to the carcasses is achieved, the carcasses shall be checked for the presence of feces (especially in the rectal area and around the hand area). The number of carcasses to be sampled must be at least 250 carcasses (which at the general working speeds of the line corresponds to about 5' with the chain practically complete) (*) A work shift is defined as the period of time between two moments of absence or stoppage in production, e.g. from the start of the working day to the first toilet stop, from the first toilet stop to the second toilet stop, and so on. The work shift maintains homogeneous production conditions throughout the shift in terms of the distribution of personnel by tasks, line speed, hygiene conditions, etc.	The quality manager (or person delegated by him/her with sufficient knowledge)	If during monitoring it is found that the operational limit is exceeded, the production manager shall be informed immediately so that corrective action can be taken: Reduction of the line speed Incorporation of more collaborators in the evisceration stations Monitoring of the preparation stages Monitoring and maintenance of work equipment If during monitoring it is detected that the critical limit is exceeded, the workers at the packaging stage shall be notified immediately. At this stage, all carcasses since the last control with correct results shall be identified as noncompliant product. This noncompliant product shall be dealt with in accordance with that laid down in the instruction for potentially unsafe products	The highest authority is the official veterinary service to take the appropriate decisions for meat (total rejection, partial rejection) The production staff may take decisions (always more restrictive than the official veterinarian, if necessary for commercial reasons) for total or partial rejection of carcasses The quality manager will have the final decision (concerning the workers of the company) for the total and partial rejection as well as for the measures to follow for the reduction of possible contaminations and the actions to solve incorrect situations

(Continued)

CCP No. 1 – Evisceration and postmortem inspection: fecal matter, tumors, parasites, other agents.							
			Monitoring				
Step	Hazard identified	Critical limit	How	Who	Correction or corrective action	Responsibility and authority	Records
Box packaging		Absence of carcasses with fecal material	All rabbits are checked at the moment they are removed from the line. If any traces of feces or contamination are found, the rabbits are identified and sorted and packed in separate boxes as reprocessing	The workers who are separating the rabbits from the line	If during packaging operations, rabbits contaminated with traces of feces are detected, they shall be removed from the line and separated and sorted into separate boxes for the packaging of reprocessed meat. These rabbits shall then be subjected to the operations necessary for the elimination of the contamination: Purging Total rejection in downstream operations (cutting operations)	The quality manager will have the final decision (concerning the employees of the company) for the total or partial rejection as well as for the measures to be taken for the reduction of possible contamination and the actions to be taken to solve incorrect situations	

Critical Limit:
Based on the experience accumulated over the years and in the absence of the development of specific or sectoral legislation or standards for the control of this operation, it was decided to establish the following limits:

Operational limit: Percentage of carcasses above 1.5. Below this value, all carcasses will be detected at any subsequent stage and, by following Good Handling Practices, will be eliminated or the defect corrected by eliminating the affected area. Between 1.5 and 3 systematic corrections at subsequent stages should be 100% effective, but this is not certain so it is advisable to warn the operators of the different posts in the slaughterhouse of the situation.

Critical limit: Percentage of carcasses with feces above 3 %. Above this value, it is not possible to remove all carcasses in subsequent stages.

The most critical area for the possible presence of feces or dirt is the area of the rectum where most of the intestinal traces of intestinal contents and consequent leakage of feces occur. Feces may be whole or in diarrheic form and range in consistency from solid to semisolid to pasty. Size is not relevant in identifying feces. However, as the size decreases, the color and texture become more difficult to discern. The dark brown color is easily identifiable.

Monitoring:
Monitoring during evisceration: during the evisceration process, operators must pay attention to the appearance of symptoms or evidence of disease or pathogenic agents in the carcasses. If detected, the carcasses must be totally or partially rejected.

In case the carcass is only partially rejected, good handling and hygienic practices must be followed in order not to contaminate other carcasses through knives and hands.
Statistical monitoring: personnel from the quality department or personnel designated by them shall be responsible for monitoring. The procedure shall be as follows: check whether a sample is greater than statistically significant (equal to the square root of the population; about 5' of time at normal line speed represents about 250 carcasses) every two hours from the start of the working day. In a stall where the carcasses are fully accessible, the rectum area in particular and the rest of the carcass in general shall be observed for the presence of feces. The number of carcasses with feces detected and the number of carcasses sampled shall be recorded. Monitoring shall be carried out before the carcasses enter the chilling tunnel.

Monitoring during packaging: if carcasses with traces of feces are detected, these carcasses shall be separated for specific crates and treated as potentially unsafe products.

Corrections and corrective actions:

Corrections during evisceration:

- Total rejection of the carcass;
- Partial rejection of carcass (contaminated parts only).
- In case of doubt, clarify with the official veterinary inspector or apply the strictest criteria.
- Observe good handling practices:
  - (i) Cleaning and disinfection of knives used in partial reject operations.
  - (ii) Frequent washing of hands after total or partial rejections if necessary.
  - (iii) Hygienic disposal of by-products. The rejected carcasses or parts all go into the by-products container used only for that purpose.

Corrections during statistical monitoring:
If the operational limit is exceeded: notify the slaughterhouse attendant to proceed with:

(i) Reduction of the speed of the line.
(ii) Incorporation of more collaborators in the evisceration stages.
(iii) Monitoring during the preparation stages.
(iv) Monitoring and maintenance of the working equipment in the preparation stages.

If the critical limit is exceeded: the quality manager must:

(i) Advise the line operators in the packaging area to identify and separate all boxes that have been prepared since the last control that have not exceeded the critical limit. You can use the traceability records to identify these products. If the product has just left the line, instruct the operators when and which channels should be separated into cartons.
(ii) All products contained in these boxes shall be identified and separated from unaffected and finished products. It shall not be allowed to carry out cutting or dispatching without first being specifically controlled.
(iii) During the correction of contaminated carcasses, the recommendations laid down in the good handling practices (see above) are respected.
(iv) Carcasses in which traces of feces or dirt are found shall be separated from the affected area using the rest of the carcass for other presentations. These operations shall be carried out under the monitoring of good production practices and under the supervision of the quality manager.
(v) If the fecal contamination is in a very extensive area or if, as its elimination by purging, it is not evident that the entire source of contamination will be eliminated, proceed to the total rejection of the carcass.

Record:
F.045 Slaughter report template.

CCP No. 2 – Carcass cooling temperature control

Step	Hazard identified	Critical limit	Monitoring: How	Monitoring: Who	Correction or corrective action	Responsibility and authority	Records
Tunnel cooling	Growth and survival of those bacteria that may have been introduced in previous stages and are present. This stage is specifically designed to prevent the growth and/or survival of microbiota	Critical limit: carcasses with a temperature in the biceps femoris muscle at the exit of the cooling tunnel < + 4 °C	Every 2 h on average and in such a way that the control coincides with the different working periods, each rabbit carcass (preferably larger than 1,200 kg) will be punched at the exit of the cooling tunnel If the temperature is within the parameters (<4°C), the CCP is controlled. If the temperature is not correct, ten more carcasses (preferably weighing more than 1,200kg) will be tested. If the temperature in any of them is higher than 4°C, the CCP is considered to be out of control and corrections or corrective actions will be established	Quality personnel or Personnel in charge of the weighing and labeling stage with sufficient and necessary training	As soon as anomalies are detected in the results of the measurements, the following must be carried out All carcasses leaving the cooling tunnel since the last control with a correct result and the conditions of the correct result shall be identified and marked as potentially unsafe product. This potentially unsafe product must be immediately placed in the cold stores The quality manager shall authorize the release of this product as long as the temperature decreases, if there is no evidence of a significant effect on the product Under no circumstances may these carcasses or batches be dispatched or cut up until the internal temperature of the muscles is below +4°C and released The person in charge of production or the quality manager must request the repair of the equipment if necessary	The highest authority is the official veterinary service to take the appropriate decisions for meat (total rejection, partial rejection) The production staff may take decisions (always more restrictive than the official veterinarian, if necessary for commercial reasons) for total or partial rejection of carcasses The quality manager has the final decision (concerning the workers of the company) for the total or partial rejection as well as for the measures to be followed for the reduction of possible contamination and the actions to be taken to solve incorrect situations	F.045 Slaughter report form Record of non-conformities, corrections, and corrective actions

Critical limit:

Temperature in the muscles of the carcass (biceps femoris (leg)) below 4°C

Larger rabbits should preferably be used for temperature measurement

Monitoring:

Monitoring after leaving the cooling tunnel: the personnel in charge of the weighing and labeling stage or the quality manager is responsible for monitoring The procedure is as follows:

1) Use a calibrated thermometer.
2) Pierce the muscle in the carcass (it may be one leg or the other).
3) Select the largest carcasses (preferably weighing more than 1,200 kg).
4) If the temperature is below +4 °C, record the value in the register. The CCP is under control.
5) If the temperature is above +4 °C, repeat the procedure on another ten (10) carcasses. If any of them shows a temperature higher than 4 °C, the critical limit is considered to have been exceeded.

Corrections or corrective measures:

If the critical limit is exceeded during monitoring, the person responsible for monitoring shall, at the time of detection, 1:

1) Immediately (without unnecessary time or delay) inform the person in charge of quality if he/she was not the one who detected it.
2) Inform the person responsible for maintenance so that the system can be checked and, if necessary, the anomalies corrected.
3) The quality manager shall request that all boxes produced since the last inspection that have not exceeded the critical limit be identified and segregated in the packaging. Traceability records may be used to identify these products.
4) All products contained in these boxes shall be identified and segregated from unaffected or finished products. These products may not be used for cutting or dispatch unless they have been immediately placed in cold storage and have reached a temperature below 4 °C. In the first instance, depending on the tests and other measures in the system, the quality manager may decide to carry out further tests before deciding to release the product. In this case, the product will continue to be identified and segregated in the cold store until a decision is made to release it.

Record:

F.045 Slaughter report form

OPRP 1 Temperature control of working premises						
		Monitoring				
Hazard identified	Control measure	How	Who	Correction or corrective action	Responsibility and authority	Records
The temperature of any premises, where orders are sorted, packed, cut, prepared, and dispatched is not suitable to ensure acceptable microbiological conditions (high temperatures favor the growth of bacteria such as *Staphylococcus*, enteric pathogens, and other spoilage bacteria). *Staphylococcus* can generate spores resistant to high temperatures	Continuous measurement and recording of the temperature of the premises where the operations described above are carried out and at the times when these operations are carried out. The temperature should be as low as possible and always less than or equal to 10 °C at these times	Temperature measurements shall be taken at different times of the day during the operations described above. Calibrated thermometers shall be used and placed in areas close to where the product is being handled or stored	The quality manager or the production manager at the time of measurement	Correction: If room temperatures exceed the set temperature: Suspend production and store products in a room with suitable temperature Proceed with repair of equipment; check the last correct temperature record Identify affected products. Study the rest of the production circumstances in order to assess the possible product affected Evaluate reprocessing/disposal/release of the affected product according to the above circumstances	The quality manager shall be responsible for taking measurements directly or reviewing measurements taken by other authorized workers The quality manager and the food safety officer shall have the authority to execute corrections and corrective actions if the ORP is found to be out of control	F.045 Slaughter Report Form Temperature charts of the premises Record of non-conformities, corrections, and corrective actions
		Temperature sensors are available, which continuously record the temperature values of the rooms. These data are recorded electronically. The data are downloaded and analyzed on a daily basis	The quality manager or the production manager	Corrective action: If the temperatures of the premises exceed the established temperature: the cause of the nonconformity shall be studied; establish if the appropriate conditions are in place so that the same situation does not occur again All these steps shall be carried out in accordance with procedure P.13 NNCC	The quality manager shall be responsible for the daily review of the temperature charts The quality manager and the food safety manager shall have the authority to execute corrections and corrective actions if the ORP is out of control	

(Continued)

OPRP 2 Cold store temperature control

Hazard identified	Control measure	Monitoring		Correction or corrective action	Responsibility and authority	Records
		How	Who			
The temperature of the storage rooms is unsuitable (below 0°C or above 4°C). These unsuitable temperatures of the storage rooms may facilitate the growth of thermophilic spoilage bacteria and the survival and multiplication of pathogenic bacteria. High temperatures favor the growth of bacteria such as *Staphylococcus*, enteric pathogens, and other spoilage bacteria	Daily punctual measurement of the temperature inside the refrigeration chamber. The temperature must be between 0°C and 4°C, and the upper limit may be exceeded up to a maximum temperature of 9 °C for a period not exceeding 1 h (defrosting, etc.)	Temperature measurements shall be taken at different times, spaced a maximum of 8 h apart during the operations described above. Calibrated thermometers shall be used and placed in areas close to where the product is being handled or stored	The quality manager or production manager at the time of measurement	Below 0 °C: noncompliant Between 0 °C and +4 °C: compliant Between +4 °C and +9 °C for a maximum of 3 h: compliant Higher than +9 °C for more than one hour: noncompliant Higher than +10 °C: noncompliant (peaks where there is a clear rise and then an immediate fall with the temperature returning below +4 °C are considered compliant) Correction: if temperatures are noncompliant: Identify the commodity stored in that store during the period in which the control measure was exceeded. These goods will be treated as potentially unsafe products Check the existing carcasses in the store with a standard thermometer and decide whether it will be necessary to change them or to apply other measures (isolation in the store for microbiological control analysis)	The quality manager shall be responsible for taking measurements directly or reviewing measurements taken by other authorized workers. The quality manager and the food safety officer shall have the authority to execute corrections and corrective actions if the ORP is found to be out of control	F.045 Slaughter Report Form Cold store temperature graphs Record of non-conformities, corrections, and corrective actions

	Continuous measurement and recording of the temperature of the storage chamber. The temperature must be between 0 and 4 °C, and the upper limit may be exceeded up to a maximum temperature of 9 °C for a period not exceeding 1 hour.	Temperature sensors are available which continuously record the temperature values of the refrigeration room. This data is recorded electronically. the data is downloaded and analyzed daily.	The quality manager or production manager	Find information on the last correct value recorded Use procedure P.03 in case the affected product is identified outside the premises Study the remaining production circumstances in order to evaluate the possible contamination of the product Evaluate the reprocessing/disposal/release of the affected product according to the circumstances analyzed and others that may be required (microbiological and organoleptic analysis). If it is concluded that the product is not fit for human consumption, it is disposed of as a by-product Proceed with the repair of the equipment Corrective action: if the temperature of the premises exceeds the established temperature: the cause of the nonconformity shall be studied; the appropriate conditions shall be established so that the same situation does not occur again All these steps shall be carried out in accordance with procedure P.13 NNCC	The quality manager shall be responsible for the daily review of the temperature charts The quality manager and the food safety manager shall have the authority to execute corrections and corrective actions if the PPRO is found to be out of control

(*Continued*)

OPRP 3 Control of transport conditions

Hazard identified	Control measure	Monitoring: How	Monitoring: Who	Correction or corrective action	Responsibility and authority	Records
Inadequate food transport temperature conditions may result in an inappropriate and uncontrolled increase of the initial microbiological load of the product Transport conditions are inadequate so that the product is transported with other products/substances that may chemically contaminate the food	Before loading: (i) Verification of the cleanliness conditions of the vehicle (ii) Verification of temperature conditions of the vehicle before loading (temperature inside the box between 0 and 4 °C for fresh and <−18 ±6 °C for frozen). Preferably by delivery of temperature ticket (iii) Verification of compliance with hygiene and clothing practices of each driver (iv) Verification of the conditioning conditions of the goods (v) Verification of the temperature conditions of the product before loading (temperature between 0° and +4 °C for fresh and > <−18 ±6 °C for frozen) (vi) Verify also what type of goods are loaded and if there are any incompatibilities	Verification of all the points set out in the control measures section for each load	The driver himself if he has the appropriate training The personnel in charge of loading the transport vehicles The person or persons responsible for quality shall verify compliance with the above points as well as their registration	If the vehicle is not in a clean condition, it must be cleaned before loading 2. If the temperature conditions are not correct, before loading, the vehicle doors are closed and the refrigeration is switched on until the temperature inside the refrigerated box reaches the specified temperature 3. Access to areas where food is handled or vehicles are loaded shall not be allowed to any personnel who do not have the appropriate clothing for that task 4. If the vehicle is transporting other types of goods, check that there are no incompatibilities with the product to be loaded (e.g. packaged product with unpackaged product, non-food products, liquid products, etc.). In case of incompatibilities, the vehicle will not be allowed to be loaded. It is checked that the goods are correctly conditioned and in such a way as to avoid falls, knocks, etc. If this is not the case, the goods will be correctly stowed	The personnel responsible for loading vehicles shall verify compliance with the control measures and make corrections to proposed corrective actions and document the entire process. The quality manager shall verify compliance with this operational prerequisite on a daily basis	F.052 Loading and unloading conditions of transport vehicles Graphical recording of vehicle temperatures. Recording of nonconformities, corrections, and corrective actions
	On return of the vehicle: (i) Proof of continuous temperature record of the transport vehicle at the time of return	At the moment each vehicle arrives at the firm on the return of the delivery, it must unload the proof of transport temperatures	Drivers and quality manager	If the temperature record is not correct, the driver shall be informed of the situation. It shall be opened as a nonconformity and shall be treated as a nonconformity		

Case Study 2

HACCP System of the Mass Catering Establishment Culinaria

Note: due to the extension of the hazard analysis and critical control point (HACCP) system as a whole, in this case study, only some of the most significant parts are included, limited to the production of meals using the cook and chill system.

1 HACCP Team

- Director
- Management manager
- Purchasing and planning manager
- Production manager
- Nutrition manager
- Coordination manager
- Picking manager
- Warehouse manager
- R+D+I manager
- Maintenance manager

Note: For each HACCP team member, a deputy is designated in case of absence.

2 Scope and References

Culinaria is a Galician specialized food service company, whose activity consists of the preparation and distribution of cooked dishes and food products, naturally prepared with top-quality raw materials, preferably Galician, without additives or preservatives. It is a company specializing in the production of dishes using the cook and chill system that provides service to communities and individuals, as well as to the restaurant, catering, and hotel sector, large supermarkets, and specialized shops through its line of ready-to-eat meals, providing products of the highest quality and safety for the consumer.

The following recommendations have been taken into account when developing the HACCP system:

– **ISO 9001:2015**: Quality Management Systems. Requirements.

HACCP System Auditing for Food Safety: Principles and Techniques, First Edition. Luis Couto Lorenzo.

- **ISO 22000:2018**: Food Safety Management Systems. Requirements for all organizations in the food chain.
- Recommended international practice – general principles of food hygiene. CAC/RCP 1-1969, Rev. 4 (2003).
- Applicable legislation and regulations identified as external documentation of the food safety and quality management system.

3 Product Description or Types

Culinaria produces the following types of products ready for immediate consumption or after heating:

- Ready meals adapted to different groups with very specific needs, both nutritional and physiopathological: very young children (in their initial phase of complementary feeding), elderly people, and people with psychomotor difficulties. These are all types of ready-to-eat dishes made from different raw materials to create nutritionally balanced menus adapted to all types of diets and textures.
- Ready-to-eat refrigerated dishes for the general public: salads, broths and creams, vegetables, rice, pasta, meat and fish stews, sauces, and side dishes.
- Poultry, pork, and beef cooked at low temperatures, refrigerated and ready to eat.
- Refrigerated ready-to-eat desserts.

4 Identification of Intended Use

The intended use for the different types of ready meals can be grouped into the following categories:

- Dishes intended for mass catering and consumer groups with specific needs, ready for consumption in mass caterers or at home.
- Dishes and desserts intended for the general public, supplied to the restaurant, catering, and hotel sector, specialized shops, large-scale distribution, and via online sales.
- Meat cooked at low temperatures, supplied to the restaurant, catering, and hotel sector, specialized shops, large-scale distribution, and via online sales.

Most of the dishes must be regenerated by sufficient heating before consumption, according to the instructions on the label of each product.

5 Flow Diagram

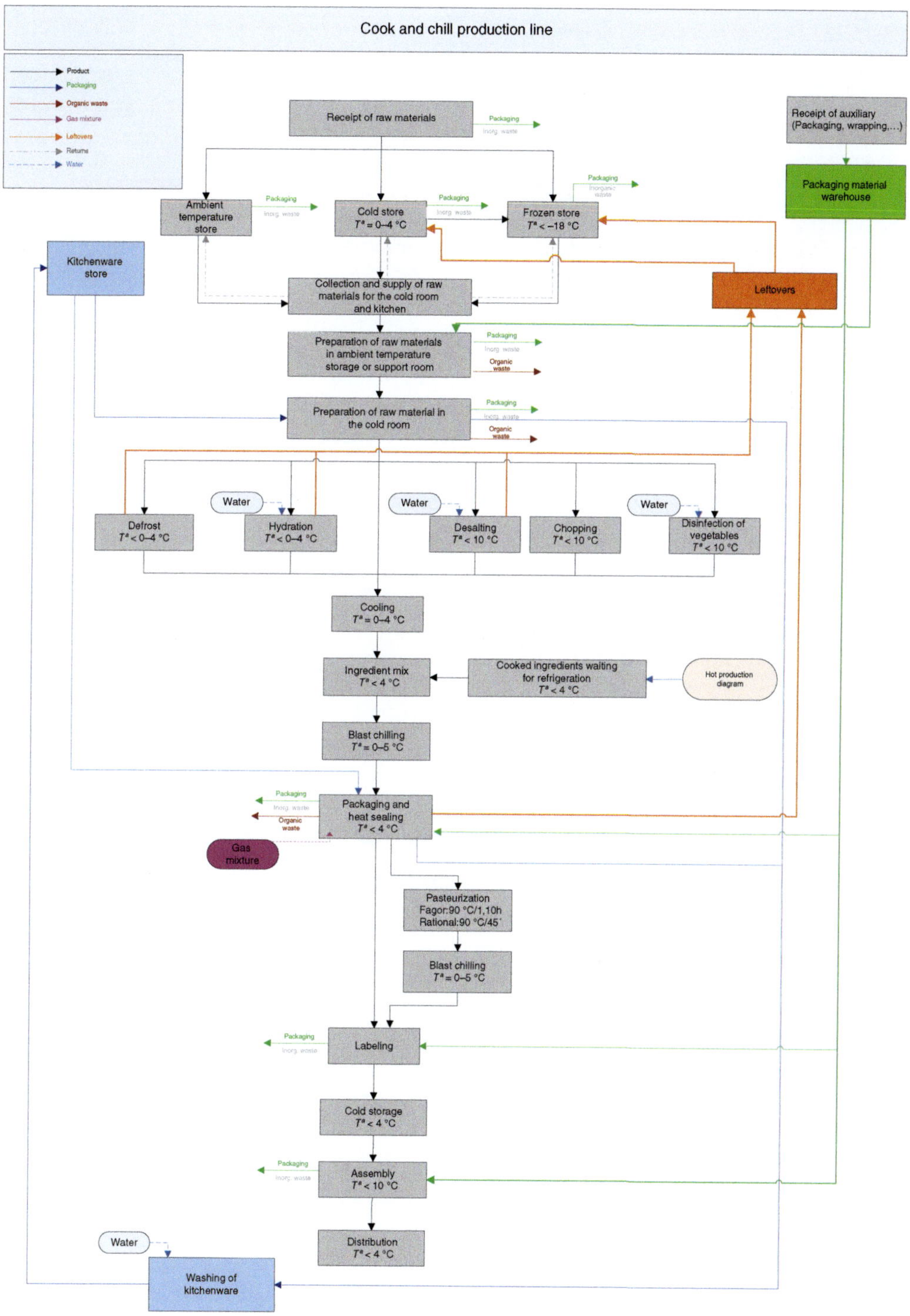

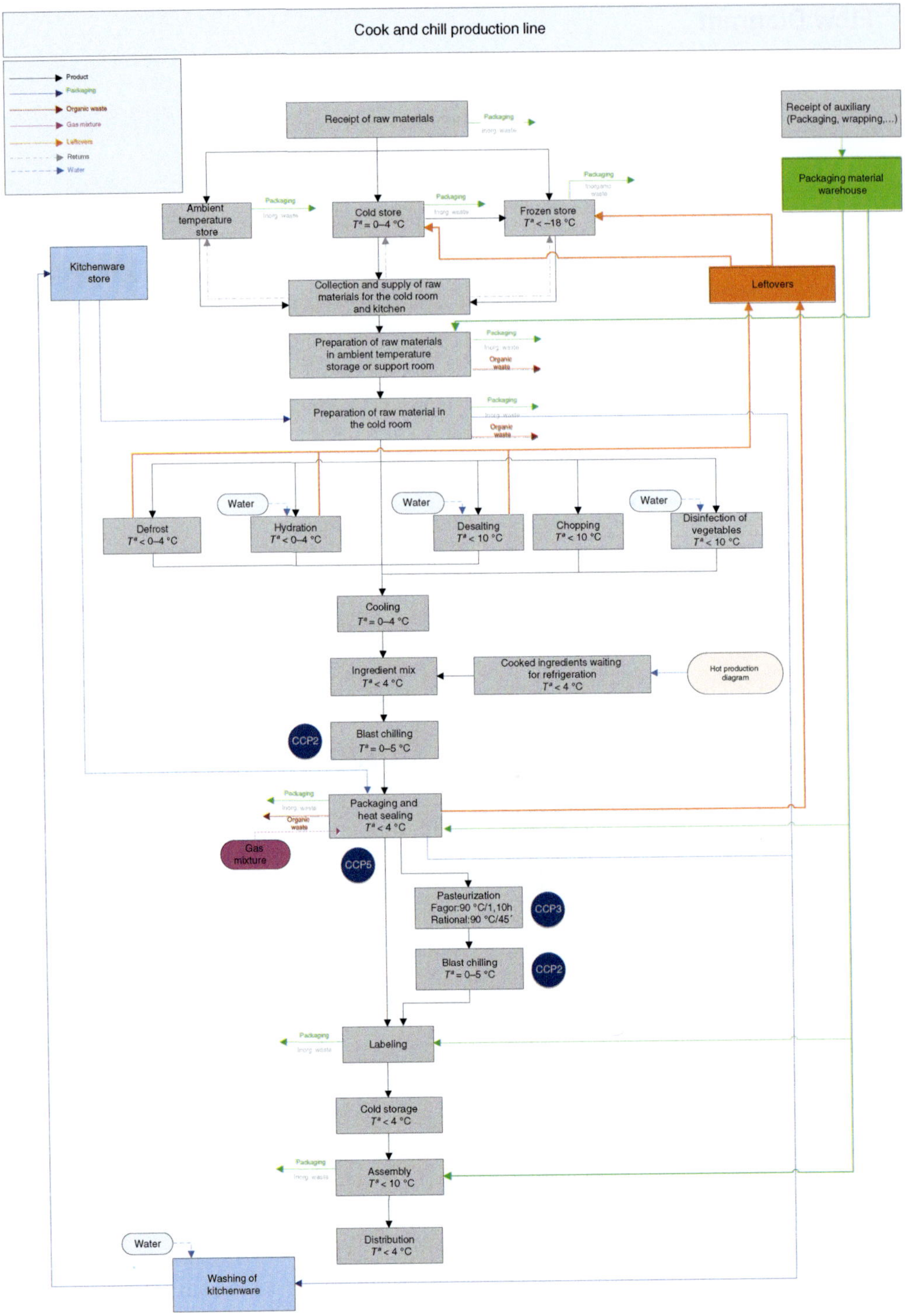

Cook and chill production line
Product
Packaging
Organic waste
Gas mixture
Leftovers
Returns
Water
Receipt of raw materials
Receipt of auxiliary (Packaging, wrapping,...)
Packaging material warehouse
Ambient temperature store
Cold store Tª = 0–4 °C
Frozen store Tª < −18 °C
Kitchenware store
Collection and supply of raw materials for the cold room and kitchen
Leftovers
Preparation of raw materials in ambient temperature storage or support room
Preparation of raw material in the cold room
Packaging
Inorg. waste
Organic waste
Defrost Tª < 0–4 °C
Water
Hydration Tª < 0–4 °C
Desalting Tª < 10 °C
Chopping Tª < 10 °C
Disinfection of vegetables Tª < 10 °C
Cooling Tª = 0–4 °C
Ingredient mix Tª < 4 °C
Cooked ingredients waiting for refrigeration Tª < 4 °C
Hot production diagram
CCP2
Blast chilling Tª = 0–5 °C
Packaging and heat sealing Tª < 4 °C
Gas mixture
CCP5
Pasteurization Fagor:90 °C/1,10h Rational:90 °C/45´
CCP3
Blast chilling Tª = 0–5 °C
CCP2
Labeling
Cold storage Tª < 4 °C
Assembly Tª < 10 °C
Distribution Tª < 4 °C
Washing of kitchenware

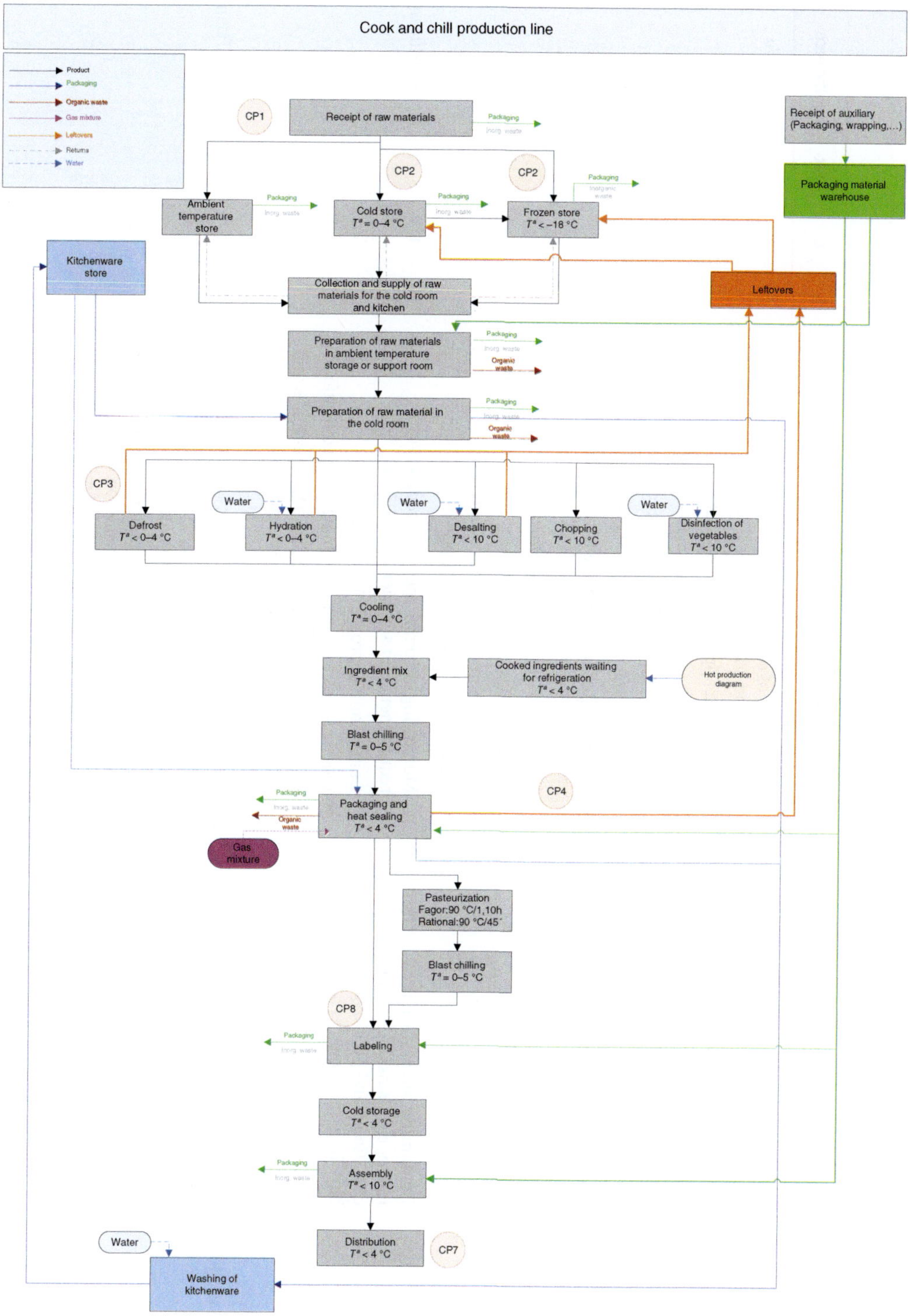

Cook and chill production line
Product
Packaging
Organic waste
Gas mixture
Leftovers
Returns
Water
CP1
Receipt of raw materials
Packaging
Inorg. waste
Receipt of auxiliary (Packaging, wrapping,...)
Packaging material warehouse
CP2
CP2
Ambient temperature store
Cold store $T^a = 0–4\ °C$
Frozen store $T^a < –18\ °C$
Kitchenware store
Collection and supply of raw materials for the cold room and kitchen
Leftovers
Preparation of raw materials in ambient temperature storage or support room
Organic waste
Preparation of raw material in the cold room
CP3
Water
Water
Water
Defrost $T^a < 0–4\ °C$
Hydration $T^a < 0–4\ °C$
Desalting $T^a < 10\ °C$
Chopping $T^a < 10\ °C$
Disinfection of vegetables $T^a < 10\ °C$
Cooling $T^a = 0–4\ °C$
Ingredient mix $T^a < 4\ °C$
Cooked ingredients waiting for refrigeration $T^a < 4\ °C$
Hot production diagram
Blast chilling $T^a = 0–5\ °C$
CP4
Packaging and heat sealing $T^a < 4\ °C$
Gas mixture
Pasteurization Fagor:90 °C/1,10h Rational:90 °C/45´
Blast chilling $T^a = 0–5\ °C$
CP8
Labeling
Cold storage $T^a < 4\ °C$
Assembly $T^a < 10\ °C$
Distribution $T^a < 4\ °C$
CP7
Water
Washing of kitchenware

Hazard analysis and assessment							
Step	**Hazard**	**Type**	**Control measure**	**S**	**L**	**Risk**	**Significance**
Reception of raw materials	Microbial growth in received chilled or frozen products caused by breaking the cold chain during storage or transport	B	Monitoring of compliance with food safety requirements for suppliers (I-FP05-01) Control and approval of suppliers Purchases according to raw material specification Carry out temperature control at reception Inspection at reception (I-FP05-02)	3	3	9	Significant
Reception of raw materials	Histamine contamination in chilled or frozen products (e.g. tuna) received, due to breakage of the cold chain during storage or transport	C	Monitoring of compliance with food safety requirements for suppliers (I-FP05-01) Control and approval of suppliers Purchases according to raw material specification Carry out temperature control at reception Inspection at reception (I-FP05-02)	3	3	9	Significant
Reception of raw materials	Contamination of fresh produce by microorganisms due to lack of cleanliness and poor hygiene, handling, or storage practices during transport	B	Monitoring of compliance with food safety requirements for suppliers (I-FP05-01) Control and approval of suppliers Purchases according to raw material specification Inspection at reception (I-FP05-02) of products and vehicle condition (cleanliness, transport conditions, etc.)	3	1	3	Not significant
Reception of raw materials	Presence of contaminants in quantities higher than the maximum quantities established in the legislation, e.g. antibiotics and other medicines, hormone residues, pesticides, heavy metals, dioxins, allergens, etc.	C	Monitoring of compliance with food safety requirements for suppliers (I-FP05-01) Control and approval of suppliers Purchases according to raw material specification. Request to suppliers for information on the analysis control plan and declaration of safety (analysis bulletins)	3	1	3	Not significant

Reception of raw materials	Presence of toxins in quantities higher than the established maximum, e.g. biotoxins in mollusks	B	Monitoring of compliance with food safety requirements for suppliers (I-FP05-01) Control and approval of suppliers Purchases according to raw material specification. Request to suppliers for information on the analysis control plan and declaration of safety (analysis bulletins)	3	1	3	Not significant
Reception of raw materials	Presence of parasites, e.g. *Anisakis*, trichinae, etc.	B	Monitoring of compliance with food safety requirements for suppliers (I-FP05-01) Control and approval of suppliers Purchases according to raw material specification Inspection at reception (I-FP05-02) **Fish**: frozen product is received (if at any time fresh fish is received, it shall be processed immediately to minimize the probability of developing the hazard) **Meat**: frozen product is received, except, in some cases, beef, chicken, pork, and turkey, and always officially inspected product from authorized establishments No preparation of raw fish or meat dishes	3	1	3	Not significant
Reception of raw materials	Presence of metallic foreign bodies (fragments of industrial origin, from handlers, etc.) or nonmetallic (stones, wood, glass, plastic, etc.)	P	Monitoring of compliance with food safety requirements for suppliers (I-FP05-01) Control and approval of suppliers Purchases according to raw material specification **Inspection at reception (I-FP05-02)**: integrity of packaging, whole pallets, free of pests, and in a correct state of cleanliness Limit the use of wooden pallets	1	3	3	Not significant
Reception of raw materials	Proliferation of microorganisms in refrigerated products due to breakage of the cold chain during reception	B	Unloading and cold storage in the shortest possible time Compliance with good practices for handling and stowage of goods on receipt	3	1	3	Not significant

(*Continued*)

Hazard analysis and assessment							
Step	**Hazard**	**Type**	**Control measure**	**S**	**L**	**Risk**	**Significance**
Reception of raw materials	Histamine contamination in chilled or frozen products (e.g. tuna) due to breakage of the cold chain during reception	C	Unloading and cold storage in the shortest possible time Compliance with good practices for handling and stowage of goods on receipt	3	1	3	Not significant
Reception of raw materials	Contamination by microorganisms due to lack of cleanliness and poor hygiene, handling, or storage practices at the receiving and storage stage	B	Compliance with good handling practices (not placing the goods directly on the floor, etc.) Full maintenance of containers and packaging Closing of gates as soon as reception is completed Correct and regular cleaning of the reception area (I-MAPPCC-05), forklifts, and pallet trucks Compliance with the established cleaning and pest control controls	3	1	3	Not significant
Reception of auxiliary material (containers, packaging, etc.)	Contamination by microorganisms due to lack of cleanliness and poor hygiene, handling, or storage practices during transport, unloading, and reception	B	Monitoring of compliance with food safety requirements for suppliers (I-FP05-01). Approval of suppliers Inspection at reception (I-FP05-02). Checking the condition of the vehicle and the material transported.	1	1	1	Not significant
Reception of auxiliary material (containers, packaging, etc.)	Presence of foreign bodies (stone, wood, glass, plastic, metals, etc.)	P	Monitoring of compliance with food safety requirements for suppliers (I-FP05-01). Approval of suppliers Inspection at reception (I-FP05-02) Check that packaging/boxes are intact and that no contamination has occurred	1	1	1	Not significant
Reception of auxiliary material (containers, packaging, etc.)	By migration of polymers in plastic material	C	Approval of suppliers Purchasing according to Material Specification. Application for a certificate of suitability for food use of the material (declaration of conformity and migration studies)	3	1	3	Not significant

Storage of raw materials at ambient temperature	By the presence of pests or foreign bodies (insects, pests, remains of packaging material, personal items, etc.)	P	Cleaning and disinfection plan, pest control good handling practices (GHP) and good manufacturing practices (GMP) (no use of jewelry, rings, watches, etc.), correct use of hats, checking the integrity of gloves, no chewing gum, no food or personal items in storage or production areas Maintain the guidelines for the use, change, and cleaning of work clothes Keep containers closed and insulated from the floor Compliance with the cleaning plan (I-MAPPCC-05) Limit the use of wooden pallets Carry out control of tools and breakable material (I-MAPPCC-10) Carry out checks of the warehouse (F-IMAPPCC-1202, F-IMAPPCC-1203)	1	3	3	Not significant
Storage of raw materials at ambient temperature	Proliferation of microorganisms due to expiry date being exceeded (raw materials)	B	Stock control according to I-FP12-01 (control of incoming and outgoing goods and control of early expiry) Control of the labeling of goods, checking that the expiry date or best-before date has been established. Stock rotation and implementation of the first in, first out (FIFO) system: placement of goods so that the first to enter (the oldest) is the first to leave	1	3	3	Not significant
Storage of raw materials at ambient temperature	Contamination of microorganisms due to incorrect handling of products or contact with dirty surfaces or tools (inadequate cleaning of stores)	B	Good hygiene and handling practices: wash hands and gloves, before and during product handling, correct condition and cleanliness of work clothes Keep containers closed to avoid contamination of the product. Correct stowage. Keep containers and products isolated from the floor Compliance with the cleaning and disinfection plan for warehouses (floors, shelves, walls, etc.) (I-MAPPCC-05).	3	1	3	Not significant

(*Continued*)

Hazard analysis and assessment							
Step	**Hazard**	**Type**	**Control measure**	**S**	**L**	**Risk**	**Significance**
Storage of raw materials at ambient temperature	Contamination of the product by cleaning product residues	C	Use of products suitable for food use Correct application of the cleaning and disinfection plan, eliminating any residues of used cleaning products Keep cleaning and disinfection chemicals correctly stored)	3	1	3	Not significant
Storage of raw materials at ambient temperature	Cross-contamination by allergens	A	Storage and specific identification of allergens in the warehouse Provide training and information for staff: Allergen typology (allergen matrix) Allergen management Correct stowage of products and keeping containers closed Periodic warehouse checks Compliance with "allergen management" instructions Apply correct cleaning and disinfection guidelines for tools, equipment, and surfaces	3	1	3	Not significant
Storage of auxiliary material (containers, packaging, labels, etc.)	Contamination by microorganisms due to incorrect handling or contact with soiled surfaces or tools	B	Good hygiene and handling practices: hand washing for the handling of materials Keep auxiliary materials correctly stored and packaged to avoid contamination of the product Keep containers and products isolated from the floor Compliance with the cleaning and disinfection plan for warehouses (floors, shelves, walls, etc.) Analysis control plan: surfaces (packaging in packaging area) (F-MAPPCC-12-01)	1	1	1	Not significant

Storage of auxiliary material (containers, packaging, labels, etc.)	Due to the presence of pests or foreign bodies (pests, remains of packaging material, personal objects, etc.)	P	Cleaning and disinfection and pest control plan GP of hygiene and handling (no jewelry, rings, watches, etc.), correct use of a hat, control of the integrity of gloves, no chewing gum, no introduction of food or personal objects in storage areas Keep containers closed and insulated from the ground	1	3	3	Not significant
Refrigerated/frozen raw material storage	Proliferation of microorganisms due to breakage of the cold chain or incorrect temperature	B	Storage of raw materials in the shortest possible time after receipt Keep freezer and cold store doors closed (open as little time as possible) Check that the product is correctly packaged (perfectly closed containers) Audible alarm in case of power failure (verify correct operation) Do not overload the cold store Maintenance plan for the cold store and calibration plan for the probes	3	3	9	Significant
Refrigerated/frozen raw material storage	Histamine formation (e.g. tuna) due to microbial growth due to breakage of the cold chain or incorrect temperature	C	Storage of raw materials in the shortest possible time after receipt Keep freezer and cold store doors closed (open as little time as possible) Check that the product is correctly packaged (perfectly closed containers) Audible alarm in case of power failure (verify correct operation) Do not overload the cold store Maintenance plan for the cold store and calibration plan for the probes	3	3	9	Significant

(*Continued*)

Hazard analysis and assessment							
Step	Hazard	Type	Control measure	S	L	Risk	Significance
Refrigerated/frozen raw material storage	Contamination by microorganisms due to incorrect handling of foodstuffs or contact with dirty surfaces or equipment (inadequate cleaning of cold stores, etc.)	B	Hygiene hands before and during product handling (especially when opening boxes, handling different products, etc.) Good hygiene and handling practices Correct stowage (avoid storing different types of products in the same column, to avoid cross-contamination by leachate, etc.). Products isolated from the floor Correct application of the cleaning and disinfection plan (cold stores)	3	1	3	Not significant
Refrigerated/frozen raw material storage	Proliferation of microorganisms due to exceeding expiry dates	B	Stock control according to I-FP12-01 (control of incoming and outgoing goods) and expiry date control Control of the labeling of goods, checking that the expiry date or best-before date has been established Stock rotation and implementation of the FIFO system: placement of goods so that the first to enter (the oldest) is the first to leave Carry out the periodic verification plan for goods (I-MAPPCC-12)	1	3	3	Not significant
Refrigerated/frozen raw material storage	By residues of cleaning products	C	Use only products suitable for food use Cleaning and disinfection operations must be carried out correctly (according to I-MAPPCC-05), applying the doses established by the manufacturer and eliminating any product residue that may remain Correct storage of chemical products	3	1	3	Not significant

Refrigerated/ frozen raw material storage	Cross-contamination by allergens	A	Compliance with “allergen management” instructions: Comply with the established instructions regarding storage and work areas due to risk of allergens (respect the defined guidelines: each product in its corresponding place according to identification and defined areas) Clean and change work tools according to the products handled; Sanitize hands prior to handling the product and in all necessary cases Check that the product is correctly packaged (perfectly closed packaging) Correct stowage of products and in closed packaging Analysis control plan: surfaces.	3	1	3	Not significant
Refrigerated/ frozen raw material storage	Due to the presence of pests or foreign bodies (pests, remains of packaging material, personal objects, etc.)	P	Keep all products protected and correctly stowed The use of jewelry, rings, watches, etc., is prohibited Correct use of hats, checking the integrity of gloves Do not chew gum, do not eat or introduce food or personal objects in storage or production areas Check that work clothes (jackets, gloves, etc.) are intact. Check tools and brittle material Keep containers closed and isolated from the floor Compliance with the cleaning and disinfection and pest control plan, review of reports.	3	1	3	Not significant
Refrigerated/ frozen raw material storage	Product deterioration due to freeze burns	C	Keep the preventive maintenance plan up to date (FP-04) Keep the product correctly stored, protected in its original container and packaging to avoid deterioration due to direct exposure to low temperatures Verify the correct state of the cold stores Periodic verification of the correct state of stored goods	1	1	1	Not significant

(*Continued*)

Hazard analysis and assessment

Step	Hazard	Type	Control measure	S	L	Risk	Significance
Refrigerated/ frozen raw material storage	Contamination of the product by contact with cooling liquid	C	Keep the preventive maintenance plan up to date (FP-04) Check the correct state of the cold stores (no leaks, low-temperature alarms, etc.)	3	1	3	Not significant
Freezing of raw materials (fresh to be frozen)	Proliferation of microorganisms due to breakage of the cold chain or incorrect temperature	B	Store the product in freezers for the shortest possible time Keep the freezer doors closed Check that the product is correctly packaged (perfectly sealed containers) Audible alarm in case of power failure (check for correct operation) Do not overload the freezing chambers Maintenance schedule for the chambers Daily temperature check (I-MAPPCC-12) and check that the probes are working correctly (I-FP04-01)	3	3	9	Significant
Freezing of raw materials (fresh to be frozen)	Histamine formation due to microbial growth due to incorrect preservation and freezing process (incorrect temperature/freezing times)	C	Storage of the product in freezing chambers in the shortest possible time Keep the doors of the freezing chambers closed Check that the product is correctly packaged (perfectly closed containers) Audible alarm in case of power failure (verify correct operation) Do not overload the chambers Maintenance plan for the chambers Daily temperature control (I-MAPPCC-12) and verification of the correct functioning of the probes (I-FP04-01)	3	3	9	Significant

Freezing of raw materials (fresh to be frozen)	Contamination by microorganisms due to incorrect handling of foodstuffs or contact with dirty surfaces or equipment (inadequate cleaning of chambers, etc.)	B	Hygiene hands before and during product handling (especially when opening boxes, handling different products, etc.) Correct stowage (avoid storing different types of products in the same column, to avoid cross-contamination by leachate, etc.). Products isolated from the floor Correct application of the cleaning and disinfection plan (chambers)	1	1	1	Not significant
Freezing of raw materials (fresh to be frozen)	Cross-contamination by allergens	A	Compliance with “allergen management” instructions: Comply with the established instructions regarding storage and work areas due to risk of allergens (respect the defined guidelines: each product in its corresponding place according to identification and defined areas) Hygiene hands before handling the product and in all necessary cases Check that the product is correctly packaged (perfectly closed packaging) Correct stowage of products and in closed packaging	3	1	3	Not significant
Freezing of raw materials (fresh to be frozen)	Due to the presence of pests or foreign bodies (pests, remains of packaging material, personal objects, etc.)	P	Keep all products protected and correctly stowed The use of jewelry, rings, watches, etc., is prohibited. Correct use of hats, checking the integrity of gloves Do not chew gum, do not eat or introduce food or personal objects in storage or production areas Check that work clothes (jackets, gloves, etc.) are intact. Control of tools and brittle material (I-MAPPCC-10) Keep containers closed and isolated from the floor Cleaning and disinfection and pest control plan	3	1	3	Not significant

(*Continued*)

Hazard analysis and assessment							
Step	**Hazard**	**Type**	**Control measure**	**S**	**L**	**Risk**	**Significance**
Freezing of raw materials (fresh to be frozen)	Product deterioration due to freezer burn	P	Keep the preventive maintenance plan up to date (FP-04) Keep the product correctly stored, protected in its original container and packaging to avoid deterioration due to direct exposure to low temperatures Verify the correct state of the chambers Periodic verification of the correct state of stored goods	1	1	1	Not significant
Freezing of raw materials (fresh to be frozen)	Contamination of the product by contact with cooling liquid	C	Keep the preventive maintenance plan up to date (FP-04) Check the correct state of the chambers (no leaks, low-temperature alarms, etc.)	3	1	3	Not significant
Arrangement of raw materials for supply to cold room/kitchen	Proliferation of microorganisms due to breakage of the cold chain or incorrect temperature	B	Handling/preparation of raw materials, and delivery to the cold room or kitchen, in the shortest possible time Keeping the product correctly packaged (perfectly closed packaging, either original packaging or internal packaging)	3	1	3	Not significant
Arrangement of raw materials for supply to cold room/kitchen	Contamination by microorganisms due to incorrect handling of products or contact with dirty surfaces or tools (inadequate cleaning of preparation areas (chambers, stores, aisles)	B	Hygiene hands before handling products (especially when it is necessary to open boxes, handle different products, etc.) Correct stowage (avoid storing different types of products in the same column, to avoid cross-contamination by leachate, etc.). Keep the products isolated from the floor Correct application of the cleaning and disinfection plan (floors, walls, tools, handling and transport equipment)	3	1	3	Not significant

			Removing any unnecessary waste or material (leftover products, packaging, pallets, etc.) in the shortest possible time				
			Filming of pallets and containers of goods until delivery and use in the cold room/kitchen				
Arrangement of raw materials for supply to cold room/kitchen	Contamination by residues of cleaning products	C	Do not place the goods directly on the floor or in contact with transport systems, pallets, etc., or in contact with walls	3	1	3	Not significant
			Use only products suitable for food use				
			Carry out the cleaning and disinfection operations correctly (according to I-MAPPCC-05), applying the doses established by the manufacturer and eliminating any remaining product that may be left				
			Correct storage of chemical products				
Arrangement of raw materials for supply to cold room/kitchen	Cross-contamination by allergens	A	Compliance with allergen management instructions: comply with the established instructions regarding storage and work areas due to risk of allergens (respect the defined guidelines: each product in its corresponding place according to identification and defined areas)	3	1	3	Not significant
Arrangement of raw materials for supply to cold room/kitchen	Presence of pests or foreign bodies (pests, remains of packaging material, personal objects, etc.)	P	Keep all products protected until use in the cold room or kitchen	3	1	3	Not significant
			The use of jewelry, rings, watches, etc., is prohibited				
			Correct use of hats; check the integrity of gloves				
			Do not chew gum, do not eat or introduce food or personal objects in storage or production areas				
			Check that work clothes (jackets, gloves, etc.) are intact. Check tools and breakable material				
			Filming of pallets or containers of goods until they are delivered and used in the cold room/kitchen				

(*Continued*)

Hazard analysis and assessment							
Step	**Hazard**	**Type**	**Control measure**	**S**	**L**	**Risk**	**Significance**
Preparation of raw materials for supply to cold room/kitchen (weighing and bagging)	Chemical contamination by migration of polymers from the plastic material (intermediate packaging bags)	C	Control and approval of suppliers Purchasing according to material specification Application for a certificate of suitability for food use of the material (declaration of conformity and migration studies)	3	1	3	Not significant
Preparation of raw materials for supply to cold room/kitchen (weighing and bagging)	Presence of foreign bodies (pests, remains of packaging material, personal items, labels, cut materials, etc.)	P	Keep all products protected until they are used in the cold room or kitchen; check thoroughly that there are no residues of any kind during handling The use of jewelry, rings, watches, etc., is prohibited Correct use of the cap, control of the integrity of gloves. Do not chew gum, do not eat or introduce food or personal objects in storage or production areas Check that work clothes (jackets, gloves, etc.) are intact. Control of tools and brittle material Keep tools and warehouse equipment (dry storage, freezer support room, etc.) in perfect condition, removing any contaminating elements	3	1	3	Not significant
Preparation of raw materials for supply to cold room/kitchen (weighing and bagging)	Cross-contamination by allergens	A	Compliance with "allergen management" instructions: comply with the established instructions regarding storage and work areas due to risk of allergens (respect the defined guidelines: each product in its corresponding place according to identification and defined areas) Clean and change work tools according to the products handled; sanitize hands prior to handling the product and in all necessary cases Check that the product is correctly packaged (perfectly closed packaging) Correct stowage of products and in closed packaging	3	1	3	Not significant

Return of raw material to chilling/freezing (raw material without handling)	Proliferation of microorganisms due to breakage of the cold chain or incorrect temperature	B	Keep raw materials refrigerated at all times until they are transferred to the cold room Plan and establish raw material requirements correctly. If it is necessary to return goods to freezing/refrigeration, do so in the shortest possible time Check correct condition and temperature prior to return to freezing/refrigeration, to ensure that these can be their destination	3	1	3	Not significant
Return of raw materials to warehouse at ambient temperature (raw material without handling)	Due to the presence of pests or foreign bodies (pests, remains of packaging material, personal objects, etc.)	P	Keep all products protected until they are used or returned to the warehouse The use of jewelry, rings, watches, etc., is prohibited Correct use of hats, check the integrity of gloves. Do not chew gum, do not eat or introduce food or personal objects in storage or production areas Check that work clothes (jackets, gloves, etc.) are intact. Check tools and breakable material Filming of pallets and containers of goods until they are delivered and used in the cold room/kitchen	3	1	3	Not significant
Defrosting of raw materials	Proliferation of microorganisms due to incorrect defrosting temperature (microbial multiplication)	B	Always perform thawing in the thawing chamber at refrigeration temperature Keep the doors of the chambers closed (open for as short a time as possible) Keep the preventive maintenance plan for the refrigeration chambers (FP-04) up to date. Check the correct state of the chambers (no leaks, alarms for temperature drops, etc.) Control temperatures and defrosting times	3	3	9	Significant
Defrosting of raw materials	Due to the presence of pests or foreign bodies (pests, remains of packaging material, personal objects, etc.)	P	Keep all products protected and correctly stowed The use of rings, watches, earrings, bracelets, etc., is prohibited Correct use of hats; check the integrity of gloves	3	1	3	Not significant

(*Continued*)

Hazard analysis and assessment							
Step	**Hazard**	**Type**	**Control measure**	**S**	**L**	**Risk**	**Significance**
			Do not chew gum, do not eat or introduce food or personal objects in storage or production areas (glasses, food, water bottles, etc.)				
			Check that work clothes (jackets, gloves, etc.) are intact				
			Control of tools and breakable material				
			Keep containers closed and isolated from the floor				
			Compliance with the cleaning and disinfection plan and the pest control plan and review of reports				
Defrosting of raw materials	Cross-contamination by allergens	A	Follow the established instructions regarding storage and work areas due to risk of allergens (respect the separation/incompatibility guidelines to avoid cross-contamination)	3	1	3	Not significant
			Carry out cleaning and change of working tools according to products handled				
			Hygiene hands prior to handling the product; especially when changing between different products				
			Thawing in closed containers				
			Inspect and check the correct state of the defrosted product on a daily basis				
Defrosting of raw materials	Contamination by microorganisms due to improper handling of food or contact with soiled surfaces or utensils	B	Hygiene hands before handling food (especially when handling different products, different tools, etc.)	3	1	3	Not significant
			Thawing should be carried out in closed containers and isolated from the floor				
			Correct application of the cleaning and disinfection plan (floors, walls, tools)				

Preparation of raw materials in the cold room: hydration	Contamination by the presence of chemical contaminants in the water, e.g. chlorine, bromine, metals, halometals, etc.	C	Compliance with water potability as established in prerequisites and I-MAPPCC-03 Consumption of tap water and request for analysis reports to the city council on the control of water quality according to reference legislation and verification plan (I-MAPPCC-12)	1	1	1	Not significant
Preparation of raw materials in the cold room: hydration	Radiological contamination of raw materials by washing and use of cold room water (drinking water) (presence of high levels of radioactivity in the drinking water used in the production process)	R	Request for information from the water management body of the city council Study of the level of radioactivity (radioactivity map) in the area (compilation of information associated with risks of radiological substances in drinking water; according to information from the University of Santiago, SERGAS, the risk is low due to the type of soil in our geographical area)	1	1	1	Not significant
Preparation of raw materials in the cold room: hydration	Contamination by microorganisms due to poor water conditions caused by inadequate maintenance of the water supply system or unsanitary water	B	Compliance of water potability according to prerequisites (internal control of water conditions according to I-MAPPCC-03)	1	1	1	Not significant
Preparation of raw materials in the cold room: hydration	Contamination by microorganisms due to improper handling of food or contact with soiled surfaces or utensils	B	Prepare the goods in the shortest possible time after entering the cold room Handle products correctly in accordance with good hygiene and handling practices: washing hands before each change of product, after handling packaging, when entering and leaving the cold room Cleaning and changing work tools according to the products handled Eliminate remains of products, packaging material, etc., from the cold room; do not allow them to accumulate Empty waste containers frequently; do not allow accumulation Correctly comply with the established cleaning plan (I-MAPPCC05) in terms of cleaning systematics, frequency, products and tools to be used	3	1	3	Not significant

(*Continued*)

Hazard analysis and assessment							
Step	**Hazard**	**Type**	**Control measure**	**S**	**L**	**Risk**	**Significance**
Preparation of raw materials in the cold room: hydration	Presence of foreign bodies (remains of packaging material, insects, pests, etc.)	P	The use of rings, watches, earrings, bracelets, etc., is prohibited Correct use of hats, checking the integrity of gloves Do not chew gum, do not eat or introduce food or personal objects in storage or production areas (glasses, food, water bottles, etc.) Check that work clothes (jackets, gloves, etc.) are intact Control of tools and breakable material (I-MAPPCC-10) Cleaning and disinfection plan and pest control plan	3	1	3	Not significant
Preparation of raw materials in the cold room: hydration	Contamination by cleaning product residues	C	Use only products suitable for food use Cleaning and disinfection operations must be carried out correctly (according to I-MAPPCC-05), applying the doses established by the manufacturer and eliminating any remaining product	3	1	3	Not significant
Preparation of raw materials in cold room: Desalting	Contamination by the presence of chemical contaminants in the desalinated water, e.g. chlorine, bromine, metals, halometals, etc.	C	Compliance with the potability of the water as established in the prerequisites and I-MAPPCC-03 Consumption of tap water and request for analysis reports to the city council on the control of water quality according to reference legislation	1	1	1	Not significant
Preparation of raw materials in cold room: Desalting	Contamination by microorganisms due to poor water conditions due to improper maintenance of the water supply network or unsanitary water	B	Compliance of water potability according to prerequisites (internal control of water conditions according to I-MAPPCC-03) Follow the guidelines set out in the desalting instructions (cold room operations manual: regular change of desalting water/control of desalting time)	1	1	1	Not significant

Preparation of raw materials in cold room: Desalting	Contamination by microorganisms due to improper handling of food or contact with soiled surfaces or utensils	B	Prepare the goods in the shortest possible time after entering the cold room Handle products correctly in accordance with good hygiene and handling practices: washing hands before each change of product, after handling packaging, when entering and leaving the cold room Cleaning and changing work tools according to the products handled Eliminate remains of products, packaging material, etc., from the cold room; do not allow them to accumulate Empty waste containers frequently; do not allow accumulation Correctly comply with the established cleaning plan (I-MAPPCC05) in terms of cleaning systematics, frequency, products, and tools to be used Regularly change the desalting water	3	1	3	Not significant
Preparation of raw materials in cold room: Desalting	Presence of foreign bodies (remains of packaging material, insects, etc.)	P	The use of rings, watches, earrings, bracelets, etc., is prohibited Correct use of hats, checking the integrity of gloves Do not chew gum, do not eat or introduce food or personal objects in storage or production areas (glasses, food, water bottles, etc.) Check that work clothes (jackets, gloves, etc.) are intact Control of tools and breakable material (I-MAPPCC-10). Cleaning and disinfection plan and pest control plan.	3	1	3	Not significant
Preparation of raw materials in cold room: Desalting	Contamination by cleaning product residues	C	Use only products suitable for food use Cleaning and disinfection operations must be carried out correctly (according to I-MAPPCC-05), applying the doses established by the manufacturer and eliminating any remaining product	3	1	3	Not significant

(*Continued*)

Hazard analysis and assessment							
Step	**Hazard**	**Type**	**Control measure**	**S**	**L**	**Risk**	**Significance**
Preparation of raw materials in cold room: Chopping	Contamination by microorganisms due to improper handling of food or contact with soiled surfaces or utensils	B	Handle products correctly in accordance with good hygiene and handling practices: washing hands before each change of product, after handling packaging, when entering and leaving the cold room Cleaning and changing work tools according to the products handled Eliminate remains of products, packaging material, etc., from the cold room; do not allow them to accumulate Empty waste containers frequently; do not allow accumulation Correctly comply with the established cleaning plan (I-MAPPCC05) in terms of cleaning systematics, frequency, products and tools to be used	3	1	3	Not significant
Preparation of raw materials in cold room: Chopping	Presence of foreign bodies (metallic debris, personal belongings, etc.)	P	The use of rings, watches, earrings, bracelets, etc., is prohibited Correct use of hats; control of the integrity of gloves Control of tools (knives, etc.) and breakable material Do not chew gum, do not eat or introduce food or personal objects in storage or production areas (glasses, food, water bottles, etc.) Check that work clothes (jackets, gloves, etc.) are intact	5	1	5	Significant
Preparation of raw materials in cold room: Chopping	Cross-contamination by allergens	A	Prepare the goods in the shortest possible time after their entry into the cold room Compliance with the "allergen management" instruction: comply with the established instructions regarding work areas due to allergen risk (respect the defined guidelines: each product in its corresponding place according to identification, defined areas, production schedules)	3	1	3	Not significant

			Clean and change work tools according to the products handled Sanitize hands before handling the product and in all necessary cases (opening of packaging, leaving the workstation, change of product, etc.) After slicing, correct storage of the product in closed containers Respect production schedules for different diets				
Preparation of raw materials in cold room: Chopping	Contamination by cleaning product residues	C	Use only products suitable for food use Cleaning and disinfection operations must be carried out correctly (according to I-MAPPCC-05), applying the doses established by the manufacturer and eliminating any remaining product	3	1	3	Not significant
Preparation of raw materials in cold room: Chopping	Proliferation of microorganisms due to noncompliance with cold room parameters (excessive time or temperature in cold room)	B	Keep the goods protected until the moment of preparation Prepare the goods in the shortest possible time after entering the cold room Depending on the destination of the goods (cooking, defrosting in refrigeration, return of leftovers to refrigeration or freezing), prepare the goods correctly: –Proper packaging and labeling Preparation on trays and pallets and in film-covered and covered bins And proceed to immediate preservation Keep cold room doors closed (open as little time as possible) Check the correct functioning of the cold room refrigeration; correctly apply the preventive maintenance plan and temperature probe calibration plan	3	1	3	Not significant

(*Continued*)

Hazard analysis and assessment

Step	Hazard	Type	Control measure	S	L	Risk	Significance
Preparation of raw materials in cold room: disinfection of vegetables	Contamination by the presence of chemical contaminants in the water, e.g. chlorine, bromine, metals, halometals, etc.	C	Compliance with the potability of the water as established in the prerequisites and I-MAPPCC-03 Consumption of tap water and request for analysis reports to the city council on the control of water quality according to reference legislation	3	1	3	Not significant
Preparation of raw materials in cold room: Disinfection of vegetables	Contamination by microorganisms due to poor water conditions due to improper maintenance of the water supply system or unsanitary water	B	Compliance of water potability according to prerequisites (Internal control of water conditions according to I-MAPPCC-03)	1	1	1	Not significant
Preparation of raw materials in cold room: Disinfection of vegetables	Microbiological contamination due to incorrect disinfection	B	Carry out cleaning/disinfection of vegetables using food-grade disinfectants Carry out washing/disinfection of vegetables following the established method (washing with plenty of cold water; application of the disinfectant product following the manufacturer's instructions; rinsing with plenty of cold water and draining)	3	1	3	Not significant
Preparation of raw materials in cold room: disinfection of vegetables	Contamination due to incorrect dosing of disinfectants	C	Use only food-grade products Wash/disinfect vegetables according to the established method: Apply the dosage recommended by the manufacturer or, if this is not indicated Apply 5 drops per liter of water or 2.5 ml per 10 l of water Leave for 5 minutes Rinse with plenty of cold water and drain	3	1	3	Not significant

Cooking (heat treatment)	Cross-contamination by allergens	A	Compliance with "allergen management" instructions: comply with the established instructions regarding work areas due to risk of allergens (respect the defined guidelines: each product in its corresponding place according to identification and defined areas) and work schedules Clean and change work tools according to the products handled; sanitize hands prior to handling the product and in all necessary cases Correctly identify the utensils used, taking special care with those corresponding to special diets Try to replace azo food colors (E-102, E-104, E-110, E-122, E-124, E-129) with natural food colors	5	1	5	Significant
Cooking (heat treatment)	Microbiological contamination due to cross-contamination of raw/cooked food	B	Clean and change tools according to the products handled Sanitize hands before handling the product and whenever necessary (opening packages, leaving the workplace, changing products, etc.) Keep the areas for handling raw food and cooked food separate	3	1	3	Not significant
Cooking (heat treatment)	Contamination by cleaning product residues	C	Use only products suitable for food use Cleaning and disinfection operations must be carried out correctly (according to I-MAPPCC-05), applying the doses established by the manufacturer and eliminating any product residue that may remain Correct storage of chemical products	3	1	3	Not significant
Cooking (heat treatment)	Contamination of microorganisms due to incorrect handling of products or contact with dirty surfaces or utensils (inadequate cleaning of preparation areas, utensils, and/or kitchenware)	B	Hygiene of hands prior to handling products (especially when handling different products, etc.) Correct stowage on oven trays (avoid placing different types of products in the same column, to avoid cross-contamination by leachate, etc.). Keep products isolated from the floor	3	1	3	Not significant

(*Continued*)

Hazard analysis and assessment

Step	Hazard	Type	Control measure	S	L	Risk	Significance
			Do not accumulate racks with products waiting to enter the oven around the kitchen. Keep refrigerated until imminent use Correct application of the cleaning and disinfection plan (floors, walls, tools, and handling and transport equipment)				
Cooking (heat treatment)	Presence of polar compounds above established limits	C	Carry out the correct control of fryer oil	3	1	3	Not significant
Cooking (heat treatment)	Presence of acrylamide in starchy foods (as a result of roasting, baking, frying, and grilling or broiling at high temperatures (especially above 120 °C)	C	Perform the correct oil control of deep fryers Cooking (potato frying and dessert preparation) at appropriate temperatures (below 175 °C when frying, and between 180 and 220 °C when using the oven; and never allow browning or burning)	3	1	3	Not significant
Cooking (heat treatment)	Microbiological growth due to incorrect temperature/cooking time	B	Perform cooking according to the times and temperatures indicated in the cooking operations manual for each type of food and according to the way it is prepared	5	3	15	Significant
Cooking (heat treatment)	Contamination by the presence of chemical contaminants in the cooking water, e.g. chlorine, bromine, metals, halometals, etc.	C	Compliance with the potability of the water as established in the prerequisites and I-MAPPCC-03 Consumption of tap water and request for analysis reports to the city council on the control of water quality according to reference legislation	1	1	1	Not significant
Cooking (heat treatment)	Contamination by microorganisms due to poor water conditions caused by improper maintenance of the water supply system or unhealthy water.	B	Compliance of water potability according to prerequisites (Internal control of water conditions according to I-MAPPCC-03)	1	1	1	Not significant
Crushing	Contamination due to the presence of foreign bodies (metallic remains of blades, work tools, personal items, etc.)	P	Check the integrity of the crushing equipment at the time of assembly after cleaning No jewelry, rings, watches, etc.	5	1	5	Significant

			Correct use of hat, check integrity of gloves. Do not chew gum, do not eat or bring food or personal items into storage or production areas Check that work clothes (jackets, gloves, etc.) are intact. Control of tools and breakable material				
Crushing	Cross-contamination by allergens	A	Compliance with “allergen management” instructions: thorough cleaning of equipment; especially after use with allergens (respect the defined guidelines: each product in its established work schedule Clean and change work tools according to the products handled; sanitize hands before handling the product and in all necessary cases Correctly identify the utensils used, taking special care with those corresponding to special diets	5	1	5	Significant
Crushing	Contamination by cleaning product residues	C	Use only products suitable for food use Correct cleaning and disinfection of the crusher, in terms of dismantling and applying the doses established by the manufacturer and eliminating any remaining product that may be left Correct storage of chemical products	3	1	3	Not significant
Crushing	Contamination by microorganisms due to incorrect handling of food or contact with dirty surfaces or utensils (inadequate cleaning of preparation areas, utensils and/or kitchenware)	B	Hygiene of hands prior to handling products (especially when handling different products, etc.) Correct cleaning of crushing equipment after use. Correct application of the cleaning and disinfection plan	3	1	3	Not significant
Preparation/ mixing of ingredients ($T^a < 4\,°C$)	Proliferation of microorganisms due to breakage of the cold chain or incorrect temperature Lack of control over leftovers (poor condition/labeling)	B	Perform the preparation (mixing) of the products in the shortest possible time Keep cold room doors closed (open as little time as possible) Do not leave the food during processing	3	1	3	Not significant

(*Continued*)

Hazard analysis and assessment							
Step	Hazard	Type	Control measure	S	L	Risk	Significance
			Sound alarm in case of power failure (verify correct operation) Maintenance plan for the chambers and calibration plan for the probes Verify the correct state of leftovers from previous preparations that are incorporated into the dish (labeling information: date of packaging/consumption; ingredients, etc.)				
Preparation/mixing of ingredients ($T^a < 4\,°C$)	Due to the presence of pests or foreign bodies (pests, material remains, personal items, etc.)	P	Keep all products protected during processing (either with the use of lids or with properly covered trays/carts/pans) No wearing of jewelry, rings, watches, etc. Correct use of hats; control of the integrity of gloves Do not chew gum, do not eat or introduce food or personal objects in storage or production areas Check that work clothes (jackets, gloves, etc.) are intact. Control of tools and breakable material Cleaning and disinfection and pest control plans	5	1	5	Significant
Preparation/mixing of ingredients ($T^a < 4\,°C$)	Cross-contamination by allergens Lack of control over leftovers (poor condition/labeling)	A	Compliance with "allergen management" instructions: comply with the established instructions regarding work areas due to risk of allergens (respect the defined guidelines: each product in its corresponding place according to identification and defined areas) Clean and change work tools according to the products handled; sanitize hands before handling the product and in all necessary cases Correctly identify the utensils used, taking special care with those corresponding to special diets	5	1	5	Significant

			Verify the correct state of leftovers from previous preparations that are incorporated into the dish (labeling information: date of packaging/consumption; ingredients, etc.)				
Preparation/ mixing of ingredients ($T^a < 4\,°C$)	Contamination and microbiological growth due to cross-contamination of raw/cooked food Lack of control over leftovers (poor condition/labeling)	B	Cleaning and changing work tools according to the products handled Sanitize hands prior to handling the product and in all necessary cases (opening packaging, leaving the workstation, changing products, etc.) Maintain separate areas for handling raw food and cooked food Verify the correct state of leftovers from previous preparations that are incorporated into the dish (labeled information: date of packaging/consumption; ingredients, etc.)	3	1	3	Not significant
Preparation/ mixing of ingredients ($T^a < 4\,°C$)	Contamination by cleaning product residues	C	Use only products suitable for food use Cleaning and disinfection operations must be carried out correctly (according to I-MAPPCC-05), applying the doses established by the manufacturer and eliminating any product residue that may remain Correct storage of chemical products	3	1	3	Not significant
Preparation/ mixing of ingredients ($T^a < 4\,°C$)	Contamination by microorganisms due to incorrect handling of products or contact with dirty surfaces or equipment (inadequate cleaning of preparation areas)	B	Hygiene of hands prior to handling products (especially when handling different products, etc.) Correctly stacking on the trays of the racks (avoid placing different types of products in the same column to avoid cross-contamination by leachate, etc.). Do not accumulate racks with products awaiting processing. Keep refrigerated until imminent use Correct application of the cleaning and disinfection plan (floors, walls, tools, and handling and transport equipment)	3	1	3	Not significant

(*Continued*)

Hazard analysis and assessment

Step	Hazard	Type	Control measure	S	L	Risk	Significance
Blast chilling	Due to the presence of pests or foreign bodies (pests, material remains, personal belongings, etc.)	P	Properly cover the blast chilling rack during the folding process The use of jewelry, rings, watches, etc., is prohibited. Correct use of the cap, check the integrity of the gloves Do not chew gum, eat or bring food or personal items into the production area Check that working clothes (jackets, gloves, etc.) are intact; check tools and brittle materials Cleaning and disinfection and pest control plans	1	1	1	Not significant
Blast chilling	Cross-contamination by allergens	A	Compliance with "allergen management" instructions: comply with the established instructions regarding work areas due to allergen risk (respect the defined guidelines: each product in its corresponding place according to identification and defined areas) Take care to ensure that the cooked product is not contaminated during its time in the blast chiller Clean and change working utensils according to the products handled; sanitize hands before handling the product and in all necessary cases Correctly identify the utensils used, taking special care with those corresponding to special diets	5	1	5	Significant
Blast chilling	Contamination by microorganisms due to incorrect handling of products or contact with dirty surfaces or tools (inadequate cleaning of racks or blast chillers)	B	Hygiene of hands prior to handling products (especially when handling different products, etc.) Correct stowage in rack trays (avoid placing different types of products in the same column, to avoid cross-contamination by leachate, etc.). Keep products isolated from the floor	3	1	3	Not significant

			Do not accumulate racks with products waiting to enter the blast chiller Correct application of the cleaning and disinfection plan (blast chillers, floors, walls, tools, and handling equipment)				
Blast chilling	Microbiological proliferation due to noncompliance with process conditions (temperature and abatement time)	B	Comply with the established guidelines for blast chilling: compliance with the established times and temperatures. Recording of temperatures and times	3	3	9	Significant
Pasteurization Fagor 1 Oven: 1.10 h at 90 °C Fagor 2 Oven: 50 min at 90 °C Rational Oven: 45 min at 90 °C.	Survival of microorganisms due to incorrect temperature or process time	B	Apply semi-pasteurization according to the times and temperatures indicated in the kitchen operations manual	5	3	15	Significant
Cold packaging/thermosealing in protective atmosphere ($T^a < 4\,°C$)	Microbiological proliferation due to inadequate temperature maintenance in the packaging process (plate and room)	B	Control of the compliance with the set packaging temperature ($T^a < 4\,°C$)	3	1	3	Not significant
Hot packaging/thermosealing in a protective atmosphere ($T^a > 70\,°C$ product center)	Microbiological proliferation due to inadequate temperature maintenance in the packaging process (plate and room)	B	Checking compliance with the set packaging temperature (>70 °C)	3	1	3	Not significant
Hot packaging/thermosealing in a protective atmosphere ($T^a > 70\,°C$ product center)	Contamination by microorganisms due to incorrect handling of the products or contact with dirty surfaces or tools (inadequate cleaning of containers, conveyor belts, packaging machines, etc.)	B	Hygiene hands prior to handling products (especially when handling different products, etc.) Immediately remove any debris produced during the packaging process Do not accumulate racks with products pending packaging	3	1	3	Not significant

(*Continued*)

Hazard analysis and assessment							
Step	Hazard	Type	Control measure	S	L	Risk	Significance
			Correct application of the cleaning and disinfection plan (floors, walls, tools, handling and transport equipment)				
			Work with masks correctly positioned so that they cover the mouth and nose				
Hot packaging/thermosealing in a protective atmosphere ($T^a > 70\,^\circ C$ product center)	Proliferation of microorganisms due to deficiencies in protective atmosphere/incorrect gas mixing	B	Control of the first plates of each packaging program and after each stop due to technical problems on the packaging line or change of packaging material (film). Recording of oxygen control according to F-IFP1004-01	3	3	9	Significant
			Ensure training/qualification of personnel to carry out controls of protective atmosphere packaging				
			Check correct heat sealing of packages				
Hot packaging/thermosealing in a protective atmosphere ($T^a > 70\,^\circ C$ product center)	By the presence of foreign bodies (remains of packaged material, personal belongings, pests, etc.)	P	Control of the correct packaging	5	3	15	Significant
			The use of jewelry, rings, watches, etc., is prohibited				
			Correct use of the cap; control of the integrity of gloves				
			Do not chew gum, do not eat or bring food or personal objects into the production area				
			Check that work clothes (jackets, gloves, etc.) are intact; control of tools and brittle material				
			Keep racks covered/filmed while waiting to be packed				
Hot packaging/thermosealing in a protective atmosphere ($T^a > 70\,^\circ C$ product center)	Cross-contamination by allergens	A	Compliance with "allergen management" instructions: Comply with the established instructions regarding work areas due to risk of allergens	5	1	5	Significant
			Clean and change work tools according to the products handled; sanitize hands prior to handling the product and in all necessary cases				

			Correctly identify the utensils used, taking special care with those corresponding to special diets				
Labeling	Incorrect information on allergen and trace identification on the label	A	Take special care in the programming/parameterization of labels	1	5	1	Significant
			Perform quality control on labeling				
			Ensure that staff are trained and informed to interpret and know the labeling rules				
Labeling	Incorrect information on allergen and trace identification on the label	B	Take special care in the programming/parameterization of labels	3	1	3	Not significant
			Perform quality control on labeling				
			Ensure that staff are trained and informed to interpret and know the labeling rules				
Refrigerated storage after packaging ($T < 4\,^{\circ}C$)	Proliferation of microorganisms due to breakage of the cold chain or incorrect temperature	B	Storage of the product in the shortest possible time after packaging	3	1	3	Not significant
			Keep the doors of the refrigeration chambers closed (open as little time as possible)				
			Audible alarm in case of power failure (verify correct operation)				
			Do not overload the chambers				
			Maintenance plan for the chambers and calibration plan for the probes				
Refrigerated storage after packaging ($T < 4\,^{\circ}C$)	Contamination of microorganisms due to incorrect handling of food or contact with dirty surfaces or equipment (inadequate cleaning of chambers, etc.)	B	Hygiene hands before and during handling of the product	3	1	3	Not significant
			Hygiene and Good Handling Practices (GHP)				
			Correct stowage (avoid storing products in badly fitted boxes)				
			Products isolated from the floor				
			Correct application of the cleaning and disinfection plan (chambers)				

(*Continued*)

Hazard analysis and assessment

Step	Hazard	Type	Control measure	S	L	Risk	Significance
Refrigerated storage after packaging ($T < 4\,^\circ C$)	Proliferation of microorganisms due to exceeding expiry dates	B	Stock control according to I-FP12-01 (control of incoming and outgoing goods) and expiry date control	3	1	3	Not significant
			Control of the labeling of goods, checking that the expiry date or best-before date has been established				
			Stock rotation and implementation of the First in - First out (FIFO) system: placement of goods so that the first to enter (the oldest) is the first to leave				
Assembly $T^a < 10\,^\circ C$	Proliferation of microorganisms due to breakage of the cold chain or incorrect temperature	B	Correct assembly of the product (use of the appropriate containers and packaging depending on the product to be transported, allowing the temperature conditions required by the product to be maintained)	3	1	3	Not significant
			Storage of the product in the shortest possible time after assembly				
			Keep the doors of the chambers closed (open for as short a time as possible)				
			Audible alarm in case of power failure (verify correct operation)				
			Do not overload the chambers				
			Maintenance plan for the chambers and calibration plan for the probes				
Assembly $T^a < 10\,^\circ C$	Contamination of microorganisms due to improper handling of dishes	B	Hygiene hands before and during handling of the product	3	1	3	Not significant
			Hygiene and handling GPs				
			Correct stowage (avoid storing different types of products in the same column)				
			Products isolated from the floor				
			Correct application of the cleaning and disinfection plan (chambers)				

Cold distribution ($T^a < 4\,°C$)	Proliferation of microorganisms due to breakage of the cold chain due to incorrect temperature or noncompliance with transport time	B	Control of the temperature of the vehicle body at the exit of the plant, halfway and at the destination Maintain vehicles in correct maintenance conditions Subject vehicles to the required checks and inspections	3	3	9	Significant
Cold distribution ($T^a < 4\,°C$) (with vehicle rest and food en route)	Proliferation of microorganisms due to breakage of the cold chain (food remains in the vehicle from one day to the next)	B	1. Verify that the resting point of the vehicle with the food has the necessary means to guarantee the necessary conditions: three-phase socket to connect the vehicle's power supply during the stop time/covered area for the vehicle/surveillance of the parking area 2. At the time of the stop, after connection to the power supply, check that the vehicle and the cooling system are in operation and record the temperature 3. First thing in the morning, prior to start of service, check vehicle temperature and record data In case of failure or verification of breakage of the cold chain, notify headquarters, request delivery of food to complete the service, and return to the plant with the food	3	1	3	Not significant
Cold distribution ($T^a < 4\,°C$)	Proliferation of microorganisms due to incorrect handling of dishes	B	Hygiene hands before and during handling of the product Hygiene and Good Handling Practices (GHP) Correct stowage (avoid storing different types of products in the same column) Products isolated from the floor Correct application of the cleaning and disinfection plan (vehicles)	3	1	3	Not significant
Cold distribution ($T^a < 4\,°C$) (with vehicle rest and food en route)	Presence of foreign body contamination (due to risk of sabotage/access to food, in cases where delivery is not completed within the day and food remains in vehicle overnight)	P	1. Verify that the resting point of the vehicle with the food has the necessary means to ensure the safety of the vehicle: preferably a covered area for the vehicle and/or with surveillance of the parking area	3	1	3	Not significant

(*Continued*)

Hazard analysis and assessment

Step	Hazard	Type	Control measure	S	L	Risk	Significance
			2. At the moment of stopping, verify that all windows and doors of the vehicle are perfectly closed and that the vehicle is locked 3. First thing in the morning, prior to the start of the service, check that everything is in order and that there are no signs of tampering or attempted access to the vehicle (doors and windows closed) If intrusion is detected, notify headquarters, ask for food to be sent to finish the service and return to the plant with the food				
Cold distribution ($T^a < 4\,°C$) (with vehicle rest and food en route)	Contamination by addition of chemical contaminants (due to risk of tampering, sabotage/access to food, in cases where delivery is not completed within the day and food remains in vehicle overnight)	C	1. Verify that the resting point of the vehicle with the food has the necessary means to ensure the safety of the vehicle: preferably a covered area for the vehicle and/or with surveillance of the parking area 2. At the moment of stopping, verify that all windows and doors of the vehicle are perfectly closed and that the vehicle is locked 3. First thing in the morning, prior to the start of the service, check that everything is in order and that there are no signs of tampering or attempted access to the vehicle (doors and windows closed) If intrusion is detected, notify headquarters, ask for food to be sent to finish the service and return to the plant with the food	3	1	3	Not significant

7 CCP Control Chart

Step	Hazard	Type	Control measures	No CCP	Critical limit	Monitoring			Corrective action		Records
						How	When	Who			
Cooking (heat treatment)	Microbiological survival due to incorrect tempera-ture/cooking time	B	Perform cooking according to the times and temperatures indicated in the cooking operations manual for each type of food and according to the way it is prepared	CCP 1	Established cooking temperature and times (according to the Cooking Operations Manual) (Baking table)	Checking the correct application of cooking times/cooking temperatures according to the applied technique	Daily	Kitchen staff	Checking the correct state of the product. Rejection of the product if it does not meet the defined parameters	Responsible for the management system	Routing Sheets Incident Record/NC F-FP07-02
Blast chilling	Microbiological proliferation due to noncompliance with process conditions (temperature and blast chilling time)	B	Comply with the established guidelines for blast chilling: compliance with the established times and temperatures. Recording of temperatures and times	CCP 2	Reach a temperature between 0 and 5 °C in a maximum of 2.5 h (except in the case of broths and stocks, which shall be kept for a maximum of 5 h)	Checking the correct application of blast chilling times/temperatures	Daily	Kitchen staff	Checking the correct state of the product. Rejection of the product if it does not meet the defined parameters	Responsible for the management system	Routing Sheets F-FP10-04 Temperature control of blast chillers F-IFP1003-01 Incident Record/NC F-FP07-02
Pasteurization	Microbiological survival due to incorrect temperature or processing time	B	Apply pasteurization according to the times and temperatures indicated in the kitchen operations manual	CCP 3	Reach $T^a > 74$ °C in the center of the product (minimum 1 second) For which: Fagor 1 oven: 1.10 h at 90 °C. Fagor 2 Oven: 50 min at 90 °C	Checking and recording of the temperature reached	At the exit of the plates from the oven (per pas-teurization shift)	Kitchen staff	Adjustment of pasteurization parameters	Responsible for the management system	Routing Sheets F-FP10-04

(*Continued*)

Step	Hazard	Type	Control measures	No CCP	Critical limit	Monitoring			Corrective action		Records
						How	When	Who			
					Rational Oven: 45 min at 90 °C Roast ham: Fagor oven: 1 h. Rational Oven: 40 min						
Packaging/ thermosealing, hot and cold, in protective atmosphere	Proliferation of microorganisms due to deficiencies in protective atmosphere/incorrect gas mixing	B	Control of the first plates of each packaging program and after each stop due to technical problems on the packaging line or change of packaging material (film). Recording of oxygen control according to F-IFP1004-01 Ensure training/qualification of personnel to carry out controls of protective atmosphere packaging Check correct heat sealing of packages	CCP 5	<1% residual oxygen $>18\%$ CO_2 Absence of bubbles in container	Check of packaging line parameters and residual oxygen measurement Immersion of container in water bath	Daily at each change of packaging program	Packaging Shift Manager	Adjustment of packaging parameters and rejection of packaged product in nonconforming condition	Responsible for the management system	Routing sheet-F-FP10-04 Oxygen Control Packaged Dish F-IFP1004-01 Incident Logging/NC F-FP07-02
						Microbiological analysis of meals	According to control plan analysis	Responsible for the management system	Rejection of the batch of product tested and withdrawal from the market if already distributed	Responsible for the management system	Analysis Report Incident Log/NC F-FP07-02.

Index

HACCP System Auditing for Food Safety: Principles and Techniques, First Edition. Luis Couto Lorenzo.

g

h

t

u

v

w

Printed and bound by CPI Group (UK) Ltd, Croydon, CR0 4YY

17/09/2026

14964334-0001